The Premise of Fidelity

The Premise of Fidelity

SCIENCE, VISUALITY, AND REPRESENTING THE REAL IN NINETEENTH-CENTURY JAPAN

Maki Fukuoka

STANFORD UNIVERSITY PRESS
STANFORD, CALIFORNIA

Stanford University Press
Stanford, California

Receipt of an Association for Asian Studies First Book Subvention has facilitated the publication of this book.

Printed and bound by CPI Group (UK) Ltd, Croydon, CR0 4YY

Library of Congress Cataloging-in-Publication Data
Fukuoka, Maki, 1972- author.
The premise of fidelity : science, visuality, and representing the real in nineteenth-century Japan / Maki Fukuoka.
pages cm
Includes bibliographical references and index.
ISBN 978-0-8047-7790-2 (cloth : alk. paper)
1. Art and science--Japan--History--19th century. 2. Botanical illustration--Japan--History--19th century. 3. Plant prints--Japan--History--19th century. 4. Photography--Japan--History--19th century. 5. Realism in art--Japan--History--19th century. I. Title.
N72.S3F85 2012
709.52--dc23 2012007601

Typeset by Bruce Lundquist in 11/14 Adobe Garamond

Contents

List of Figures

Acknowledgments

It would be an understatement to say that this project came easily. It would be even more inaccurate to characterize this project as my solo endeavor. The generosity and support of institutions, colleagues, friends, and family members have been instrumental to this publication.

Since the "genesis" moment in a graduate seminar on the intellectual history of modern Japan in the fall of 1999, Professor Tetsuo Najita's insight and criticism helped tremendously in modeling this project. In the Department of Art History at the University of Chicago, where I completed my graduate work, Professors Tom Gunning and Joel Snyder advised me with care and skill to steer my ideas from a (sometimes useful) misguided direction back to an appropriate path. Tom's generosity in sharing his knowledge and ideas of modernity continue to shape my interest and research. Working with Joel has opened many closed doors of inquiry about photography and representations and guided me into unexplored territories. Professor Susan Burns stepped into this project at the final stage of dissertation, and her suggestions brought into light a constructive perspective on the broader historical landscape that became essential in revision.

In Tokyo, Professor Kinoshita Naoyuki at the University of Tokyo kindly accepted me to his *kenkyūshitsu* in November 2001. His eye for objects and history, and his uncanny way of navigating the strange continue to mesmerize and inspire me. Kinoshita-sensei's humor gave me a profound sense of comfort that lasts to this day.

Since my first archival research trip to Nagoya in 2001, Endo Shōji has been more than generous in sharing his collection and knowledge of materials related to the Shōhyaku-sha. I only hope that I did some justice to the wealth of the Shōhyaku-sha material. Also in Nagoya, Yokoyama Susumu of the Higashiyama Botanical Garden welcomed me with great kindness during my repeated visits to the archives. In Tokyo, Sato Mona of the University of Tokyo gracefully helped me relearn classical Chinese and Japanese, and her willingness to decipher cultural connotations was indispensable in many of the close readings of the primary text. In Leiden, Professor Matthi Forrer's initial encouragement convinced me to take the first step in this endeavor. Since 2006, Christine Guth, Katsuya Hirano, William Schaefer, and Timon Screech have given me invaluable comments and suggestions on versions of the manuscript drafts.

This project was supported by several generous fellowships. The initial dissertation research in the Netherlands was supported by a Lipman travel grant and a Smart Foundation travel grant, both from the Department of Art History at the University of Chicago. The archival research in Japan in 2001–02 was made possible through a Japan Foundation Dissertation Fellowship. A dissertation write-up grant from the Center for East Asian Studies at the University of Chicago in 2003–04 and a Whiting Fellowship from the Division of Humanities at the University of Chicago between 2004 and 2005 provided me with the critical financial support to continue and finish my dissertation. For post-dissertation revisions, a Faculty Research Grant from the Center for Japanese Studies at the University of Michigan allowed me to examine objects in the UK that had otherwise been difficult to inspect. In 2008–09, a Robert and Lisa Sainsbury Fellowship from the Sainsbury Institute for the Studies of Japanese Arts and Cultures offered me a tremendous opportunity and time to concentrate on revising the manuscript.

The collegial support of the Department of Asian Languages and Cultures at the University of Michigan has been instrumental, and I am particularly thankful for the support of Micah Auerbach, Ben Brose, Deirdre de la Cruz, Ken Ito, Christi Merrill, Youngju Ryu, and Jonathan Zwicker. Nancy Florida and Donald Lopez, the chairs of the department, have offered me realistic and experienced guidance since my arrival in Ann Arbor. In the larger community of the University of Michigan, Kevin Carr, Joan Kee, Christian de Pee, Leslie Pincus, and Sarita See have been there when I needed a sounding board. For its financial support of this publication, I am grateful to the Faculty Grants and Awards Program of the University of Michigan's Office of the Vice President for Research. Kerry Boeye and Jenn Wright kindly copyedited my draft with such grace and intellect, and I am indebted to their refined editing skill and friendship. The support of Stacy Wagner and Jessica Walsh, editor and editorial assistant at Stanford University Press, has proven vital in the last phase of this project.

Many family members and close friends who shall remain unnamed in Japan, the Netherlands, the UK, and the United States made writing this book easier through their extraordinary friendship and compassion. Their companionship and willingness to engage in dialogue was critical to navigating, both literally and metaphorically, the webs of life that presented themselves, unexpectedly and delightfully. You know who you are. I am profoundly indebted to my parents, Fukuoka Kanefusa and Yamaguchi Hiroko, who first instilled in me the pleasures of traveling to unknown cities and museums and seeing the world with my own eyes.

Despite all of the generous and constructive advice I have received, all errors in this book are entirely my own.

Note on Names

Japanese names in this book follow the Japanese order with family name followed by given or literary name. It is standard practice in the study of Tokugawa Japan to refer to historical figures by their given or literary name rather than surname. As a result, I have adopted this convention in this book. For the sake of consistency, I also use the given or literary name to refer to historical figures of modern Japan, though use of surname would otherwise be the convention.

The Premise of Fidelity

Introduction

An unpublished manuscript entitled "*Honzō sjasin*," compiled in 1826 by the scholar Mizutani Hōbun, first revealed to me that a vital part of the history of nineteenth-century Japanese visual culture had yet to be written. The term *shashin*, I discovered later, would be the keyword to apprehend the interconnected fields of visual culture and scientific studies in Tokugawa Japan.[1] The manuscript resides today at the Special Collection of Leiden University Library as part of the collection of books and maps amassed by Phillip Franz von Siebold, a German physician who worked in Japan at the Dutch factory from 1823 to 1829. I had initially traveled to Leiden in search of a Japanese zograscope, a viewing device for *vue d'optique* (*uki-e*) brought back to Europe by Siebold. The *uki-e* prints, which gained popularity around the end of the eighteenth century, employ one-point linear perspective, and when seen through a zograscope they create an optical illusion that gives the viewer a sense of receding and expanding space. This device is one of the few extant examples of its kind produced in Japan before the latter half of the nineteenth century. As transparent transmitters of light, lenses are often overlooked in studies of visuality, yet lenses for *uki-e* were often opaque and warped, unintentionally calling attention to the mediating mechanism of the lens itself. The way the materiality of the *uki-e* lenses underscores the refracting of light—the lenses' alteration rather than mere transmission of the light—provides a metaphor for how conceptual categories and frameworks of knowledge similarly shaped practices of representation and the reception of photography in Japan.

The manuscript "*Honzō sjasin*" crystallizes in its very title the question of the impact of concepts and discourses upon Japanese visual culture. The first part of the title, *honzō*, refers to a discourse on *materia medica*, the medicinal properties of plants and minerals. *Honzōgaku*, the study of *materia medica*, was integral to medical training in early modern Japan. The other word in the title, *sjasin*, is a nineteenth-century rendering of the word now Romanized as *shashin*. In contemporary Japanese *shashin* means "photography." But the medium did not exist when "*Honzō sjasin*" was compiled. The invention of photography was announced years later on another continent (in 1839 in France and England) and did not reach Japan until the 1850s.[2] Exploring what the term *shashin*—"transposition (*sha*) of the real (*shin*)"—meant to the manuscript's author and his circle in 1826 opens a window onto the sophisticated relationship posited between seeing and knowing. This relationship rested upon a premise of fidelity in visual representation that would shape artistic practices, including photography, for decades.

Around 1828, a few years after compiling "*Honzō shashin*,"[3] Mizutani Hōbun founded the Shōhyaku-sha, a small group of scholars with an interest in *materia medica* (*honzōgaku*) based in Nagoya, the castle town of the Owari domain. Many were practicing physicians and pharmacologists, while others worked in the employ of the government bureaucracy. Still other members were directly engaged in agriculture as farmers and gardeners. The Shōhyaku-sha (the Society of One Hundred Tasters) took its name from the story of *Shen Nong*, a mythological Chinese god of agriculture and medicine who was said to have tasted one hundred plants in order to find the correct natural preparations to cure a given disease.[4]

"*Honzō shashin*" exemplifies a group of materials—some printed, others left as manuscripts—produced by Shōhyaku-sha, which remained active through the Meiji Restoration of 1868 and into the mid-1880s. But "*Honzō shashin*" is also only one part of an extraordinary archive comprising objects and images ranging from specimens of dried plants and minerals to numerous botanical illustrations in etching, ink rubbing, and woodblock. The activities of the Shōhyaku-sha were many and varied: they made excursions to collect botanical specimens, organized exhibitions of *materia medica,* and printed and published their findings. In addition to their public activities, the group also got together to share and examine ideas, and exchange texts, objects, and scientific instruments.[5] None of the members of the Shōhyaku-sha, including those most directly responsible for production of the extraordinary quantity and array of extant images, fits easily into a discursive field of art history, the discipline in which I am trained. As art historians Kitazawa Noriaki and Satō Dōshin have demonstrated, the categories of "art" (*bijutsu*) and "art history" (*bijutsushi*) were not readily available frameworks before the Meiji period; rather, it was the concerted effort of the Meiji bureaucrats who actively applied and refined these concepts through their exhibition practices and institutionalizations of artistic production in Japan that cemented these categories as meaningful and useful sources for their nation building.[6] The textual and pictorial records produced by the Shōhyaku-sha do not fall within this trajectory of art and its history. Indeed, the diversity of repositories in which their work is held today—art museums, libraries, natural science museums, and botanical gardens scattered across Japan and Europe—attest to the difficulty modern disciplines face in attempting to capture the interconnected and interpenetrating aspects of the group's activities.[7]

"*Honzō shashin*" consists of seventeen botanical images, fifteen of which were drawn in great detail, one a mere sketch, and one an inked impression of a plant (see Fig. 1.1 in Chapter One). Each image is accompanied by the name of the plant in Japanese and Chinese, with no other text. This stark juxtaposition of image and text in two languages underscores one of the central concerns for the group: naming, particularly with matching the Chinese names found in imported texts such as *Compendium of Materia Medica* (*Bencao gangmu*) to the local flora of Japan.[8] Indeed, many of the major publications by members of the Shōhyaku-sha, such as *Clarification on the Names of Things* (*Buppin shikimei,* 1809) by Mizutani Hōbun and *Nominal Differentiations in Western Materia Medica* (*Taisei honzō meiso,* 1829) by Itō Keisuke, are records of their efforts to concretize the relationship between names and things by reexamining the Chinese canon of *materia medica* within the context of the local environment and the importation of Western texts on natural history. The representational practices that the Shōhyaku-sha engaged in throughout the nineteenth century hinge on securing the relationship of fidelity among textually rendered knowledge, illustrations, and the existent plants native to Japan. The recursive probing in their study of *materia medica* that entailed correlating published knowledge with their own experience and observations required that all three spheres of their knowledge function together and mutually reinforce one another.

The Shōhyaku-sha emerged from a line of inquiry that began over a century earlier with Kaibara Ekken, an influential Neo-Confucian scholar and an avid practitioner of traditional pharmacology. In his *Materia Medica in Japan* (*Yamato honzō*), published in 1709, Ekken noted the wide disparities between the plants and minerals catalogued in the Chinese canon of *materia medica* and his own environment. As a result of this observation, he urged a refocusing within the study of *materia medica* from faithful memorization and recitation of the Chinese text to a more flexible approach that reconsidered Chinese knowledge within the context of a Japanese ecology. Ekken thus sought to recalibrate the relationship between texts and reality. In the century that followed, a variety of scholars such as Inō Jakusui, Ono Ranzan, and Hiraga Gennai began a turn toward an encounter with things—plants, trees, minerals, animals, birds, fish, and insects—as well as noncanonical texts of natural history—Dodonaeus's *Cruijdeboeck* (1544) or Johannes Jonstonus's "Zoological Album" as it was known (1660),[9] for instance—to address what they saw as a widening gulf between the ossified world presented in the Chinese canon and the living environment to which they increasingly redirected their attention.[10] These scholars were trying to "make sense" of the world by correlating the two, an endeavor that resonates with Peter Dear's deeply historicist conviction that "the world pictures that we believe in owe much more to what we find plausible than to the way the world 'really' is: their acceptance, rather than being determined by the natural world itself, depends on the ways in which we choose to live in the world."[11]

The members of Shōhyaku-sha were distinguished by an emphasis on the medical utility of this body of knowledge. What was at stake for the eighteenth-century predecessors of the group was a discrepancy between representation and reality, but their concern still remained largely theoretical and abstract. The rectification of names was here not tied to the specific efficacy of the *materia medica* in treatment but rather to a broader reordering of classificatory systems and reimagining of natural world. For the Shōhyaku-sha, the concerns were more immediate and more practical: the medical efficacy of the materials they studied, and their use in treating the patients they saw daily as practicing physicians.

These underlying conceptual and pragmatic motivations inform how the imagery in Hōbun's manuscript participates in the process of naming and knowing alongside the Chinese and Japanese texts, a process that is most effectively explored through the term *shashin* in the manuscript's title. The manuscript lacks a preface or any other written indication explaining the use of this term. The mixture of drawing and inked impressions of plants in the manuscript indicates that whatever specific meanings *shashin* had for the manuscript's author, they were not tied to one specific medium. Art historians who have traced the shifting meanings of the term *shashin* argue for a close connection between the history of this concept and the emergence of the practices of pictorial realism in Japan beginning in the late eighteenth century.[12] Originating from discourses on portrait painting in the late Han period, the term came into common use in Japan during the eighteenth century among literati painters influenced by theories and images of Ming visual culture.[13] While in China the term *shashin* (Ch: *xiezhen*) was distinctly used for portraits, and *shasei* (Ch: *xiesheng*) for flower and bird paintings, in Japan *shashin* came to encompass both subjects. Two decades later, those scholars who took to Western learning (*rangaku*) began to reappropriate the term as a concept that designated Western pictorial methods. Shiba Kōkan, a late-eighteenth-century painter and printmaker and one of the early exponents of Western theories of pictorial representation, deployed the term to describe a "method" absent in the representational traditions of China and Japan, in his 1799 *Discussion of Western Pictures* (*Seiyō gadan*). For him, the method of *shashin* would "give an illusion to the viewers that the depicted subjects are about to move out of the picture."[14] This *shashin* method, in his estimation, made indigenous modes of representation appear "similar to children's play."[15] Six years later, Kōkan would pair this *method* with a *device* when he referred to camera obscura as *shashin kyō* (literally, "lens/mirror that does *shashin*") in *Dutch Navigation* (*Oranda tsūhaku*, 1806).[16]

Kōkan's approach to picture making and his uses of the term *shashin* have come to be accepted as standard for the entirety of the early modern discourse on *shashin*. Historians of Japanese photography perceive the association Kōkan made among the pictorial techniques of Western pictures, the device of the camera obscura, and the term *shashin* as a kind of originary moment that would lead more than fifty years later to the equation and conflation of photography and *shashin*.[17] Historians of paintings and prints, on the other hand, use the same term to refer to the emergence of an early modern discourse of realism in Japan. As a result, the advent of photography in Japan becomes a telos in scholarly accounts of the history of medium toward which the term *shashin* invariably points.[18] In such a narrative, the question of whether and how a photographic image corresponded to what was in front of the camera, and how viewers perceived the images, become obscured by an overgeneralized teleological conception of the term *shashin* as already embodying pictorial realism and an affinity for Western pictorial traditions. In other words, the conflation of the pictorial method with the device in Kōkan's writings is taken as the self-evident statement that *explains* why photography would later be called *shashin*.

Yet Kōkan's own attitude toward representation and his practices as a maker of images can account neither for the diverse ways in which this term *shashin* was used from the late eighteenth century onward nor for the array of surviving pictures that were considered by their makers and viewers as *shashin*. Indeed, this obsessive focus on a single genealogy of *shashin* running through the figure of Kōkan, which was only part of a larger discourse on the transposition (*sha*) of the real (*shin*), has flattened out of the complexities and ambiguities that inhered in discourses on representation that cross a number of modern disciplines, ranging from art history proper to the histories of science, medicine, and technology.

The "*Honzō shashin*" evinces precisely this problem with the prevailing scholarly understandings of *shashin* and the histories of visual representation in nineteenth-century Japan, since the illustrations in the manuscript neither accord with Kōkan's notion of illusionistic effect nor were produced with Western devices for picture making such as the camera obscura. It was a meeting in Nagoya in 1826 of some of the future members of Shōhyaku-sha and Phillip Franz von Siebold that led to the German physician's possession of the manuscript now kept in Leiden. Siebold has long been understood to be one of the central figures in the introduction of Western scientific and medical knowledge to Japan. The Shōhyaku-sha occupies a complementary role in accounts of the history of science in Japan, and the group is now primarily remembered and celebrated as a liaison between the "premodern" science of the Tokugawa era and the practices of "modern" science institutionalized by the Meiji government. Yet the process through which Hōbun and the Shōhyaku-sha came to understand and even to transpose "Western science" was very much guided by practical concerns, and dependent on a language and conceptual framework that they brought to, rather than took from, their encounters with "Western science."

The future members of Shōhyaku-sha had requested the 1826 meeting with Siebold in part to inquire after the proper Linnaean names for the plants that would later be represented in "*Honzō shashin*." This meeting superficially validates conventional narratives of Western scientific progress that attach primary importance to the understanding and implementation of scientific theory—whether Linnaean nomenclature or the Copernican heliocentric model of the solar system—in constructing a teleological account of the advance of knowledge.[19] Underlying this narrative of movement from premodern to modern is the assumption that characterizes the study of *materia medica* prior to the introduction of "Western science" as an activity incommensurable—and antithetical—to modern science. These narratives construe instances such as the meeting between Siebold and the Japanese physician-scholars as moments of active transmission by Europeans and passive inculcation of a theory by the Japanese that signifies the "leap" necessary to the development of proper science. Hōbun and Keisuke's investigation of Linnaean nomenclature is thus interpreted as an indication of their farsightedness in anticipating "modern science," rather than a historically driven curiosity that was grounded in their studies of *materia medica.* This understanding obscures the vital roles of practice and representation in mediating and forming knowledge of the world, and it overlooks the consequences of mingling scientific ideas with discourses on visual representation for the practice of science and Japanese visual culture. Methodologically, my approach in this book emphasizes the material as an integral part of scientific *practice* within which linguistic and conceptual understanding of the scientific theory was articulated and secured.[20]

Thus we return to the key term, *shashin,* with which I opened the introduction. At the center of the Shōhyaku-sha's translingual negotiation—and the intersection of epistemologies that it represents—lies this term, which traces its origins back a millennium to Han China. But the term is not easily reducible to any of its constituent parts; it is of Chinese origin used to represent a collection of pictures of Japanese plants. Yet *shashin* is not peculiar either to this manuscript or to this encounter; rather it serves as part of the bedrock of the Shōhyaku-sha's ongoing struggle to reconnect names and things. Indeed, for the Shōhyaku-sha *shashin* referred not to the effects of conjuring an illusion of reality but to attempts to quite literally create a residue of the real itself.[21]

In the Shōhyaku-sha's constant negotiation among the canonized knowledge of Chinese *materia medica,* Western botanical nomenclature, and the actual environment in which members lived, the concept of *shashin* emerges at the very intersection between visuality and knowledge. For Hōbun and other members, the role of direct observation began to gain more currency and value in this process of synthesizing their text-based understanding and their observed physical conditions. In the various types of pictorial representations they produced and studied, the group questioned not only what is represented but also *how* it is represented and *where*. The problem of method carried intellectual implications that were informed by the desire to make the image knowable and intelligible as the representation of something real. The group incorporated and experimented with numerous methods of picture making, including copper etching, hand sketches, copying, woodblock printing, and ink rubbing—many of which we see grouped together in Hōbun's manuscript under the associated meanings of *shashin*. Thus images occupy an extremely important place in the work of the Shōhyaku-sha as these various techniques of pictorializing their subjects helped them visualize, question, and articulate their knowledge in ways that language alone could not.

This is not to say that textual renditions lost significance in the work of the Shōhyaku-sha. Quite the contrary, the pictorial aspect of written language itself became an issue of visuality. The group worked with four languages (classical Chinese, Japanese, Latin, and Dutch) and four scripts (Chinese characters, *hiragana*, *katakana*, and Roman alphabet). While translation served to introduce particular bodies of knowledge, the very mode of writing within which the knowledge was expressed came to play a role as a visual code. The strategic choice to represent a name in a particular script played a decisive role in stabilizing the relationship between the plant and its nominal expression. Thus both pictorial and linguistic representation functioned as visualizations of a relationship between seeing and knowing that became a running theme of their work.

Significantly, some of the members of the group decided to take up photography when it became available to them in the 1860s, and they unhesitatingly called the technology *shashin*.[22] In France and England a variety of new terms were coined to describe and comprehend the new medium of photography. In Japan, something rather different happened: an existing term with a dense semantic history related to representation was co-opted. This raises an intriguing question, especially considering that it has been called by other names, such as *in'ei kyō* (literally, "impressing shadow lens") and *rui'ei kyō* ("arresting shadow lens").[23] To understand the early history of photography, therefore, we need to understand both the semantic history and the history of the very process through which it would become conflated with the new medium. The Shōhyaku-sha's choice of *shashin* to describe photographic technology attests that they identified a continuity between the representational possibilities offered by photography and their own representational practices that supported their studies.

The Shōhyaku-sha emerges, therefore, as a group of key practitioners of seeing, representing, and knowing in nineteenth-century Japanese visual culture. The group offers a chance to reimagine the early history of photography and the visual culture of nineteenth-century Japan in a way that does not rely on the uncritical dichotomy between Japan and the West that has played such a crucial role in many prevailing narratives of visual modernity. Rather, analysis of the oeuvre of the Shōhyaku-sha emphasizes the role of *practice*. In the last twenty years, historians of photography have distanced themselves from a modernist account of the medium—an approach championed by Beaumont Newhall in *The History of Photography* in 1937—and dispensed with the idea of photography as a unified subject for which an internally coherent narrative can be written.[24] At the same time, historians have also begun to examine histories of photography in non-Western contexts. Turning away from universalizing histories of photographic technology, scholars such as Christopher Pinney, Karen Strassler, and Rosalind C. Morris have revealed specific histories of photography in Asian contexts that defy the modernist approach.[25] In these pioneering works, photography is not an autonomous technology that produces consistently hegemonic experience and representations, but rather a parasitic *medium* that latches on to a historical and cultural context to transform preexisting local practices and needs, while also creating new hybrid visual expressions. Rearticulations of photographic history through the lens of intricate social, cultural, and economic relations have produced historical insights that could not be reasoned through the assumption of a Cartesian formation of subject and object as an essentialized tenet of photography. These new scholarships on photographic history share the emerging conviction, best summarized by Pinney, that "photography is a cultural practice with no fixed outcome."[26] Yet because of the elusiveness of Shiba Kōkan's treatise on Western pictorial representations, coupled with the persistent influence of the distorting dichotomy of West versus Japan, writings on the early history of photography in Japan continue to avoid serious engagement with contemporaneous representational and intellectual issues that occupied the minds of the very people who took up the technology.

The practices of the Shōhyaku-sha with a particular attention to their understanding of *shashin* enable us to write a new history of Japanese photography and nineteenth-century visual culture that more accurately captures the complex shifts in which the meanings and practices were embedded in the deeper negotiations between epistemologies of seeing and knowing that characterized the era. Like the lens Siebold brought back from Japan, the lens formed by the term *shashin* marks a specific history. Just as a physical lens alters visual perception in particular ways, so too does the Shōhyaku-sha offer a "lens" that filters our understanding of a visual culture in the past and opens a new perspective on history.

What I offer in my exposition of the genealogy of *shashin* is an account of photographic history in Japan that differs from—and thus questions—the received narrative. *Shashin,* originally a compound word and a concept from China, was more than a term associated with the camera obscura or Western pictorial method. Its multiple uses and significance in the activities of the Shōhyaku-sha reveal how the group deployed the term to negotiate, rearticulate, and most importantly concretize the relationship of plant names, pictorial representations, and the existence and availability of the actual plants. The conceptual history of *shashin* thus reveals that the challenges in representational practices in the Owari domain were not confined to naïve effort at mimetic representation assisted by application of Western pictorial devices and techniques. Rather, the representational challenges that faced the Shōhyaku-sha emerged from the persistent questioning and probing of their knowledge of *materia medica.* The term *shashin* occupied a pivotal position in this particular epistemological discourse as a primary arena for negotiating the conceptual relationship among representations, represented objects, and names. When photographic technology became available, the Shōhyaku-sha had already established *shashin* as a term that designated the fidelity of a representation to the name of the represented object, and to the object itself. This history begs us to reflect upon the implications of conceptual appropriations in studying a history of imported technology. Moving beyond the history of photography as a history of technological adaptation, serious consideration of the very language through which photography was understood in nineteenth-century Japan allows us to recapture the densely entangled meanings and roles of representations.

Each of the chapters that follow addresses one arena where the notion of fidelity played out in the intellectual endeavors of the Shōhyaku-sha: in their profession as physicians, in translations of texts from China and the

West, within organized public exhibitions, in the production of pictorial representations, and finally in the governmental institutions in which certain members of the Shōhyaku-sha were employed. In each chapter, the notion of fidelity is examined in relation to the real (*shin*) and transposition of the real (*shashin*) by examining a set of overlapping practices of the group. The succession of chapters thus builds a cumulative view of the intellectual genealogy of *shashin* from a variety of key perspectives.

Chapter One, "The Eye of the Shōhyaku-sha: Between Seeing and Knowing," opens by situating "*Honzō shashin*" historically on two levels: within the local context of Owari, and within the discursive context of medicine and *materia medica*. Central to this chapter is an account of the role of visuality in medicine and *materia medica*, which foregrounds how a particular mode of processing knowledge was formed: a triangular relationship among textually rendered knowledge, the actual object, and illustrations.

In Chapter Two, "Ways of Conceptualizing the Real: Scripts, Names, and *Materia Medica*," I turn to the issue of translation in the study of *materia medica*. In 1829, Itō Keisuke, a leading member of the group, published *Nominal Differentiations in Western Materia Medica* (*Taisei honzō meiso*), the first published attempt to apply the Linnaean botanical nominal system to Japanese flora. I focus on the process through which Keisuke articulated the Linnaean system in textual representations to show how the visuality of the script itself played a significant role in this publication. The centrality of the concept of the real (*shin*) in Keisuke's articulation of the Linnaean nomenclature is explored by attending to how various scripts employed in this publication solve the epistemological and representational issues that faced the Shōhyaku-sha, and more broadly the study of *materia medica* in nineteenth-century Japan. Rather than evaluating the group's activity against what is assumed to be a transcendent set of values such as scientific accuracy and objectivity, this chapter grapples with the particular local and historical stakes in introducing and understanding the Linnaean nomenclature and the effort to best ensure the transplantation of this abstract system into textual representations.

In the following chapter, "Modes of Observation and the Real: Exhibition Practices of the Shōhyaku-sha," I situate the group within the discursive history of *materia medica* and detail the process in which the faculty of vision began to influence this discourse. Approaching the practice of public exhibition as an innovative response to reevaluate the content of *materia medica*, this chapter focuses on the significance of direct observation and how the exhibitions served as a site of contention and reevaluation of previously shared knowledge by privileging the actual object. This chapter throws into relief specific observational modes, tactics, and strategies explored by the group, a process of negotiation among what and how they saw, what they knew previously, and what they could discern on site.

In Chapter Four, "Picturing the Real: Questions of Fidelity and Processes of Pictorial Representation," I consider two techniques of representation used by the group: copper etching and ink-rubbing prints. Central to the group's understanding of these prints was the idea of the real (*shin*, the second character of the compound *shashin*). The Shōhyaku-sha differed from other contemporary scholarly groups by continuing to use the ink-rubbing method into the Meiji period, referring to these images as shadows of the real (*shin'ei*). I assert that the modalities of existence that these prints articulate emerged out of the group's collective desire for access to a particular plant. Compared to other pictorial media that the group employed, such as woodblock and copper etching, ink-rubbing prints guaranteed and authenticated the fidelity between the object and images with an indisputable certitude.

Finally, in Chapter Five, "*Shashin* in the Capital: The Last Stage of Metamorphosis," I give an account of the Shōhyaku-sha's use of photography itself from the 1860s onward and situate the group's interest in photography within a broader discourse of art history and photographic history. Central to this chapter are the figures of Itō Keisuke and Yanagawa Shunsan, two members of the Shōhyaku-sha who occupied important posts in the scientific establishment of the last years of the Tokugawa period and the early years of the Meiji period; and Takahashi Yuichi, an artist who worked at the same government institutions as Keisuke and Shunsan. By recovering the process through which the Shōhyaku-sha's discourse on the transposition of the real became linked with the new technology of photography, this chapter reopens the problem of what photography meant to its earliest practitioners in Meiji Japan. In this way, I delineate the process through which the meanings of *shashin* merged during this critical period of transition.

Although the Shōhyaku-sha began as a small provincial group, their concern with the relationship between names and things came to occupy a central place within the scientific establishment of modern Japan. As members of the group left Nagoya and took up posts first in the Academy for Western Studies (*Bansho shirabejo*) and then in the Academy of Developing Intelligence for Successful Enterprises (*Kaiseijo*), the forerunner of Tokyo

Imperial University, the discourse on *shashin* that had developed out of the particular concerns of practicing physicians assumed a central role in how the modern world came to be apprehended. This book, then, is not simply a biography or portrait of the Shōhyaku-sha but a study that highlights the interdependence of the history of ideas, scientific practices, and visual culture. By returning to the moment of photography's introduction to Japan, my hope is to show not the inevitable triumph of a new technology in transforming a static visual field but rather the historically contingent ways in which the new technology was understood within an established epistemology and set of representational practices, which unsettles the very equivalence that came to be established between *shashin* and photography. Indeed, what the case of Japan allows us to see is that the history of photography is also a history of photographic technology and practices being understood and used in particular places at particular times. It should allow us to return to the various sites of photography's "origins" to reexamine the history of photography not only as a technology but also as a particular site of negotiation between seeing and knowing.

ONE

The Eye of the Shōhyaku-sha

Between Seeing and Knowing

> Prioritizing a theory over an actual object is similar to ignoring wise advisors. By experimenting from the start with objects, any average person will be able to come to a fuller understanding of an object.
>
> Yamawaki Tōyō, *Thoughts on Internal Organs* (*Zōshi*, 1758)

The interest of the members of the Shōhyaku-sha in the world around them was governed by a set of practical concerns related to the medical efficacy of the local flora. But this practical bent in the approach of such figures as Mizutani Hōbun and Itō Keisuke was neither inevitable nor even natural. Indeed the medical world of early modern Japan, at the very moment at which the Shōhyaku-sha emerged, was marked by a tension between two dominant approaches: one reliant on the deep textual tradition of Chinese medicine, and the other an emerging approach indebted to Western anatomical science. What both of these approaches shared was an attempt to understand the living world through recourse to the world of printed books. In this context, experimental human dissections—initiated by both traditions of Chinese and Western medicine—in the late eighteenth century radically shifted the channels of medical understanding by adding the actual and the real human body as the source of their medical knowledge. Where the Shōhyaku-sha departed from these practices was in an abiding concern that emerged in this group to reinterrogate the textual record in relation to lived experience and the environment, a concern that itself emerged from the daily demands placed on its members as practicing physicians. In this light, recording and preserving what they deemed to be the real (*shin*) encompassed varied activities that included their study of textual resources, excursions to nearby mountains and villages, experimentation, and comparisons of their findings.

To understand the thought of the Shōhyaku-sha and the place that the transposition of the real (*shashin*) assumed within it, we need first to grasp the social world that its members occupied. "*Honzō shashin*," the Shōhyaku-sha manuscript that Siebold brought with him when he left Japan in 1829, was a product of this world and serves as a point of entry for reimagining the intellectual and social environment within which Hōbun and his contemporaries operated.

Mapping "Honzō shashin"

On March 29th, 1826, the core members of the Shōhyaku-sha—Mizutani Hōbun (1779–1833), Itō Keisuke (1803–1901), and Ōkōchi Zonshin (1796–1883)—met with twenty-eight-year-old Philipp Franz von Siebold at the local Atsuta shrine in Owari. Trained as a physician, Siebold was accompanying Dutch factory chief Johan Willem de Sturler to pay the obligatory visit to Shogun Tokugawa Ienari in Edo castle.[1] The Edo castle served as the headquarters of the shogunal government, known as *bakufu,* which reigned from 1603 to 1868. For their meeting with Siebold, the Shōhyaku-sha members prepared a variety of samples that included plants, fruits, minerals, and dried fish as well as illustrations. During their meeting, the Shōhyaku-sha representatives asked Siebold to examine the objects they brought and identify them, affixing to them Latin and Dutch names. The Shōhyaku-sha were fully aware of the Dutch investment in collecting and understanding Japanese fauna and flora and wanted to understand the place of these local objects within a larger classificatory system.[2]

In May of the same year, the Shōhyaku-sha met with Siebold again when the Dutch envoys were returning to Dejima and exchanged more information as well as samples with the German physician. Siebold describes this encounter in "Journey to the Court of the Shogun in 1826" ("Reise nach dem Hofe des Sjogun im Jahre 1826").[3]

> I met there Mizutani [Hōbun] with whom I have been exchanging letters from Dejima, and a few others who do not specialize in medicine but whom I have asked to collect specimens on my behalf before. It is then that I first met Itō Keisuke and Ōkōchi Zonshin, both of whom later came to contribute greatly to my research. [Hōbun] ... brought a wide range of objects to

> the meeting. Among them was a collection of plants, mainly from the area of the Shrine, which had been made into a unique pressed specimen, accompanied by their names in Japanese and Chinese. There are some fruit from China and Japan included in these illustrations as well.[4]

The collection of "pressed specimens" that Siebold describes here is clearly the manuscript "*Honzō shashin*" now at Special Collection of Leiden University Library, which is described in the catalogue of that Siebold collection, compiled in 1845, as "Honzō shashin, botanical sketches according to nature drawn by Mizutani Sukeroku [Sukeroku is Mizutani's common name]" ("Honzō sjasin, adumbrationes botanicae ad naturam del. *Midsutani Sukerok*").[5] The manuscript came to Leiden as a part of the first installment of the Siebold collection, donated after he was expelled from Japan in 1829 and returned to Europe, finally settling in Leiden in 1831.[6]

The manuscript includes fifteen illustrations of native Japanese flora, including poisonous plants (such as *kashioshimi Lyonia ovalifolia*, subspecies *Neziki*) as well as edible ones (*kuzu Pieraria lobata* and *nadeshiko Dianthus superbus L. var. longicalycinus*; Figs. 1.1 and 1.2; see also color plate 1).[7] Of the fifteen images, four depict fruit—apple, loquat, fig, and apricot—with their branches. Seven illustrations portray plants valued as *materia medica* for their medicinal properties, and two others depict plants that were not included in canons of *materia medica*.[8] The size of the paper approximates *Oban* (39 x 26.5 cm) and is rather large for the period, when most published illustrations of *materia medica* plants were printed in *Chūban* (26 x 19 cm), approximately half of Oban. In addition to these hand-painted illustrations, the manuscript includes an ink-rubbing print and a pencil outline of flowers and leaves of *nishikiran* (*Goodyera schlechtendaliana*), one of the plants that caused a gardening frenzy at the beginning of the nineteenth century. Each illustration is accompanied by inscriptions giving the Japanese name of the plant in *katakana* script and the Chinese name in Chinese characters.

At first glance these appear to be fully hand-painted illustrations. Close inspection, however, reveals that they were, in fact, created by a rather unusual method. Deep indentations, clearly visible under natural light, surround each illustrated plant like an outline. In some cases, the paper even bears the fine impressions of the stamen and pistil, details best seen in the spidery network of ridges on the reverse of the pages (Fig. 1.2, reverse side of Fig. 1.1). These illustrations look markedly different from illustrations

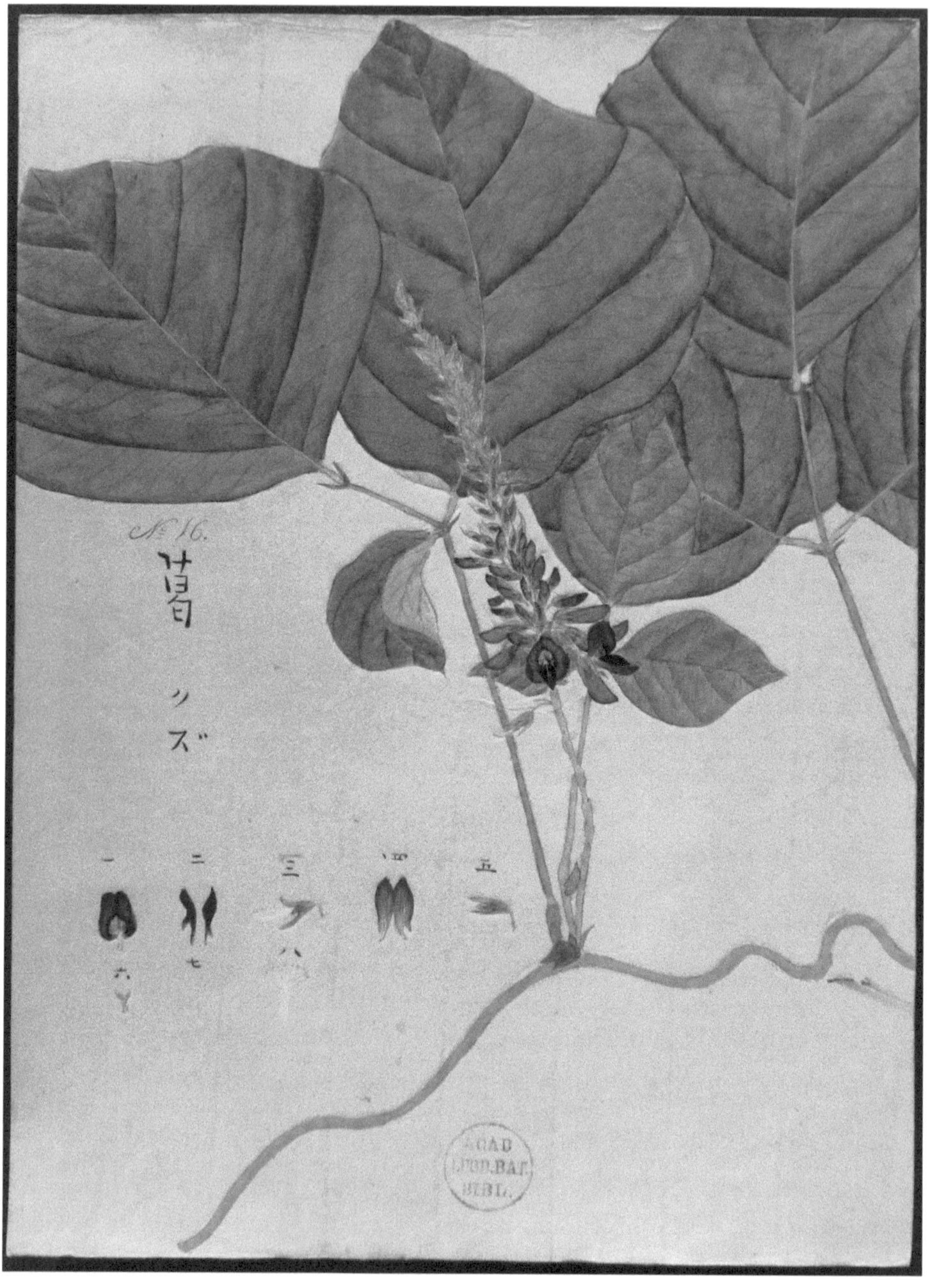

Figure 1.1 Kuzu, from "*Honzō shashin*," before 1826, ink on paper, Leiden University Library

used and trusted in *materia medica* around this time (Fig. 1.3, from *Ka'i* [1765], by Ono Ranzan, Mizutani Hōbun's teacher). Traditional depictions of *materia medica* plants had relied heavily on outlining to delineate the contours of the plant, effectively flattening the depicted plant. Comparatively, the illustrations in "*Honzō shashin*" (Fig. 1.1 and color plate 1) use little outlining and instead employ gradational hues to sculpt the plant. The monochromatic woodblock prints necessarily relied on two colors to differentiate the front and back of the leaves, which further creates a stark tension in the crowded areas of leaves, but multicolored depiction delineates the two sides more ambiguously. In fact, in some areas the front and back of the leaves are difficult to discern. The rectangular frame that traditionally marked the pages of published illustrations demarcates the representational space and gives a sense of confinement and separation. The two representations in Hōbun's manuscript, though, suggest a rather different sense of space. The gradual differentiations in hue create less visual contrast,

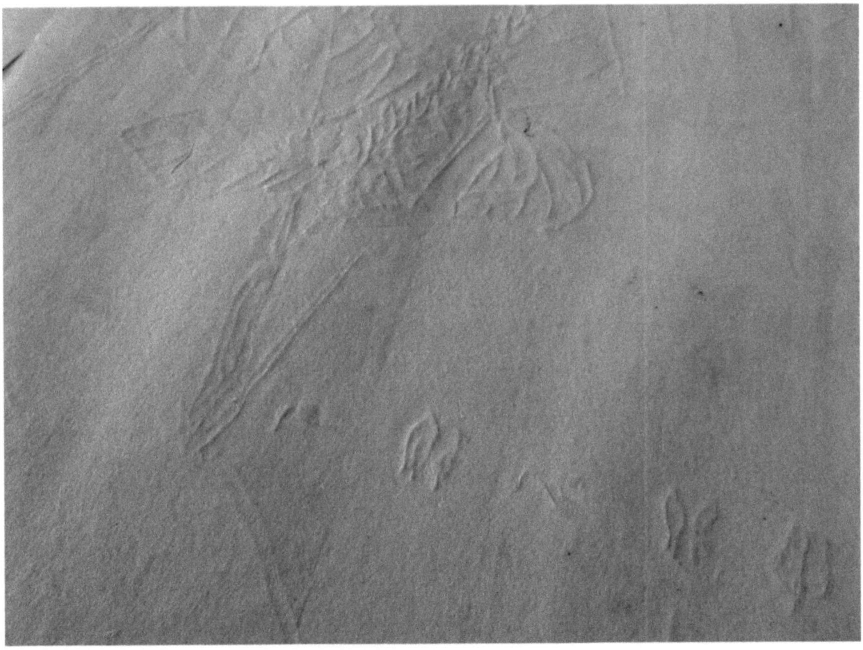

Figure 1.2 Recto, *Kuzu*, from "*Honzō shashin*," before 1826, ink on paper, Leiden University Library

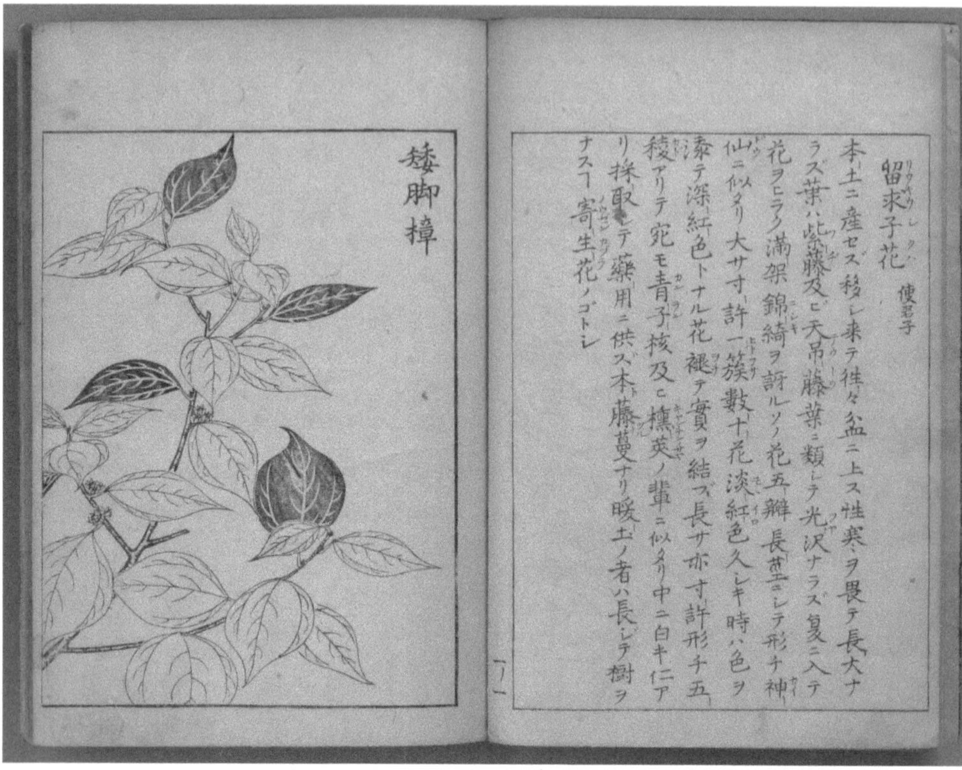

Figure 1.3 Ono Ranzan, *Waikyakushō*, from *Ka'i*, 1759–1765, woodblock print, Waseda University Library

deemphasizing the difference between the front and back of the leaves. The lack of framing allows the representations to appear less controlled and confined. The impressions visible on the reverse of the pages (Fig. 1.2) indicate that these unusual qualities of the illustrations stem from an innovative attempt to transpose visually living plants onto the pages. These impressions, together with the fact that the fifteen illustrations closely follow the size of the plants, suggest that the outlines were produced from actual specimens, a process of illustration unknown in any other Japanese botanical manuscript.[9] For each image, the plant seems to have been impressed directly into the paper first, providing an indented outline that the colorist then painted inside with the appropriate colors. Although none of the illustrations bear signatures or seals, the entry for the manuscript in the 1845 catalogue of the Siebold collection indicates that Hōbun, perhaps in collaboration with others, was the maker of these singular images. The details of his exact process remain obscure. If he used dried specimens and a pressing machine to create the indentation on paper, he would not have been able to attain the fine details from fragile parts such as the stamen and the pistil. Yet if Hōbun had used fresh plants with a press, the paper would have absorbed some of the water and nutrients seeping from the stem of the plant from the applied pressure. Alternatively, he might have copied another illustration and traced the outline by hand with enough pressure to make indentations on the blank paper, but if so, the illustrations that they would have used have not been found.

This unusual method of illustration resembles, and was perhaps even inspired by, the technique of ink-rubbing prints, a medium to which, as we will explore in Chapter Four, the members of the Shōhyaku-sha devoted much energy and time. The technique involved inking a pressed specimen and taking an impression of it on paper; an example is included in the manuscript "*Honzō shashin*" (Fig. 1.4). Considering their ongoing interest in incorporating the actual plant as a way of documenting and accumulating knowledge of *materia medica*, it is highly likely that they used a previously prepared specimen and put the plant into a press machine to produce these illustrations. The minute details of the illustration, however, would have to be drawn in with high pressure separately by hand.

The "pressed specimens" that Siebold describes in his account of the 1826 meeting with Hōbun and other members of Shōhyaku-sha affirms the material evidence of the pictorial process found in "*Honzō shashin*": the illustrations were made with a unique method of impression using plants common in Owari at that time. The descriptions of the inscribed names in Siebold's account from the first meeting also correspond to the illustrations in the manuscript. Correlating Siebold's accounts of the meeting in 1826, we begin to see "*Honzō shashin*" as an object embedded in a particular moment and location. This manuscript, then, served to facilitate transmission and reevaluation of knowledge between the Shōhyaku-sha and Siebold, and thus it offers us a window through which to examine what was at stake for the Shōhyaku-sha in their pursuit of *materia medica*. In other words, "*Honzō shashin*" carries with it a historical residue that sets itself at the intersection of inveterate and new epistemological frameworks and practices.

The two meetings with a German physician in 1826 embody, both figuratively and literally, three interrelated dimensions of the broader historical climate that conditioned scientific discourse in early-nineteenth-century Japan. The first is an awareness of the diversity of ways of thinking about

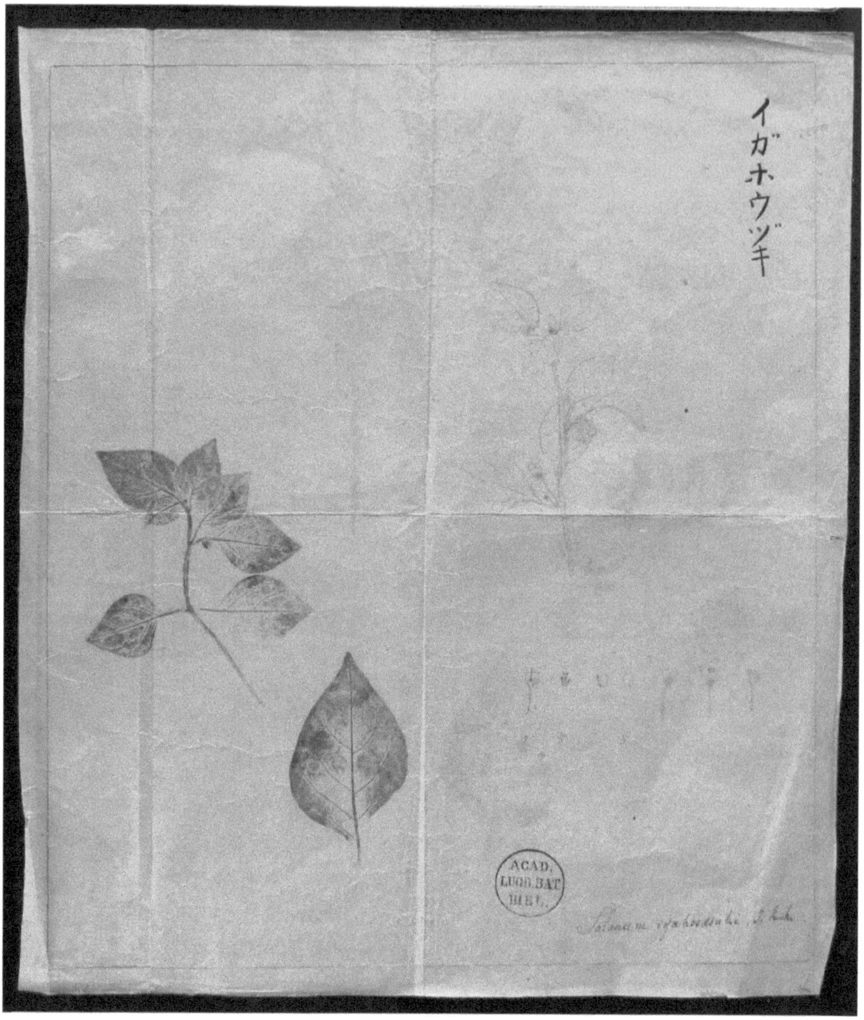

Figure 1.4 Igahōzuki, from "*Honzō shashin*," before 1826, ink rubbing and pencil on paper, Leiden University Library

the natural world, exemplified by Linnaean taxonomy. The second is an understanding that situating local Japanese flora within the broader systematic approach offered by Western botanical taxonomy might grant insights into the medicinal properties of native plants. And the third is an intense interest in physical objects as a means of reexamining and recalibrating textual knowledge derived from the canon of Chinese *materia medica* but removed from the living world around them.

According to Siebold's account of his meeting with the Shōhyaku-sha, the group presented him with a series of illustrations that already bore plant names based on Linnaean taxonomy. Further evidence of the group's familiarity with the Linnaean system prior to their encounter with Siebold comes from the fact that of the 102 examples they presented, only four were wrongly classified.[10] But what would be the benefit of knowing a plant's place within the Linnaean system for someone who already knew its Chinese and Japanese names?

To understand what was at stake for the Shōhyaku-sha in their meeting with Siebold and their interest in Linnaean taxonomy, let us turn to one particular example from the first meeting, Hōbun's inquiry into the Linnaean classification of a plant known as *hashiri dokoro* in Japanese.[11] For physicians trained in the Chinese traditions, the plant known in Japan as *hashiri dokoro* was understood to be a local variant of the plant known as *rōtō* (Ch: *langdang*) in Chinese *materia medica*.[12] By the time of the 1826 meeting, however, some Japanese scholars identified *rōtō* with the tobacco plant.[13] To investigate the relationship between *rōtō*—and hence its local variant *hashiri dokoro*—and tobacco, Hōbun obtained tobacco seeds from Siebold and began an experiment that allowed him to compare and note the differences between the plants.[14] This exchange of seeds and the study of mature plants complement the role of the illustrations in "*Honzō shashin*" by underscoring the decisive role that vision and observation played in the pursuit of medicine and pharmacology in nineteenth-century Japan. For the Shōhyaku-sha the encounter with Western natural history was critically mediated not only by text but also by object and image.

Another facet of Hōbun's questioning of the identification of *hashiri dokoro* and *rōtō* reveals that this emphasis on observation was intimately paired with the desire to verify the names of plants. Scholars familiar with the traditions of Western medicine argued that the *rōtō* of Chinese texts such as the *Compendium of Materia Medica* (*Bencao gangmu*) in fact corresponded to the plant known as *belladonna* in imported Western texts. Accurate linking of names and physical specimens, a process mediated through text, constantly preoccupied the Shōhyaku-sha, since these linkages facilitated coordination of and access to knowledge. For physicians like Hōbun and Keisuke, however, this was not simply an abstract question of classification. In Chinese *materia medica*, *rōtō* was identified as a poisonous plant, consumption of which could lead to death. In contrast, the translations of Western herbal books described *belladonna* as a plant used for mydriatic treatment of the eye (Figs. 1.5 and 1.6).[15]

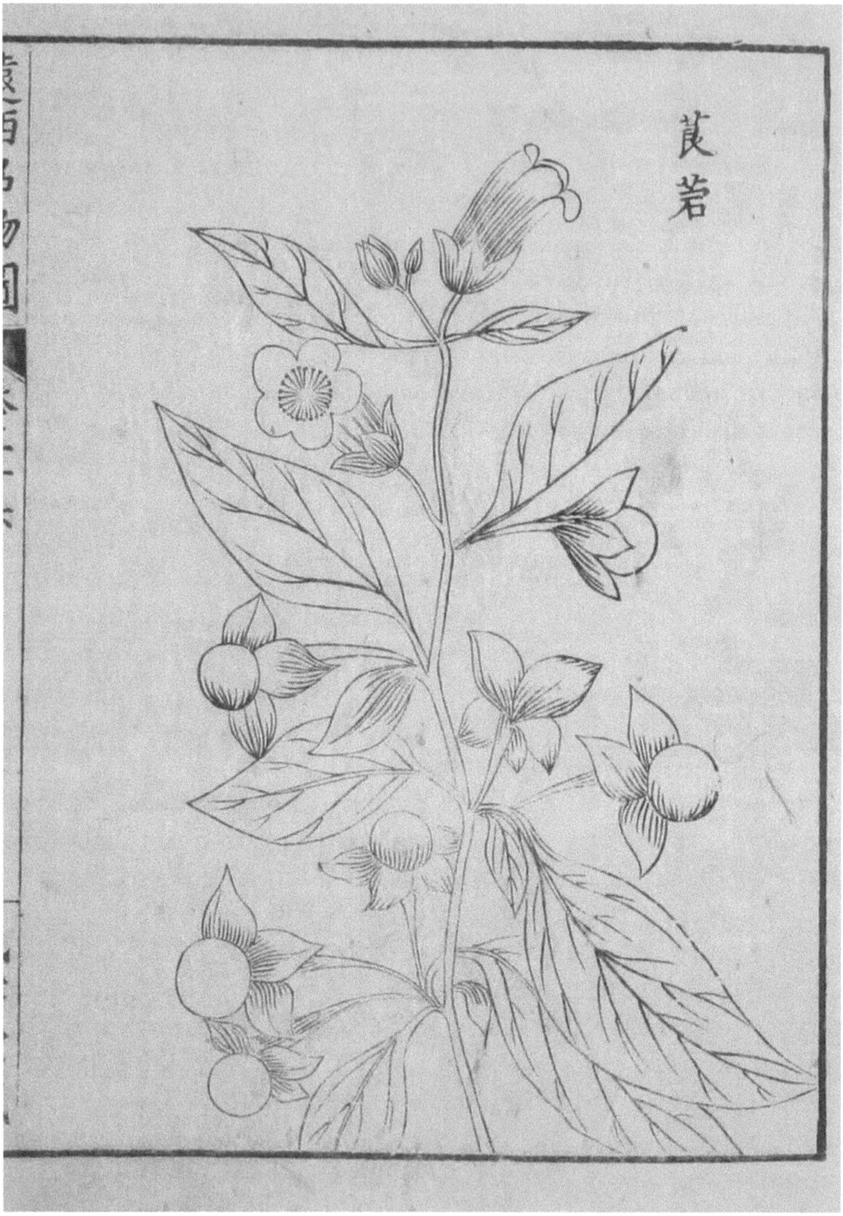

Figure 1.5 Udagawa Genshin, *Rōtō*, from *Ensei ihō meibutsu kō*, 1822, woodblock print, Waseda University Library

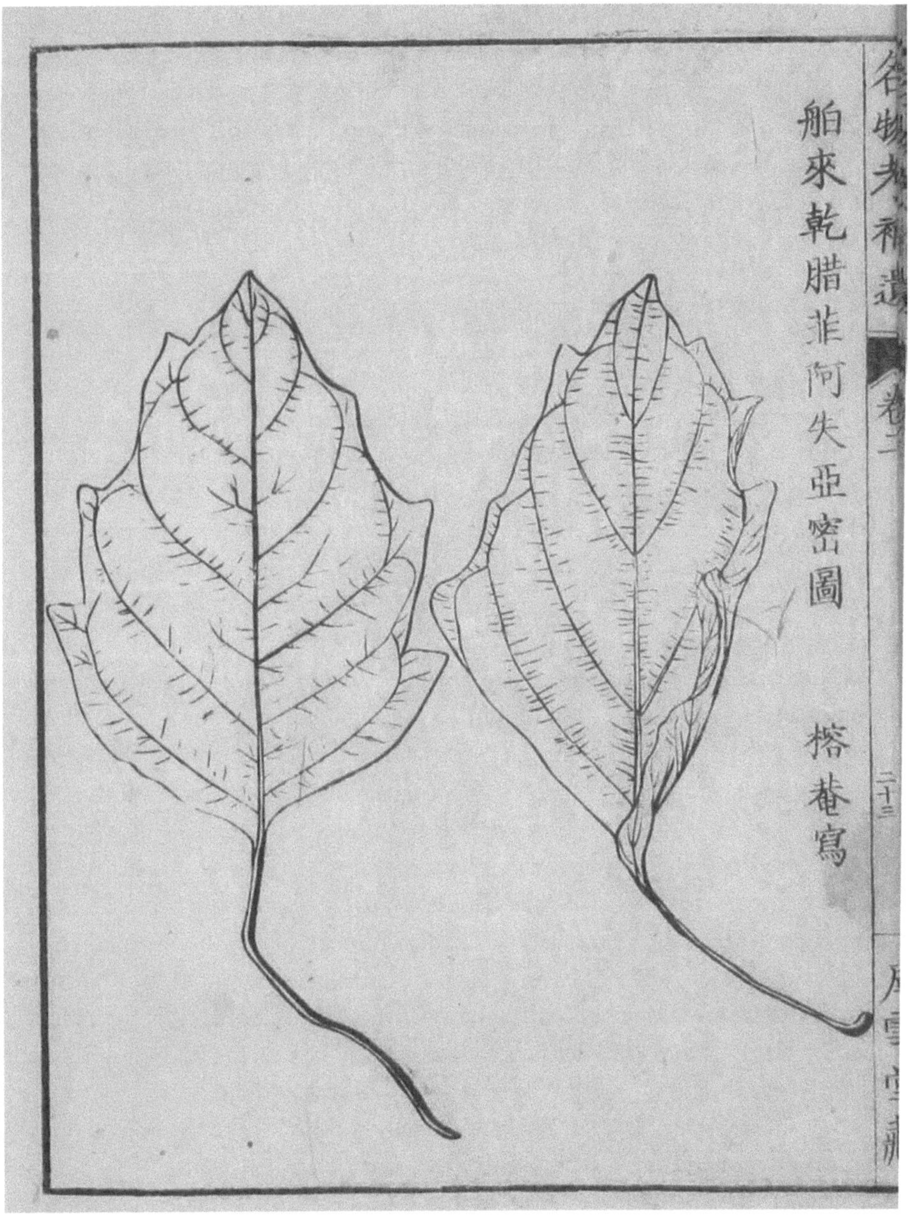

Figure 1.6 Udagawa Yōan, *Hiyosu*, from *Ensei ihō meibutsu kō hoi*, 1834, woodblock print, Waseda University Library

Proper naming thus had very practical consequences for the physicians: if *hashiri dokoro* was correctly identified as *belladonna*, then it could be used medicinally so long as the remedy was prepared properly.[16] As this example demonstrates, the allure of Linnaean taxonomy for Shōhyaku-sha rested ultimately not in the theoretical and abstract systematization of plant life that constituted the novelty of its approach but in the practical, almost pedestrian, knowledge that the taxonomy brought to treatment.

Medical Knowledge and "the Real"

It was around the time of the 1826 meetings that the Shōhyaku-sha began to gather regularly and adopted their collective name. Their scientific exchanges with Siebold, moreover, indicate the questions and issues that would guide the group's activities for the next half century. Along with Hōbun's inquiry into the nature of *hashiri dokoro*, another illustrative example of the group's concerns comes from their interest in a number of wild grasses commonly found in the environs of Nagoya. The "*Honzō shashin*," for instance, includes two illustrations of local plants, known in Japanese as *sawagiku* and *igahōzuki*, that existed entirely outside of the canon of *materia medica* and whose medicinal properties were consequently unknown. Both plants appear in Hōbun's 1809 *Clarification on the Names of Things* (*Buppin shikimei*), a compilation of names that correlated Chinese names with their Japanese counterparts. But both were listed under only their local Japanese names and lacked a Chinese analogue, suggesting that the plants existed essentially outside of medical knowledge. What Hōbun sought from Siebold was a way around such lacunae that enabled him to investigate the medicinal properties of these sorts of common grasses that held potential for pharmaceutical use. Hōbun and his colleagues were interested not in abstract placement of these grasses within a larger classificatory system for its own sake but in understanding the practical uses of the plants in medicine.[17]

The Shōhyaku-sha's interests were rooted in a practical drive to find and apply knowledge that served a curative purpose. In this sense, one needs to situate the group in relation to the emergence of Practical Learning (*jitsugaku*) in the late seventeenth century and the crisis of knowledge that permeated eighteenth-century thought.[18] The advocates of Practical Learning stressed the importance of engaging with the actual world and found themselves "in opposition to false learning [*kyogaku*]," targeting the metaphysical doctrines of Buddhism and Daoism in particular. Bound up with the Neo-Confucianism of Chu Hsi, which had been central to early modern Japanese epistemology, Practical Learning derived from the desire to establish a position from which to ascertain moral certitude for the self in the world.[19] Within the thought of Practical Learning, nature and history were the two main sources by which an individual could establish her or his own position in the world. Those who looked into history, exemplified by Ogyū Sorai, read and reinterpreted ancient Chinese texts; those who turned toward nature as their guiding source proceeded with an assumption that nature provides "principles" that in turn affirm what Tetsuo Najita calls "natural ontology."[20] In this view, nature was identified as the ultimate and unlimited source of knowledge, and practical engagements with nature, such as the pursuit of farming, fishery, and astronomy, served as experimental practices that helped extract and clarify natural principles. In this light, the interest of Shōhyaku-sha in the Linnaean system appears more as an extension of the practices and concerns of Practical Learning than a desire to mimic "Western science."

The core members of the Shōhyaku-sha were trained physicians; their concerns with questions of medical efficacy need to be understood in relation to the intellectual impulses of Practical Learning. Of the twenty-seven members whose names appear in the record of a gathering in 1835, nine worked as physicians and one was a pharmacist.[21] Equally important to our understanding of the group, moreover, is the fact that these physicians were trained in different medical traditions. The founder of the group, Mizutani Hōbun, for instance, was trained in Chinese medicine and then later studied Western medicine with Nomura Ryū'ei in Nagoya in the 1790s. Itō Keisuke, too, first studied and practiced Chinese medicine and then began studying Western medicine in 1821 with Fujiwara Fuzan in Kyoto. Some members of the group, on the other hand, were trained in only one tradition or the other: Shibata Dōgen and Ōkōchi Zonshin, Keisuke's biological brother, practiced Chinese medicine, while Yoshio Jōsan studied Western medicine.[22]

The members of Shōhyaku-sha navigated the differences between the two medical traditions of China and the West with surprising ease, however unbridgeable the differences may seem in historiographical hindsight.[23] Despite each tradition's distinct approach to fundamental aspects of medical knowledge and practice—identification and exploitation of medicinal plants, mapping of the human body, and diagnostic method, for instance—Shōhyaku-sha worked comfortably in practice between Chinese and Western medicine. Indeed, the members of the group never highlighted or commented on the two traditions as systematically different or irreconcilable. In their publications, letters, and diaries, the members discuss how these traditions differ, but there is no substantial elaboration of how the differences are conflicting or problematic in their professional activities.[24] The organization of knowledge, the observation of the real, and the exploration of new methods of representation exemplified by "*Honzō shashin*" provided the necessary foundations for Shōhyaku-sha's negotiating and merging of traditions in their medical practices. In their eyes, the two approaches to medicine seem simply to coexist as two modes of treating illness.

The Medical Environment in the Owari domain

The case of Keisuke, Hōbun's younger colleague, demonstrates the fluid nature of medical knowledge in the Owari domain in the early nineteenth century. In the 1860s, Keisuke would become a central figure in the scientific establishment of the shogunate and then the new Meiji state. But in 1824, two years prior to the Shōhyaku-sha's meeting with Siebold, Keisuke was a young physician of twenty-two attempting to make sense of the various medical discourses and practices circulating in Japan.

Keisuke had been trained in the school of "latter-day" medicine practiced by his father and had spent a year in Kyoto learning Western medicine. When he returned from this training, three schools of medicine existed: the school of "latter-day" medicine *(goseihō*, literally, methods of later generations), the school of "old way" medicine (*ko'ihō*, literally, the method of classic medicine), and Western medicine (*ranpō*). The school of "latter-day" medicine had been practiced in Japan since the later thirteenth century and was expanded and institutionalized by Manase Dōzan during the sixteenth century. The traditions of the school of "old way" medicine looked back to classic texts of Chinese medicine, specifically *On Cold Damage* (*Shanghanlun*) written by Zhang Zhongjing in the late Han dynasty (25–220 C.E.). The school of "old way" medicine emerged in the mid-Tokugawa period in opposition to the school of "latter-day" medicine. Whereas the school of "latter-day" medicine involved multifaceted theories based on natural phenomena explored and articulated in China during the Jin (1115–1234 C.E.) and Yuan (1128–1368 C.E.) dynasties, the school of "old way" medicine was marked by a move away from the school of "latter-day" medicine and a return to the classic texts aiming to simplify the theoretical framework.

In the 1824 draft of a farewell address he delivered on the occasion of the departure from Nagoya of his friend Asai Tokuin, Keisuke wrote of the difficulties facing physicians who were working at the intersection of different medical traditions and epistemologies. In his farewell, Keisuke first outlines how these three medical traditions with which he was acquainted had their own limits and disadvantages. He then explains that his own approach was "recently to read a myriad of different books and to incorporate my experiences of Japanese as well as Western medicine."[25] Keisuke continues: "When one reaches the point where one efficiently practices measures flexibly in any given situation, whether or not life is restored back to a nonbreathing body or a withering body is reinvigorated, is entirely reliant on the physician."[26] In other words, in Keisuke's mind the role of the physician is to concentrate on matters of life and death. The origin of the particular medical intervention in one tradition or another is irrelevant.

In the 1824 draft, Keisuke laments how some physicians immediately accept the new methods of Western medicine and discard classical Chinese teachings altogether. Those who agree with the popularized Western medical teachings, he writes, try to find substitutes for the actual Western medical remedies that were difficult to mix. As a result these practitioners end up concocting remedies that are informed only by their own conjectures. But equally problematic to uncritical and dangerous absorption of Western medicine, Keisuke continues, is a situation in which the adherents of each tradition insist on their own inherent correctness to the exclusion of all others. The students who study medicine, as a result, experience nothing but confusion.[27] For Keisuke, the world of medicine around him had lost focus on its professional goals and instead had devolved into a petty debate about intellectual superiority totally divorced from questions of medical efficacy.

Keisuke's own approach to medicine needs to be situated within the medical practices of the Owari domain in the eighteenth and nineteenth centuries. Situated within the broader historical tension that surrounded the profession, we begin to see the ways in which the Owari domain would give rise to a heterogeneous group like the Shōhyaku-sha. The domainal medical school, *Igakkan*, was dominated by the Asai family beginning in 1725, when the Owari *daimyō* Tokugawa Tsugutomo recruited Asai Tōken from Kyoto and appointed him as a domainal physician and awarded him a government stipend (*han'i*).[28] Although the lectures at the medical academy drew from foundational texts of the school of "latter-day" Medicine, examinations also tested students' knowledge on *materia medica*, using *Compendium of Materia Medica* (*Bencao gangmu*; Jp: *Honzō kōmoku*) as the primary textbook.

When Keisuke became a physician in the early nineteenth century, the Owari domain possessed a flexible system for evaluating physicians that distinguished it from other regions of Japan. For instance, the domain did not privilege a hereditary system in selecting and training physicians, as was common in other domains. The Owari, moreover, had implemented an internal infrastructure to promote physicians to higher ranks regardless of their family background or the tradition of medicine that they practiced. By the beginning of the nineteenth century, a biannual examination was administered by the domain-appointed physician, Asai Tei'an (1770–1828), which ranked more than five hundred physicians in the domain so that the Asai family could make appropriate recommendations to the *daimyō*.[29] The institutionalization of medical practice in Owari recognized physicians on the basis of their practical medical skills and knowledge, not their family ties or allegiance to a particular medical tradition.[30]

During his tenure as the head domainal physician, Tei'an not only permitted but encouraged the activities of the Shōhyaku-sha. Tei'an himself wrote the introduction to the 1818 *Critical Methods of Western Internal Medicine and Surgery* (*Oranda naigai yōhō*) by Yoshio Jōsan, a physician trained in Western medicine and a member of the Shōhyaku-sha. In this introduction, Tei'an ridicules those physicians who renounce Western medicine with only a superficial understanding of its methods; he writes that, as long as one grounds oneself in the teachings of the ancient sages in China, there is much to be learned from Western medicine.[31]

But as the Shōhyaku-sha began its activities in earnest—the collective gatherings, excursions, and exhibitions that will be discussed in the following chapter—the institutional climate of the domainal medical academy changed radically after the death of Tei'an in 1828, merely two years after the open intellectual exchange between Siebold and the members of Shōhyaku-sha, and appointment of his son Shizan in place of Tei'an. Under Shizan the academy adopted a different stance toward Western medicine. As new head domainal physician, Shizan criticized Western medicine as a barbaric and harmful (*gaiari*) method. Knowledge from the West became increasingly politicized in the latter half of the 1830s, especially after 1838, when the U.S. merchant ship *Morrison* came to Japanese shores to deliver Japanese castaways in exchange for a commercial treaty.[32] Following the *Morrison* incident, a fear of Western presence spread among the ideologues of the shogunate in Edo, encouraging tighter control over the spread of Western books and teachings. In particular, the Neo-Confucian Hayashi family, the official tutors to the shogunate, pressured the shogunal office to limit dissemination of Western learning (*rangaku*), including Western medicine.[33]

In Owari domain, the denunciatory statements of Asai Shizan acutely politicized the practice of Western medicine in Edo. What had been a flexible and inclusive atmosphere in the domainal medical system under Shizan's father Tei'an was transformed into a polemical and contentious one by his son. Physicians now became segregated along the lines of the main medical traditions. It is within this politicized state of medical practice that Keisuke published *A Rare Book on Vaccination from England* (*Igirisukoku shutō kisho*) in 1841 and began to offer monthly vaccination service to the public in 1850.

The Quest for Medical Efficacy

Keisuke's promotion of vaccination exemplifies his conviction that efficacy—not allegiance to a particular medical tradition—was of paramount importance to the work of the physician. In practicing vaccination, Keisuke implemented the heterodox approach he had advocated early in his career in his farewell address, which would continue to guide his career as a physician until the 1860s, when he moved to Edo as a scholar of natural history and stopped seeing patients. Many of Keisuke's publications are dedicated to an effort to intervene at critical times of famine and epidemic: *Listing of Edible Plants in the Times of Poor Crop* (*Kyūkō shokumotsu benran*, 1837) during the Great Tenpō Famine (1833-1839), *A Rare Book on Vaccination from England* (1841) for small pox vaccination, and *Information for Nonspecialists in Regard to Cholera* (*Bakushabyō teate shirōto kokoroe sho*, 1863) during the cholera epidemic, are vivid examples of this *modus operandi*.[34]

In 1841 Keisuke revised a Chinese book on vaccination by providing *kunten* reading marks to a Chinese text, thereby rendering it readable in Japanese.[35] The original Chinese text was itself a translation from English to Chinese of a series of lectures by the British orientalist George Stauton, published in China in 1805.[36] Keisuke published the translated book, titling it *A Rare Book on Vaccination from England*, at his own expense. It was one of the few books on smallpox vaccinations published in Japan at this time before an effective vaccination was available.[37] Keisuke undertook this work at precisely the moment when Western medicine became politicized and its practices attacked as barbaric within the official medical establishment of Keisuke's own domain.[38] The publication and Keisuke's perseverance in implementing vaccinations demonstrate a foundational focus on practical health care that dominated Keisuke's approach to medicine and that was broadly shared by members of the Shōhyaku-sha.

By 1849, the shogunate had officially denounced Western medicine except for the purposes of surgery and curing eye disease.[39] The shogunal medical academy argued that the methods and cures offered by Western medicine were inappropriate for Japan, since the natural environments of the two regions differed greatly.[40] This sentiment was echoed, even magnified, by Asai Shizan in the Owari domainal school, when he claimed:

> What can vaccination be other than mudding of the sacred land and disrespecting the sages of medicine! Inserting cowpox into the people of our sacred country proves nothing other than the fact that finally people have fallen into the category of brute animals. Those [who practice vaccination] are nothing other than insurgents who want to take over the country. They are dogs and pigs.[41]

With the attacks by Shizan still reverberating in the background, Keisuke continued to practice his medicine as he had before, as well as his study of *materia medica* as a member of the Shōhyaku-sha through exhibitions and excursions. In late 1849, Shibata Hōan, a physician trained in Western medicine, was traveling from Nagasaki to his home domain of Mito to deliver the cowpox virus in the form of scab. He stopped in Owari to vaccinate Keisuke's daughter, and through him Keisuke obtained the vaccine.[42] He subsequently tried to promote vaccination of small children, but his effort met with obstacles. Despite Shizan's hostility, coupled with public suspicion over the safety of the vaccination, Keisuke paid poor families to vaccinate their children, while also sending assistants to nearby villages in search of children who needed to be vaccinated.[43] Immediately after he opened a monthly vaccination service at his own home, Keisuke was questioned by domainal officials about his vaccination practice and in 1852 was asked by the Owari domainal lord Tokugawa Yoshikatsu to head an officially sanctioned vaccination center. The success and spread of Keisuke's vaccination practice persuaded the domainal officials, and it gave him the stamp of approval from the domain. Indeed, what runs throughout Keisuke's medical activities, here exemplified by his role in promoting vaccinations, is a commitment to incorporate new and effective methods regardless of the political cost, a stance he advocated for early in his career.[44]

The importance of efficacy also guided Keisuke's approach to knowledge. In 1829, only a few years after the meetings between Siebold and Shōhyakusha's activities that produced the "*Honzō shashin*," Keisuke published *Nominal Differentiations in Western Materia Medica* (*Taisei honzō meiso*), the first printed book to introduce the Linnaean taxonomy in Japanese. In the preface he writes:

> Scholars in Japan have begun studying Western theories in areas from astronomy, calendar making, and geography to practices of internal and surgical medicine. [The knowledge of the Western theories] has become sufficient because of these efforts. However, the area of *materia medica* is different because available plants differ depending on the region. Therefore, the previous studies [of *materia medica*] are predominantly filled with distortions, discrepancies, and misinformation. It follows that although there have been innovative studies and publications, our study [of *materia medica*] is insufficient. This shortcoming is the fault of us, the medical professionals. Illness is most urgent for us. Following Western studies, our primary job lies in curing and helping those with serious illness and nowhere else. If we hope to cure illness, there is no choice but to master this Western theory. In cases of astronomical and geographical theories, even if there were differences of opinion, these [differences] do not result in injuring a living spirit. In cases of *materia medica*, however, even a small misunderstanding can immediately lead to harm.[45]

Keisuke recognizes and accepts that the ultimate stake in the practice of medicine is human life. This deeply ethical concern trumps any ideological allegiance or claims to truth of a particular medical tradition in the quest to find appropriate and effective methods to sustain human life.[46]

The classifying categories, such as school of "old way" medicine, school of "latter-day" medicine, and Western medicine, are useful to the extent that they work as signposts to permit a generalized view of a given historical landscape. However, as Keisuke's career demonstrates, these categories did not necessarily function as the defining categories through which Keisuke and other physicians identified themselves. While Shizan reduced the category of Western medicine to a stereotype that he wielded as a polemical device to further his own position, Keisuke worked beyond the confines of a particular tradition with a fluidity anchored by ethical conviction to help save lives. The content of Keisuke's unfinished manuscripts demonstrates this commitment. His unpublished papers include five volumes of lecture notes on the Han dynasty text *On Cold Damage*, a classical textbook of the school of "old way" medicine, as well as a draft of a translation of a work of Western anatomy. Also included are notes of Asai Tei'an's lectures on *I-ching* and a translation of theories of Western medicine.[47] And Keisuke's diaries, finally, which cover a span of more than sixty years, witness his unabated and rigorous study of *materia medica*.

Triangulating the Sources of Knowledge: the Real Object, Illustration, and Texts

To better understand the particular concerns and intellectual problematic faced by Keisuke and his colleagues, it is important to look briefly at the genealogy of inquiry to which they were indebted, extending back into the middle decades of the eighteenth century. Around 1750, a seminal shift occurred in the acquisition of medical knowledge that would lead to new interest in visuality, which would later inform the work of the Shōhyaku-sha. New practices of human dissection motivated this shift in both the school of "old way" medicine and Western medicine, as physicians moved away from accepting knowledge transmitted through apprenticeship and classical texts toward an emphasis on the actual experience of witnessing the human interior vis-à-vis the practitioner's own understanding of the human body.

Two dissections carried out in the second half of the eighteenth century underscore the role that observation came to play within the discourse at the time on the human body. The first case was in 1754, and Yamawaki Tōyō (1706–1762), a practitioner of the school of "old way" medicine observed and recorded the preceding. The second was observed in 1771 by Sugita Genpaku (1733–1817) and Maeno Ryōtaku (1723–1803), the pioneers of Western medicine in Japan. Previously, the Tokugawa *bakufu* had prohibited human dissection. These dissections are often linked to the role that experimentation played within early modern Western discourse. But they are more properly situated within the broader discursive field of Practical Learning, since Tōyō and Genpaku perceived the human body as part of a seemingly inexhaustible nature from which principles could be derived. These two physicians harbored deep suspicion toward ancient texts as sources of knowledge.

The dissections attended by Tōyō and Genpaku triggered a major shift in the role of observation in medical discourse, sparking a reassessment of received understanding of human anatomy in particular medical traditions and highlighting the role played by the visual in anatomical illustrations.

The pictorialized image of the interiors of the human body came to serve as a verified and agreed standard against which accepted anatomical knowledge could be measured. Put differently, an awareness emerged that observational comparison between the illustration and the actual body amounted to a meaningful and productive reckoning of one's own knowledge of human anatomy. Thus the body itself served as a verifying and affirming medium of knowledge capable of contradicting theoretical classical anatomy. These new approaches to medical knowledge paved the way for an intense interest in vision as a source of knowledge, which would mark the work of the Shōhyaku-sha in the 1820s.[48]

Tōyō's publication of his autopsy observations in *Thoughts on Internal Organs* (*Zōshi*, 1758) seems to have inspired Genpaku's own interest in attending a dissection in 1771. To overlook this genealogy of dissection and the various intersections of the school of "old way" medicine and Western medicine is to risk misrepresenting the intellectual and epistemological formulation of medicine in eighteenth-century Japan.[49] More importantly, the common interests of these practitioners from different medical traditions foreshadows a heterodox approach to knowledge among the members of the Shōhyaku-sha, who saw no conflict in engaging with these fields.

The overlapping goals of these dissections centered on identifying human organs by their visually verifiable and quantitative attributes such as shape and size, prioritizing these aspects over otherwise important qualitative aspects such as the functionality of organs. A visual understanding of the human interior was beginning to supersede the functionality articulated by particular branches of medicine.

As Yamawaki Tōyō writes in *Thoughts on Internal Organs*, the fifty-year-old physician attended the first *bakufu*-authorized dissection of a human body in Kyoto in the third month of 1754.[50] The school of "old way" medicine in which Tōyō was trained based its practices on classical texts of Chinese medicine. Tōyō, however, increasingly doubted the foundational organizing principles and pedagogical assumptions of his own tradition (especially the distinction of small and large intestines), and he decided to study human organs with the naked eye to resolve his doubt:

> I began my pursuit of medicine at an early age, and dedicated myself to my profession as a healing medical physician. As I carefully studied these canonical texts, however, a large cloud of doubt began to emerge in my mind....One day, I visited the home of my teacher Gotō Kōzan and we dis-

> cussed the theories of viscera. My teacher said, "of course, there is nothing better than opening the human body to see for yourself. However, attendance at a dissection is controlled by and granted only though the *bakufu*. Would you perpetrate a crime [by organizing a dissection]? Since you have no choice, why don't you try to dissect an otter? There are many who say that the internal organs of a person are similar to those of an otter.... A glance is worth listening to a hundred accounts. Why don't you see [the similarity] for yourself?" I humbly accepted my teacher's advice and carried out dissections of otters. The area covered by lungs, the position of the heart, the way in which the liver and *dan* are connected, and the ways that *hi'i* and *boucho* converge, all nine viscera were certainly there in fact, and in that respect the observation by my teacher that otters and humans share similarities in the internal organs was correct. I realized that those theories put forth by talented individuals should not be exaggerated. However, I have not been able to distinguish the small intestine from the large intestine in otters. I have tried to look for the differentiating elements several times, only to become more doubtful that such a distinction exists at all.[51]

Tōyō then cites numerous classical texts, specifically pointing out the discrepancies in the number of viscera and organs thought to exist within the human body. He follows by listing those facts taught in the tradition of the school of "old way" medicine that seemed dubious. To Tōyō, it became clear that witnessing a dissection was a chance to obtain experiential proof that could correct, once and for all, the discrepancies found in the various textual sources of the school of "old way" medicine.

What Tōyō came to understand through the process of witnessing the autopsy was the interrelationship between knowledge derived from books and knowledge obtained through direct observation. And what is critical is the fact that as a result of his observation, Tōyō displaced his understanding of the human body as taught in the "old way" medicine with an understanding that was part of a broader epistemology of the human body that had previously stood apart from his own:[52]

> Some time ago, I obtained a book on anatomy written by a Westerner. I was unable to understand the book as I myself was confused. However, I see this book now, and realize that various organs and bones [of an actual body] are just like the illustrations of the book. I find it utterly impressive that those who witness the actual interior of a human body share the same *fu*.[53]

The evocation of *fu*, a method of verification and authentication, is particularly relevant and meaningful. When several parties made an agreement, *fu* was practiced by breaking a piece of wood into smaller pieces and distributing them among those present. When the parties met again in the future, they would match their pieces to identify the other parties and verify the authenticity of the agreement they had made. The practice of *fu* functions as a seal that possesses the capacity to verify an agreement visually. The *fu*, that is, was a physical remnant of the shared past invested with the power to document authenticity through its visuality and materiality. Tōyō extends this metaphor here by characterizing the anatomical illustrations in the imported books as a kind of *fu*, thereby recasting the illustrations not only as a picture of a human interior that resembles what he saw with his eyes but as a physical object that serves to prove and authenticate a particular agreed and shared principle of human anatomy. Tōyō's association between *fu* and anatomical illustrations conjures an image in which each copy of the illustrations serves to uphold the knowledge of human anatomy. When Tōyō saw the anatomical illustrations in the imported anatomy book, it strengthened the understanding he derived from his own observations at the dissection, and it confirmed his suspicions that the understanding of human anatomy in the school of "old way" medicine was wrong. Furthermore, by linking Western medical illustrations to *fu* as a physical object of shared agreement, Tōyō comes to understand his recently acquired knowledge of human anatomy as part of a larger whole.

Here, dissemination of knowledge became tripartite: whereas classical Chinese texts or direct transmission had previously been the basis of knowledge, in this case these sources are replaced by illustrations, and the actual human body serves as the third source for understanding human anatomy. It is precisely this triangulated relationship that "*Honzō shashin*" captures in the Shōhyaku-sha's meetings with Siebold in 1826. The group's concern was not human anatomy; they exercised and indeed expanded on this process of inquiry by making physical specimens of plants and minerals—real objects—central to their conversation. In this light, the illustrations included in "*Honzō shashin*," produced by impressing actual specimens into the pages (a direct transference of the real), can be projected on this triangular relationship. These illustrations as well as the one ink-rubbing print included in the manuscript sit closer to the real plants they represent precisely because these prints involve physical contact with the actual, whereas depictions in woodblock illustrations do not—and could not—carry the physical impression. The seed of the tobacco plant that Hōbun received from Siebold during their meeting in 1826, then, served as a physical manifestation of the real that provided an experimental case with which to compare and contrast the received canonical knowledge associated with the plant.

In Tōyō's case the organs represented in the illustrations came to function as a prioritized and shared understanding that superseded the teachings of his own medical tradition. The severe criticism by fellow physicians of the school of "old way" medicine targeting his participation in the dissection reinforces the unimportance such textually based physicians attached to looking inside a body; the perceptible qualities of organs were irrelevant within their tradition. Thus Sano Yasutada attacked Tōyō's findings in *Thoughts* in the polemically entitled *Denunciation of Zōshi* (*Hizōshi*, 1759). Yasutada asserted that ". . . the appearance and shape do not make viscera viscera. Rather, a person's breath and *qi* constitute viscera. Without breath and *qi*, viscera are nothing but empty vessels."[54] He continued to assert that the distinction between large and small intestines is determined by the content that passes through them, and not by their physical appearance, thus denouncing the visual attentiveness that underlies *Thoughts* as heretical.

Although Tōyō was unable to distinguish the small intestine in this particular dissection, he was able to find one erroneous aspect of the school of "old way" medicine in relation to the concept known as five viscera and six repositories (*gozō roppu*). Tōyō asserted confidently:

> While the perplexing theories [of medicine] confuse us, actual objects will not deceive us. Prioritizing a theory over an actual object is similar to ignoring wise advisors. By experimenting from the start with objects, any average person will be able to come to a fuller understanding [of an object]. In other words, if one starts with the actual body, he will understand that the intestines of Yao [an ancient king known for his benevolence], those of the tyrant ruler Jie, as well as those of Westerners are all similar. A pine tree fashions itself as a pine tree, and an oak tree becomes an oak tree by itself. Those beings that can fly can fly naturally, and those that can run can run without intervention of others. These rules have not changed since the ancient times, and remain the same regardless of the regions. How could this fact not be more clear and evident? However, not everyone is enlightened [by this fact]. There are those who define and ostracize people who experiment with actual objects as worthless and meaningless, and criticize those engaged with classical texts as being behind the times. How unfortunate our present situation has become.[55]

Here, Tōyō stresses the importance of experimentation as a way to not only complement but challenge the theoretical and textual knowledge of the human body. What is "clear" to him is the visible similarity that links figures as disparate as Yao, Jie, and Westerners. In other words, observation served as a way to foreground the natural principle that unites these individuals whose historical import and time differ greatly. Thus even though Tōyō's own misgivings began from differing historical exegeses of classical Chinese medical texts, what the dissection itself allowed him to see was the existence of an essential natural principle that makes those beings that can fly able to fly and those that can run able to run.

More than seventy-five years later, the Shōhyaku-sha followed this process of disentangling text from reality through observing and illustrating actual objects during their meeting with Siebold. The source of nominal identification of a particular plant in Chinese and Japanese, as well as the knowledge associated with it, was no longer limited to the classical Chinese canon. Indeed, by including the actual plant itself as a part of their process of understanding, the Shōhyaku-sha, too, sought to decouple the local flora from the canons of *materia medica*, and resituate them within their actual environment and observational knowledge.

Between Tōyō's dissection and the Shōhyaku-sha's exchange with Siebold would come the second important encounter with the insides of the human body. It was a similar suspicion of existing medical orthodoxy that led the practitioners of Western medicine (*ranpō*) themselves to participate in a human dissection. In 1771, Sugita Genpaku, Nakagawa Jun'an, and Maeno Ryōtaku sought permission to witness an autopsy performed at Kotsugahara in Edo. What distinguished this event from Tōyō's previous observation was that these Western-trained physicians brought an imported anatomical treatise, J. A. Kulmus's *Anatomische Tabellen* (1734), to the autopsy itself.

In *The Beginnings of Rangaku* (*Rangaku jishi*, 1812), an eight-three-year-old Sugita Genpaku recalled the transformative experience thirty years earlier:

> The old fellow who cut open the body told us that he had done several dissections since his youth. Until today, dissection was conducted by relying completely on him to cut out and identify each organ, such as the lungs and kidneys, for observers. The observers, in turn, passively watched him cut and display the parts. Then they would say at the end, "we have thoroughly and directly observed the interior mapping of the body." Needless to say, the actual organs inside the body do not have nametags attached to them for easy identification. Therefore, whatever the old fellow would say [the organ] was, the observers would acquiesce to its identification and [this

> structure of communication] became the routine practice at past dissections. Likewise, on our day of dissection, the old fellow pointed to the heart, kidney, liver, and stomach. He pointed to unidentifiable organs, and said "I don't know the names of these organs, but in my experience since youth, inside every body, there is an organ like this in this position, and an organ like that in that location." We compared what we saw with the illustrations [of *Anatomische Tabellen*], and later identified the organs the old fellow was referring to as two trunks of blood vessels (an artery and a vein) and the adrenal gland. The old fellow also said that "in every dissection I have performed in the past, I have displayed organs as this or that, but no invited physician ever raised a question." When Maeno Ryōtaku and I compared [the interior of the body] to the illustrations included in the Dutch book, we realized there is no difference [between what we saw and what was illustrated].[56]

What is particularly striking in this dramatic narrative is the hierarchical relationship that Genpaku and his colleagues saw among the actual dissected body, pictorial representations in their *Anatomische Tabellen*, and the information provided by "the old fellow," an outcast assigned to carry out the "unclean" activity of cutting open a human body.[57] This triangulated relationship for formulating knowledge shares a similarity to the operative methodological assumption found also in the reasoning of Tōyō's conclusion. In both cases, the process of verifying involved the dissected human body, illustrations of Western anatomical books, and their act of observation. What differed in Tōyō's case was the role of illustration as a retrospective affirmation of knowledge gained from observation. In Genpaku's case, illustration and direct observation worked together simultaneously. As Genpaku remarks, the assumption that there would be name tags on the actual organs is absurd, yet the comment that such tags are obviously lacking points succinctly to the critical role of ascribed names that immediately confirm observation, an issue that preoccupied Genpaku and other physicians who sought to further their study of Western medicine.

Genpaku would go on to participate in the Dutch translation of Kulmus's *Anatomische Tabellen* (1734), and his prefatory remarks to the 1774 *New Book on Anatomy* (*Kaitai shinsho*) demonstrate the important place of visuality within Genpaku's conception of scientific inquiry:[58]

> —The original [Dutch] book included footnotes. However, for this translation, we have selected only those footnotes that relate to appearance and omitted others.

—The most significant aspect of books on anatomy is to read the text while looking at the illustrations.

—When the details of the illustration of the original imported texts were hard to observe, we placed the illustration under a microscope and copied it.[59]

The decision of Genpaku and his colleagues to use the "original" illustrations and their choice to focus on the visible features of organs are particularly revealing of their interest in the internal mapping of a human body rather than in the functions of the internal organs themselves. Indeed, for Genpaku these illustrations were more significant for understanding the human body than the accompanying text. Two common elements connect the arguments put forth by Tōyō and Genpaku: the emphatic and pivotal role ascribed to observation, and the conviction that through observation otherwise unknown facts can be deduced.

The interrelationship implicit in Genpaku's writing on dissection echoes the pivotal underlying logic in Tōyō's account: pictorial representations and the actual body serve as evaluative yardsticks for determining the veracity of pre-existing knowledge, a measurement guide against which the degree of fidelity was determined. In both cases of dissection, the illustrations in the imported anatomical books were posited as the verifying standard in their understanding, over and beyond the knowledge the physicians possessed. It is within this conjunction, moreover, that the question of fidelity surfaces. Both Tōyō and Genpaku chose to observe dissections because of grave doubts each harbored about knowledge of the human body based primarily on textual sources. Each claimed, furthermore, to have found faithful reciprocity between the pictorial representations in imported Western books on anatomy and the dissected body they had seen with their own eyes. Thus in each account text is subordinated to illustration as fidelity inheres in the pictorial representation itself. The experiences of actually seeing the interior of a body shifted the status of illustrations in such a way that the anatomical pictures were assumed to uphold higher fidelity to the real than the textual renditions, a tradition that had shaped the dissemination of medical knowledge.

Here we can recognize the tripartite relationship among questioned knowledge, actual objects, and pictorial representations that would form the mode of questioning that characterized the Shōhyaku-sha. As we can see in their meetings with Siebold, members of the Shōhyaku-sha used objects and illustrations as anchoring devices for their inquiry. All information external to the object itself, whether derived from Chinese, Japanese, or Western sources, was subordinate to the real object; all knowledge was thus contingent and affirmed only through verification with one's own eyes.

Following publication of *New Book* in 1774, the reputable physicians of Western medicine in Edo became completely preoccupied with translations of Western medical books. Yet the extent to which they actually practiced or implemented their understanding of Western medical techniques is uncertain. At the Dutch Academy (*Shirandō*, literally the "Hall of Holland in Shiba," an area in Edo), founded in 1789, Ōtsuki Gentaku produced only thirty students by the end of the century, and more efforts seemed to be placed on translation and publication. In the early nineteenth century, the Ōtsuki family published the four-volume *Shōi shinsho* (a translation of L. Heister's book on surgical medicine, 1825) and *Yōjutsu chishin* (a translation of sections drawn from L. Heister, 1824, details tools used in surgical medicine). The members of the Udagawa family, a hereditary family of physicians in Edo, similarly spent much of their time deciphering and translating the instructional textbooks on internal medicine, ophthalmology, and Western medical remedies: *The Names and Things of Western Medicine* (*Ensei ihō meibutsu kō*, 1822–1825), *Selected Western Theories of Internal Medicine* (*Seisetsu naika senyō*), and unfinished translations of a book on pediatric care by Nils Rosen von Rosenstein and Joseph Jacob Plenck's *Doctrina de Morbis Oculorum* (1779) are just a few examples. At these sites of knowledge diffusion, translation became a focal point of understanding the Western medical traditions and the actual acquisition of and experimentation with techniques and remedies seem to have remained secondary.[60]

In the decades before the emergence of the Shōhyaku-sha, the role of observation and engagement with real objects had thus brought about a series of interrelated transformations in these approaches to medicine: Tōyō in the school of "old way" medicine, and Genpaku in "founding" Western medicine. It was within this lineage that stressed the potential of perception as a means to ascertain reliable knowledge that Shōhyaku-sha emerged, and it is this fidelity among the actual objects, illustrations, and texts that they would pursue in the decades to come. The Shōhyaku-sha, however, would extend this problem into new areas of scientific inquiry, with significant consequences for the place of visuality within late-nineteenth-century cultural and scientific discourses.

"Kōsei shinpen" and shashin

The central activity of the Shōhyaku-sha took place in Nagoya, but the members actively communicated with scholars of *materia medica* in distant cities, exchanging specimens, illustrations, and ideas. One such example is a relationship between Keisuke and Udagawa Yōan, who lived in Edo as a member of a famous Udagawa family of Western learning. In 1827, Keisuke stayed with Yōan for a month before traveling to Dejima to study with Siebold, an intense period of absorption and tutelage that resulted in Keisuke's publication of *Nominal Differentiations* (1829; see Chapter Two). Further evidence of their contact comes from an undated letter (before 1846) that Yōan sent to Keisuke in which he laments that Edo lacks the flourishing rigorous study of *materia medica* found in Nagoya.[61] A brief look at how Yōan and his circle deployed the concept of *shashin* in a distinctly different way from other scholars of Western learning brings to the fore the specific connotation that this term carried in the study of *materia medica*. This exploration also brings attention to another way in which concern for transposition of the real (*shashin*) emerged out of the triangulated relationship of received knowledge, illustration, and the actual entity. The Shōhyaku-sha would continue to place a localized emphasis on the structure of recursive triangulation throughout the nineteenth century. The analysis here sheds light on how the relationship between the actual world and the illustrations came to be grasped through the concept of *shashin*.

The unfinished translation project "Encyclopedia for Household Use" ("*Kōsei shinpen*") began in 1811 when a multilingual translator, Baba Sajūrō, was asked by the *bakufu* to translate into Japanese the Dutch revised edition of Noel Chomel's *Dictionnaire œconomique, Algemeen/huishoudelijk-, natur-, zedekundig-en konst-/woordenboke*.[62] The *bakufu*'s effort to translate Chomel's encyclopedia owes much to a series of Western ships arriving on the shores of the archipelago.[63] These continued and unforeseen visits by foreign ships and the resulting demands made of the *bakufu* that began in the late 1780s led the *bakufu* to secure their shorelines, but also to realize that acquisition of foreign languages was vital to their effective defense and negotiations with the threatening forces.[64] Perhaps the most physically threatening to the *bakufu* was the confrontation in 1808, when a British Naval ship HMS *Phaeton* entered the port of Nagasaki while showing a Dutch flag and ambushed Dutch fleets that were scheduled to enter the port. The British captured two Dutch sailors and demanded water and supplies in return for freeing the captives. The translation project of *Dictionnaire œconomique* administered by the *bakufu* shows an attempt to understand and acquire knowledge of the West beyond the military might.

Of the team of seventeen translators who worked on this project, Yōan is of particular importance here. Along with his adopted father, Genshin, Yōan joined the translation project around 1813 at the *bakufu*'s newly established Office Dedicated to Translating Western Books (*Waranshoseki wage goyō*).[65] According to historian Sugimoto Tsutomu, Genshin translated 836 articles, and Yōan 224 for the "Encyclopedia" project, approximately half of the total number. Sugimoto examined two manuscripts of the translations of *Dictionnaire œconomique* and ascribed translators to sections of the encyclopedia by correlating the overlapping sections and archival materials related to the translation project. I am relying on the chart of the structure, content, and translators that Sugimoto compiled to designate the roles of Udagawa Genshin and Yōan in translating and revising the entries.[66]

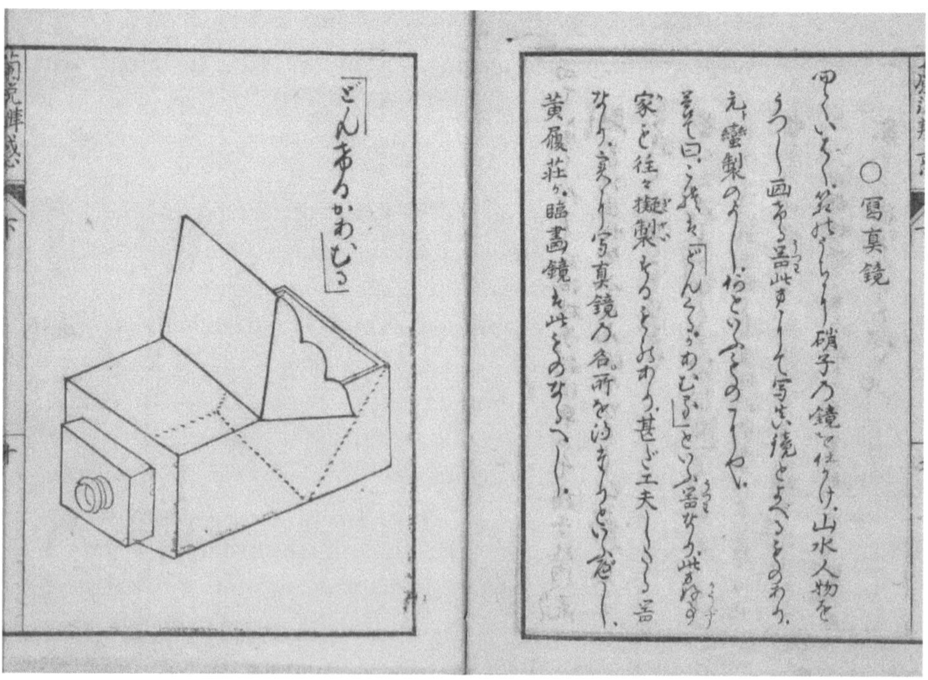

Figure 1.7 Camera obscura, from Ōtsuki Gentaku, *Ransetsu benwaku*, 1799, woodblock print, Waseda University Library

In their translations, the father-and-son team applied the concepts of *shashin* and *shin'ei*, which came to be used by the Shōhyaku-sha for ink-rubbing prints. Udagawa's use of *shashin* departs dramatically from previous and contemporary applications of the term exercised by scholars of Western learning (*rangaku*), who used the term to translate camera obscura as *shashin kyō* (*shashin* lens): Morishima Chūryō in his *A Lexicon of Barbaric Words* (*Bangosen*, 1798), Ōtsuki Gentaku in *Correcting Errors About the Dutch* (*Ransetsu benwaku*, 1799; Fig.1.7), Shiba Kōkan in 1811, and Sugita Genpaku in 1815.[67] Udagawa Yōan, for instance, used the term *shashin* in translating the section on herbarium:[68]

> *Oshiba chō*; Latin: *Herbaricum vivum*; Dutch: *Leendif Kruidboek*
>
> It refers to a collection of pasted dried pieces of plants in a pamphlet. In general, in order to record and pursue the principle of plants, published shashin pictures [*shashin zu*] are indispensable. The herbarium, too, is vital [to achieve these goals]. Even when the artisans do their best to depict the shape and form of the plant in detail, there are always parts [of a plant] that perception cannot reach, and are therefore not represented. The benefit of the herbarium is limited not only to these purposes. It is also highly advantageous in terms of understanding medicinal plants and their applications thoroughly. The following is an abridged version of how to make a herbarium.[69]

That Yōan used the term *shashin zu* (literally, "pictures of transposition of the real") to translate *Kruidboek* (herb books) alludes to Yōan's perceptive comparisons between the herbarium and the concept of *shashin*.[70] On the nominal level, Yōan understood the published *Kruidboek* as the equivalent of a Chinese compound *shashin*.

It is useful here to recall that the compound of *shashin*, originally an aesthetic term that came to Japan, was used in the discourse of landscape paintings in Japan in conjunction with and in contrast to the concepts of transposition of will (*sha'i*), transposition of the living spirit (*shasei*) and transposition of the actual (*shajitsu*). The ways that art historical studies valorize the pervasive term *shashin* present several issues in analyzing the activities of *materia medica*, exemplified in the case in translation of *herbarium*.

The historiographical tendency to formulate two dichotomies, between East and West and between art and science, and to situate the illustrations of *materia medica* from Tokugawa Japan at the intersection of these two dichotomies, presumes the definitive boundaries of each category—art, science, East, West—without clear explication of the historical significance of these notions.[71] Particularly in regard to the interchangeability ascribed to the four terms *shashin*, *sha'i*, *shajitsu*, and *shasei*, I argue these methodological presumptions cannot grasp the significance of the process and shifting epistemology of *materia medica* and the prevalence of the concept of *shin* and *shashin* within the discourse of *materia medica*.

Naruse Fujio, a contemporary art historian of the Tokugawa period, investigates the relationship between theories of Western pictorial studies and the intellectual history of the Tokugawa by constructing two categories of pictorial style: utilitarianism (*jitsuyō shugi*) and prioritizing the actual (*shajitsu shugi*).[72] Naruse takes into account the limited exposure of Western oil paintings in the Tokugawa period, and contends that most encounters between Japanese artists and Western art were facilitated through illustrations in imported books on natural history. He centers his analysis on the treatises on Western painting of Satake Shozan and Shiba Kōkan, both of whom, according to Naruse, used the language of "utilitarianism" and "prioritizing the actual" and critiqued the dominant traditions of picture making. Naruse argues that although tendencies of utilitarianism exist in imported Western illustrative books, the historical tracings of the theories put forward by Shozan and Kōkan reveal that their critique of Eastern traditions of picture making originates not with Western ideas but with the Eastern ideas they claim to critique.

In regard to prioritizing the actual, Naruse asserts that both Shozan and Kōkan understood the traditions of painting in Japan and China as prioritizing the interiority and subjectivity of the painter, and it is from this understanding that they claim "objective expressions are deterred in pictorial traditions in Japan."[73] Naruse then posits the concept of *shashin* that permeates the treatises of Shozan and Kōkan but that would come to be equated with photography as an intrinsic aspect of the "rational" and "objective" Western pictorial tradition. He does this by placing *shashin* in opposition to the traditional practices of picture making. His argument, however, falls down a slippery slope when he claims that Shuzan and Kōkan's move away from classical picture making and toward direct observation is "precursory to the modern idea of art that values the eye of an individual."[74] Without specific explorations of "the modern," Naruse's argument conflates key terms on multiple levels. His rendering of prioritizing the actual (*shajitsu shugi*) as "rational" and "objective" and the Western artistic tradition as "scientific," inadvertently characterizes "Eastern artistic tradition" as "subjective."[75]

In this regard, Kōno Motoaki's work on the concept of *shasei* stands out. As if to resist the simplifications entailed by the formulated dichotomy of East and West and the concepts of utilitarianism and prioritizing the actual in previous scholarship, Kōno explores the multivalence of the concept of *shasei* (Ch: *xiesheng*) in the history of Chinese art and interprets *shasei* in Tokugawa artistic discourse as one that is influenced by but also diverges from the complexities of historical Chinese *xiesheng*. His work thus brings focus to *shasei* as a concept enveloped within the predominant influence of the Chinese artistic discourse during the Tokugawa period.[76]

Kōno's exploration of *shasei* is illuminating because it substantiates the limitations of fixed East-West comparisons in regard to the concepts of *shashin*, *shasei*, and *shajitsu* by showing the ways in which the meaning and significance of *shasei* itself deviated from the Chinese original within the aesthetic discourses of Japan. However, I am hesitant to project this model over the works of the members of the Shōhyaku-sha or scholars of *materia medica*. The most difficult and complex task in accommodating these specific discursive models to the works related to *materia medica* lies in formulating concepts such as subject and object, and their attendant notions, subjective and objective.[77]

In this light, art historian Satō Dōshin provides a compelling argument for using the concepts of *shashin*, *shasei*, and *shajitsu* as openings into further historicization and meta-criticism. In his study on the history of the establishment and proliferation of the category "fine art" in the early Meiji period, Satō considers both the discursive establishment of the category "artist" as well as the concepts of *shasei*, *shajitsu*, and *shashin*.[78] In focusing on the prevalence of these terms throughout the discursive formulations of Japanese artistic practices, Satō appropriates a model from statistical linguistics that views the Japanese language not through a genealogical model but rather through a model of appropriations. Summarizing previous works on the concepts of *shasei*, *shajitsu*, and *shashin*, Satō articulates the art historical issues concerning these terms:

> Previous investigations indicate that the three terms *shasei*, *shashin*, and *sha'i* all bear double meanings pointing on the one hand to perceptual actuality and on the other to the interior truth. However, in producing pictorial works, these terms create seemingly contradictory situations because the art-

> ists' ultimate goals were to unify the subject-object relationship by combining exterior truth with interior truth and objectivity with subjectivity. Depending on which aspect is emphasized, even when they employ these terms to articulate their ideas, the concepts appear to signify completely opposite meanings. Rather, it may be more fitting to describe these terms by qualifying them as ones that embody a certain "ambiguity" that refuses absolute definitions. Therefore, the subjective understandings of the picture maker played an expansive role.... In other words, each of these terms is employed in "fluctuating" positions, and the volition of the cognitive subjectivity of the picture maker or the audience is largely responsible for this "fluctuation."[79]

By considering the use of and need for illustrations, Satō's thoughtful characterization of *shasei* and *shashin* as always in transitional movement provides a way to dispense with the traditional dichotomies, and instead find the relevance of these concepts within historical overlaps in materials that may seem unimportant in traditional disciplinary discourses.[80] I follow Satō's assertion that these three terms are best understood by envisioning them in constant flux, and it is my aim to articulate the cultural landscape that surrounded the discourse of *materia medica* and the role of *shashin* within it.

Returning to Yōan's entry on herbarium in the drafts of "Encyclopedia," we note that on the conceptual level he associated a particular mode of visualization with the imbricated meaning of the term *shashin*. The Dutch entry itself notes that there are two types of herbarium, living (*leevendig*), and printed (*gedrukt*). The fact that the entry begins by "Kruid-boek (leevendig)" affirms the difference between the two. The Japanese translation, however, does not make this difference textually apparent. Rather, for a translation of "leevendig Kruidboek" Yōan employed "*sakuyōchō*" (literally, "pressed leaf notebook"), and for "gedrukt Kruidboek" he used "*shashin zu*."[81] However, his choice of "*shashin zu*" does not make recourse to the then-common Japanese translation of camera obscura as *shashin kyō*.

The meaning Yōan imputed to the term *shashin* comes to the fore in comparison with another section of the "Encyclopedia" project in which the concept of the shadow of the real (*shin'ei*) makes an appearance. The section describes the camera obscura:

> A device that makes the *shin'ei* of scenes and objects appear in a darkened room. Latin: *camera obscura*; Dutch: *Donkere Kamer*
>
> This is a device set up by using one completely darkened room during the day, save for one small hole on one wall about the size of a one-*yen* coin,

> which allows for sunlight to enter the room. Through this hole, shadows [*ei*] that accurately correspond to the objects outside are projected onto a white wall or a piece of white paper located at an adequately distanced position [from the hole] in the darkened room. Furthermore, the projected image is represented with the natural colors of the *shin* forms of the objects outside. This device makes it possible for the shapes of objects to emerge and appear completely. The finer the weather is, the more clearly the objects appear. However, the projected shadow [*ei*] is reversed horizontally and vertically. In order to correctly project the inverted shadow [*ei*], one needs to place a mirror in the hole. There is also a portable device made to take it anywhere one desires called a "donkerekamur." It comes in a variety of sizes. This device is used particularly by astronomers to observe and measure sunspots. The portable version is particularly indispensable and vital for artists and amateur picture-makers to *shashin* notable places in local areas, and particularly to represent [*egaki*] the geographic landscape of each area in detail and to develop one's picture-making skill [*gajutsu*].[82]

In the original Dutch translation of Chomel's encyclopedia, the entry is simply named "Camera Obscura," which suggests that the Udagawa father-and-son team deviated from the original text and added the lengthy description.[83] Here we witness the decision to render the device differently from the well-known translations of camera obscura as *shashin kyō* (*shashin* lens). The suggestive gesture sheds light on two aspects that will be vital to the Shōhyaku-sha in the following three decades: (1) difficulty in translating and describing devices and images that were not familiar at that time, and (2) keen interest in the issues of representation as a process that hinges on the fidelity of imagery to the actual world. The established translation of camera obscura, *shashin kyō*, places emphasis on the apparatus itself, particularly through the word *kyō*, or "lens," and foregrounding a particular interest in the mechanical structure of the apparatus. Translation of the same device in "Encyclopedia" places the emphasis on representation, rather than the instrument. The concept of *shin* was used to convey the process of representation, specifically the unusual relationship between the representations made with the aid of camera obscura and what they represent. Here, Yōan apprehends the Dutch term *waar gedaante* (true shape) as "*shin* form" in Japanese and translates *zijgen* in Dutch as a verb, *shashin*, to describe the process of representation entailed in using a camera obscura. The resonant meaning of *shin* implied in "*shin'ei*," "*shin* form," and "*shashin*" in this translation reverberates strongly with the use of *shin* in *shin'ei* to refer to ink-rubbing prints by the Shōhyaku-sha (see Chapter Four). The *shin* form in these uses refers to the perceptible and observable shape of the subject of the representation. In all three cases of deployment of *shin*—*shin'ei*, *shin* form, and *shashin*—the relationship between the existent subject of the representation and the representation itself is foregrounded as the primary value of the resulting image. The added description of the device by the Udagawas reinforces this estimation of the value of camera obscura as an aid in producing representations (*shin'ei*). The conceptual resonance implied in these uses of *shin* as pointing to an *already existing* entity becomes even more pronounced when we compare the entry for camera obscura with the entry for the device known as a "magic lantern."

> *Matō*; Latin: *lantern magica*; Dutch: *Toverlantaarn*
>
> *Matō* is a form of lighting that by shining a light on regular objects makes everyone laugh at the strangeness of the projected image. Children and the common people are particularly delighted and entertained [by *matō*].
>
> *Matō* was discovered by the German priest Kircher. Its method involves drawing a picture with translucent pigment on a small piece of glass and placing the piece of glass in front of a candle. Then, on either a white wall or white piece of cloth, the shape from pictures [*ga no zō*] is projected in enlarged form.[84]

Here, we witness how the Japanese translation of this entry tries to render the distinction between the projected image from a pictorial representation in a magic lantern and the projected image originating directly from the existing physical entity in the camera obscura. While a camera obscura is a device that enables one to trace a projected image from the actual world as the basis for pictorial representations, a magic lantern takes preexisting representations and projects them outward. The connotation carried by *shin* in the section on camera obscura as an already existing entity is irrelevant in the case of magic lanterns. In referring to the projected images of the magic lantern as "shapes from pictures" (*ga no zō*), the Japanese translation makes it explicit that the magic lantern projects outward an already represented image that may or may not originate in the actual physical world. Comparing the explicit uses of "shapes from pictures" (*ga no zō*) to *shin'ei*, which Yōan used in translating the entry on camera obscura, we see the marked differentiation in the concepts of *shin* (the real), *ei* (shadow), *ga* (pictures), and *zō* (shapes) come to the fore. While the term *shin'ei* embodied the distinct relationship to the physical outside world, *ga no zō* does not concern itself with the relationship between representation and reality. The images projected using a magic lantern in Yōan's time, indeed, included imaginary scenes based on Hiroshige's work, as well as fictionalized narratives of murders and ghost stories.[85] The marked conceptual differentiation designated by the terms *shin'ei* and *ga no zō* not only acknowledges but also instantiates textually the explicit differences between a pictorial representation that has a direct relationship with actuality and representations that are not trying to account for actuality.

These instances of translation in "Encyclopedia" point to keen awareness of the relationships between various kinds of visual representation and the actual world. This awareness was no doubt triggered by experiments with human dissection, renewed interest in direct observation, and the available translations of Western anatomical books, which employed pictorial representations that accurately followed reality as a superior means of transmitting knowledge. In the case of "Encyclopedia," the challenges faced in translating and describing the uncommon device of camera obscura led Udagawa Genshin and Yōan to deploy the concept of *shin* to mark the aspects of both image making and the images themselves that possessed a direct relationship to the existing actual world. As we will explore in the next chapter, the issue of translation that the Shōhyaku-sha faced extends the difficulty that confronted Udagawa father and son in "Encyclopedia" and rearticulates and reshapes the notion of *shin* that they developed in their translations.

TWO

Ways of Conceptualizing the Real

Scripts, Names, and Materia Medica

> The descriptions [of the plants] given by the Chinese are clear, and their assigned Chinese characters are also clear and understandable. However, when it comes to their practical application in Japan, [using their texts] would be like falling into fissures and holes.
>
> Ishiguro Sei'an, introduction to *Nominal Differentiations in Western Materia Medica* (*Taisei honzō meiso*, 1829)

> Perhaps the thing to do is go beyond the deconstructionist stage of trying to prove that equivalents do not exist and look, instead, into their *manner of becoming*.
>
> Lydia H. Liu, *Translingual Practice* (1995)

As the medical traditions practiced in Tokugawa Japan began to incorporate and prioritize the triangulated relationship among the actual body, text, and illustration, the means of accessing anatomical knowledge shifted. In this process of finding fidelity among the three entities, physicians were faced with various languages and scripts. The *lingua franca* of the early modern Japan was classical Chinese, conveyed solely in Chinese characters (Jp: *kanji*). Vernacular Japanese, on the other hand, incorporated Chinese characters in addition to two Japanese alphabets, *hiragana* and *katakana*. At the same time, the imported Western books—for instance, the *Anatomische Tabellen* that Sugita Genpaku had taken to a dissection in 1771—were Dutch translations of German books rendered in the Roman alphabet.

Within the broad scope of their practices and interests, it was in their research into *materia medica* that the Shōhyaku-sha confronted directly the challenges of securing medical knowledge through negotiation of languages. These problems were paramount, for instance, in their 1826 meetings with Siebold, when the Japanese physicians collaborated with their Western counterpart to match Chinese with suggested Western names, and to identify plants in their local region by correlating Chinese, Japanese, and Western names. The task of integrating different bodies of knowledge about medicinal plants and connecting them to the real (*shin*) depended fundamentally upon decoding and translating various languages, scripts, and illustrations. Names and naming, in particular, posed the crux of the translational problem. As practicing physicians, unmistaken identification of plant specimens was pivotal in healing their patients. Mizutani Hōbun's investigation of the names for *rōtō* demonstrates the pragmatic goals driving both the search for a name of a specimen whose properties are only known from Chinese texts and the integration of knowledge from diverse medical traditions. This chapter focuses on the Shōhyaku-sha's process of establishing and strengthening faithful relationships among names, their representation as texts, and actual specimens. The following pages thus illuminate one facet of fidelity in the triangulated relationship of text, illustrations, and the real that underpinned their approach to knowledge and representation, which proves vital to our understanding of nineteenth-century visual culture in Japan.

My interest lies not in determining whether the discourse surrounding *materia medica* fits into theoretical ideations of "translation" per se but rather in seeing polyglot *materia medica* as a multilayered intellectual arena that operated beyond the common set of dichotomies (Japan versus the West, and feudal versus modern) that have colored historical studies of *materia medica* in early modern and modern Japan. Rather than exploring the field through paired notions that are already differentiated, approaching the practices and concerns of *materia medica* through a translational and translingual engagement allows us to witness and take note of the "manner of becoming," to use Lydia Liu's phrase.[1] The practice of translation in this chapter explores how the Shōhyaku-sha tackled the issues of naming and names in response to their particular needs through an active and strategic act of translation, with an ultimate goal of textually adapting a different abstract nominal scheme to the traditional Chinese and Japanese naming of plants.

What tools were available to the Shōhyaku-sha to secure and stabilize a relationship between actual plants and names rendered in four types of script—Chinese characters, Japanese *hiragana* and *katakana* scripts, and Roman alphabet? In what way did Linnaean nomenclature contribute to organizing the disarray? The Shōhyaku-sha sought to resolve the epistemological incongruities among Chinese *materia medica*, Linnaean nomenclature, and native Japanese flora through close attention to representation of language as text. Publications by the Shōhyaku-sha evince dawning awareness of the graphic and visual nature of scripts, and eventual reliance on the visuality of the Roman alphabet to solve the challenges of translating names that the group confronted. This nominal problem for the Shōhyaku-sha was thus also an issue of representation, one of ensuring fidelity between names expressed in various scripts without furthering the confusion in the field of *materia medica*.

Michel Foucault's work on the stakes of natural history in Europe helps delineate precisely the ways in which the problem of naming is inextricably related to issues of representation. His insights into the discourses of natural history underscore that the question of names and naming faced by scholars of *materia medica* concerned not only languages but also the visuality of scripts in which particular languages are embedded and expressed.

Foucault describes the discourse of natural history in Europe as "nothing more than the nomination of the visible."[2] He argues that the emergence of natural history should not be attributed to and understood as a sudden rise of interest in the world of nature as such; rather, it is intimately related to formation of a certain *episteme* that "finds its locus in the gap that is now opened up between things and words."[3] He emphasizes here that the act of naming that constitutes natural history is intimately related to and takes root in practices of representation:

> We must not see the constitution of natural history, with the empirical climate in which it develops, as an experiment forcing entry, willy-nilly, into a knowledge that was keeping watch on the truth of nature elsewhere; natural history—and this is why it appeared at precisely this moment—is the space opened up in representation by an analysis which is anticipating the possibility of naming; it is the possibility of seeing what one will be able to say, but what one could not say subsequently, or see at a distance, if things and words, distinct from one another, did not, from the very first, communicate in a representation.

Asserting the importance of historicity and structural contingency in formulating an *episteme*, Foucault points to a space of possibility for natural history between naming and seeing. In this newly available space, things and words exist *as* and *within* representation. The names of plants represented in Roman letters and the illustrations that accompany these texts in the books of the Shōhyaku-sha manifested what needed to become visible for the given names to make sense under the new classificatory scheme. Thus the discursive features of natural history necessitated matching the seeable and the sayable in forms of representation. The problem of names in the work of Shōhyaku-sha on *materia medica* presented a nebulous discursive space for navigating the increasingly heightened tensions among various representational systems. The case of the plant *rōtō* demonstrates this succinctly: negotiations among a name in Chinese characters (Ch: *langdang*; Jp: *rōtō*), the Japanese regional name (*hashiri dokoro*), and the Latin name written in the Roman alphabet (*belladonna*) entailed matching the sayable—the names—with the seeable, the actual plant to which the names properly correspond.

Elaborating on the dichotomy of the seeable and the sayable, Foucault explains the role of language vis-à-vis the constitution of natural history:

> The theory of natural history cannot be dissociated from that of language. And yet it is not a question of a transference of method, from one to the other; nor of a communication of concepts; nor of the prestige of a model which, because it has "succeeded" in one field, has been tried out in the one next to it. Nor is it a question of a more general rationality imposing identical forms upon grammatical thinking and upon *taxinomia*. Rather, it concerns a fundamental arrangement of knowledge, which orders the knowledge of beings so as to make it possible to represent them in a system of names.[4]

In nineteenth-century Japan, the tension was seated in the confluence of *material medica*, an *episteme* originating in China that was based on the region's native species, and the indigenous flora of Japan, which departed in many ways from the received knowledge from China. The wide variety of plant names in Japanese regional dialects introduced further complications. The Shōhyaku-sha's research and publications intervened in this amorphous discursive space of *materia medica* and attempted to rearrange the available knowledge in different orders and positions *in* and *as* representational systems.

This chapter begins with Kaibara Ekken's *Materia Medica in Japan* (*Yamato honzō*, 1709), which first outlined the problem of naming to the field of *materia medica*. The Shōhyaku-sha's struggle with names and naming engaged with and extended the concerns that emerged in *Materia Medica in Japan* in new directions. The key work in the group's explorations was Itō Keisuke's *Nominal Differentiations in Western Materia Medica* (*Taisei honzō meiso*, 1829), which introduced the new system for naming—the Linnaean binominal scheme—for the first time in publication form. Analyses of how Keisuke carefully mobilized the graphic and visual aspect of alphabetic script in this book reveals Keisuke's overriding concern in his seminal publication with maintaining fidelity between Latin names expressed in Roman letters and the scheme of Linnaean taxonomy.

In reorganizing the schema by which nature was understood, Keisuke foregrounds the concept of the real (*shin*) as the essential component underpinning the very practices and study of *materia medica*.[5] Detailed examination of *Nominal Differentiations* reveals the process of sifting and filtering that the new Linnaean taxonomy enabled in relation to a concept of the real. Examining the interdependency of translation, names, and *materia medica* opens critical perspectives on the historicity of the concept of the real (*shin*)—recognizing in the concept the "fluid sense of etymology" endorsed by Lydia Liu.[6] These perspectives illuminate nuances, differences, and changes often overshadowed in discussion of the introduction of Linnaean taxonomy in Japan but vital to the construction of knowledge and practices of representation. The centrality of the real (*shin*) illuminates the historical contingencies at stake for the Shōhyaku-sha, but also how the practices of representation and translation merged effectively at the moment when the relationship between names and plants was critically questioned. The attached meanings and connotations of the concept of the real points to the ways in which the Shōhyaku-sha's use of *shashin* (transposition of the real) would come to indicate a particular relationship of fidelity.

From Translation to Observation in the Study of Materia Medica

Translation of foundational Chinese texts presented the first hurdle faced by Japanese scholars seeking knowledge of *materia medica*. As a body of learning founded in China, successful communication of this knowledge in Japan depended to a significant extent on translingual representations of classical Chinese texts. Beginning with *Compendium of Materia Medica in Japan* (*Wago honzō kōmoku*, 1689), in which physician Okamoto Ippo'shin translated the text of Li Shizen's *Compendium of Materia Medica* (*Bencao gangmu*) from classical Chinese into vernacular Japanese, scholars made multiple efforts to render Chinese texts more accessible to Japanese scholars, many of whom lacked complete mastery of Chinese. In 1769, Ōkuchi Mimei, a scholar of *materia medica* in Osaka, published a ten-volume work *Punctuated and Annotated Compendium of Materia Medica* (*Honzō kōmoku jiben*), which provided the glossary and annotations necessary to read *Compendium* for those without fluent reading knowledge of classical Chinese. At the same time, versions of Li's *Compendium* were published repeatedly throughout the Tokugawa period with various aids to help readers decipher the classical Chinese: *okurigana*, which added kana suffixes to stems of Chinese characters, and *kunten* marks, punctuation that helped readers grammatically construe the text. These editions aimed to facilitate the reading process of comprehending *Compendium* by allowing the Chinese text to be read in Japanese.[7] In this sense, these efforts prioritized comprehension and absorption of *Compendium* and the Chinese tradition of *materia medica*. The central concern in these efforts was to overcome apparent linguistic difference. Significantly, many scholars assumed that the alleged transparency of translating Chinese to Japanese extended to the actual availability of *materia medica* contained in the texts: the presumed equivalence of words led to perceived equivalence of content. The shared script of Chinese characters in Chinese and Japanese, and the centrality of texts from China in the field of *materia medica*, compelled those translators in Japan to assume *Compendium* possessed a unified and transparent authority.

While many scholars continued to translate and accept the authority of such works well into the nineteenth century, there emerged another group of scholars in *materia medica* who questioned and challenged the transparency of *Compendium* and the presumed availability of items described in these translated works. From the early eighteenth century on, there emerged a concern for not only translating the Chinese text but also accounting for how it might be used to understand and classify the plant life of Japan, which often differed markedly from the world of medicinal items explained in *Compendium*. Ekken's *Materia Medica in Japan*, Ono Ranzan's *Dictated Compendium of Materia Medica* (*Honzō kōmoku keimō*, 1803), as well as Mizutani Hōbun's sixty-volume manuscript "Compendium of Materia Medica, Heard and Seen" ("*Honzō kōmoku kibun*," before 1833) testify to an epistemological transmutation. The process of transmutation both challenged and moved beyond the translingual practices that assumed linguistic and contextual equivalence, and indeed homogeneity, between *Compendium* and the Japanese natural environment. The discursive shift instigated by Ekken rests precisely in recognizing the specificity of Japanese flora, and thus the uselessness of translating imported knowledge as if it were universally applicable to any context. Ekken and his followers instead assumed that the specific names given to plant specimens existed independently of the knowledge contained in *Compendium*, conditioned by the local linguistic practices in which the plants grew. Unlike the translators of Chinese texts, these scholars began their study by abandoning the presumed equivalence between Chinese knowledge of *materia medica* and their native environment; they accepted the possibility of difference between

Japanese flora and Chinese *materia medica*. The notion of heterogeneity was thus at the root of epistemological transmutation. Redirecting their gaze from the imported texts to their immediate and directly accessible environment, Ranzan, Hōbun, Keisuke, and the Shōhyaku-sha built on Ekken's articulation of difference between textual tradition and Japanese flora and attempted to bridge the canonized knowledge of *materia medica* with their own surroundings through observational verification. The relationship of text, illustration, and the actual object consequently became interwoven in this epistemological transmutation by a process of insinuating, rejecting, and modifying existing knowledge.

Those engaged with epistemological transmutation were not surprised to find disparities between Chinese learning and their Japanese context. When they found discrepancies between the canon and their local environment, they continued to modify the received knowledge so as to accommodate their own epistemological needs. The *Compendium* provided a framework of nomenclature by which the Japanese scholars could begin to name, organize, and know native plant life, but differences between what they learned in the Chinese texts and what they saw around them demonstrated the need for nominal uncoupling if more accurate knowledge was to be attained. Similarly to the motivations that led Yamawaki Tōyō to attend human dissection, the lingering suspicion of canonized knowledge and the applicability of this knowledge propelled the scholars of *materia medica* to seek alternative methods to vet and match the received knowledge with actual plants. As the case of *rōtō* hinted, there were ample instances in which a Chinese name could not be matched to its alleged Japanese counterpart. In this sense, epistemological transmutation necessarily went hand in hand with transvaluation, and the Shōhyaku-sha's engagement with *materia medica* operated at both these levels: using *Compendium* on one hand as a framework, and recognizing on the other hand that the same Chinese canon would not contain the plants found in their region.

In 1709, Kaibara Ekken (1630–1714), a philosopher and ardent practitioner of agricultural endeavors living in the Fukuoka domain in the southern island of Kyūshū, published *Materia Medica in Japan* at the age of seventy-nine.[8] In this book, which consolidates fifty years of his studies, he dramatically shifted the focus of *materia medica* from passive recitation of *Compendium* to active verification of the content of *Compendium* based on one's own sensory experience.[9] And it is through Ekken's publication that observational verification became a significant practice for the Japanese study of *materia medica*.

The publication was seminal because it first called attention to the incongruity between the plants included in *Compendium* and the available plants in Japan. Second, it addressed the gap between the Chinese text and Japanese flora by promoting exploration and use of one's own senses as reliable means for accumulating and verifying knowledge.[10]

Almost one hundred years after publication of *Compendium*, Ekken's work inverted its fundamental presumption, still adhered to by followers of the canon, by calling attention to the uncertainty of whether specimens included in the imported Chinese text actually existed in the local environment. Moreover, the fact that Ekken included plants having only Japanese names attests to the recognition of the differences between specimens available in China and Japan.[11] In an introductory remark, Ekken asserted his frustration with the issue of names in *materia medica* and how he envisions his book would alleviate his frustration.

> Among the books on *materia medica* and others from China, many are written about plants, trees, animals, insects, fish, minerals, and soil. Although the knowledge included in these books has been disseminated for a long time, each of the various works contains differing information. Repeatedly, they are misapprehended and attribute confusing names and facts. One should know that books on *materia medica* and other books from China, too, may include contradictions. One should not recklessly believe what any text says. As Mencius says, believing every word in a book is tantamount to the nonexistence of the book.... There are most likely some misunderstandings and errors in my text as well. I would appreciate it if readers would point them out to me.[12]

Accepting the disarray of names and natural items in books from China, Ekken here reiterates an underlying epistemological conviction that continued to be adapted by the Shōhyaku-sha: one is incapable of understanding the "whole" of *materia medica*.[13] In his mind, it would be impossible to imagine a full knowledge of *materia medica* that covers all of Japan; equally impossible was a book on *materia medica* that would never require an update or a correction. Rather, he encourages readers to assume that every book on *materia medica*, however thoroughly researched, is bound to contain erroneous information. Ekken imagines the discourse of *materia medica* as infinitely expanding and shifting, as a never-contained and static sphere of knowledge. The publications on *materia medica* thus inform and direct the discourse so as to make room for modifications to be acknowledged and contested. Ekken then offers new approaches whose standard of accurate information is not rooted in canonical textual knowledge:

> Those who study *materia medica* must have a wide range of knowledge and hear and see as much as possible; this will then lead to the elimination of doubts. It is not actual knowledge so long as one only refers to other works and only roughly understands the information. One must not carelessly reckon what he has seen and heard as correct or make the mistake of insisting that differing opinions are wrong. The following four traits are always erroneous: (1) not seeing and hearing enough, (2) believing whatever one sees and hears, (3) applying one's own theory, and (4) making reckless conclusions.[14]

Ekken warns scholars who absorb the information in *Compendium* that they must be reflective and critical. He urges his readers to integrate observation into their shaping of knowledge. In Ekken's epistemological constellation, the experience and knowledge gathered and verified through one's own senses must work hand in hand with the information provided in Chinese texts. The role of suspicion in practicing *materia medica* requires doubting the applicability of the received knowledge, while searching for ways to affirm or reject canonized information. In his emphasis on correlating knowledge with observation, however, Ekken does not completely renounce *Compendium* and recitation and memorization of texts. Rather, he renders the method of *materia medica* into a discourse best characterized by heterogeneity. The new method incorporated various senses, while expanding the scope of inquiry to include objects unknown to *Compendium*. Renewed interest in the act of observation will be explored fully in the next chapter, but for our purposes here it is sufficient to note that Ekken's methodological intervention in the study of *materia medica* cultivated room to doubt canonical knowledge vis-à-vis one's own experience and knowledge.

In the local domain of Owari, the ripples of the methodological shifts instigated by Ekken became particularly manifest in a series of publications on names dating from the end of the eighteenth century: in 1776, the domainal Neo-Confucian scholar Matsudaira Kunzan published *Accuracy and Flaws of Materia Medica* (*Honzō seigo*) to which Yamaoka Kyoan responded with *Response to Accuracy and Flaws of Materia Medica (Honzō seiseika)* in 1778, which in turn received a rebuttal from Kunzan's student Sugiyama Ikei, *Accuracy and Flaws of Materia Medica: Revised Edition* (*Honzō seiseika kango*) in 1779. In addition, in 1809 the founder of the Shōhyaku-sha group, Mizutani Hōbun, published *Clarification on the Names of Things*

(*Buppin shikimei*), another work that aimed to pin down the names associated with pharmacological items by prioritizing and indexing them according to their Japanese names. Hōbun continued his effort to fill in the gaps between Chinese and regional Japanese names by publishing a supplementary volume to this book, *Supplement to Buppin shikimei* (*Buppin shikimei shui*) sixteen years later. Shibata Dōgen, a physician in Owari and a member of the Shōhyaku-sha, published in 1810 *Medicinal Items for Everyday Use* (*Nichiyo yakuhin kō*), which further elucidated the Chinese and Japanese names associated with various commonly found medicinal plants. All these local publications emerged out of a shared concern to find a match between plants in Owari and Chinese canons, a concern founded on Ekken's recognition that the efficacy of a particular plant elucidated in a Chinese text did not—and could not—immediately translate into knowledge of the native plants of Japan.

Hōbun's introduction to *Clarification* attests to the complexity and the convoluted process entailed in reorganizing and altering the received knowledge of Chinese *materia medica*. The challenges were especially acute in the process of renaming, which involved suspending the established pairing of a Chinese name and a plant, finding a concrete nominal relationship between names in Chinese and Japanese, and then reintroducing the names into the arena of *materia medica*.

—This book lists things from our country with the corresponding Chinese names. Lately, great advances have been made in what is known as *materia medica*. However, despite the number of books published to this day by scholars on Chinese names and Japanese names, such as *Wameishō* and *Honzō tashiki*, the opinions put forth by them fail to agree in many cases. I am not in any position to determine what is right and wrong [in these differing opinions]. Most of my understandings rely on Ono Ranzan's interpretations.

—There are enormous amounts of natural objects in our country and it is impossible to know every one of them. I am jotting down only what I have heard and seen, in addition to what is commonly known. In particular, I was not able to accumulate many regional names, since the range of dialects in Japanese is rather copious.[15]

Hōbun echoes Ekken's frustration expressed in *Materia Medica in Japan* with the need to stabilize the system of nomenclature, and he reiterates the earlier writer's underlying skepticism about the possibility of apprehending the whole world of *materia medica*.

Whereas Ekken could introduce through his observations only discrete alterations and additions to the received Chinese knowledge of *materia medica*, the Shōhyaku-sha directly addressed the fundamental problem of furnishing a stable system of nomenclature. Motivated by the desire to create a basis for knowledge and practice—something akin to Ekken's "principle"—Hōbun and Keisuke pioneered implementation of the Linnaean taxonomic system. Their meetings with Siebold in 1826, from this perspective, were more than an attempt to translate a new epistemological framework by correlating Linnaean names with the Chinese and Japanese equivalents. The Shōhyaku-sha knew from their experience with Chinese *materia medica* the pitfalls of such an approach. Introducing a new set of intellectual practices and processes from the West necessitated remapping the world of plants, and it expanded the conceptual and representational possibilities for rebuilding a view of the natural world. For the Shōhyaku-sha, the study of the Linnaean system not only involved acquisition of linguistic skills; the representational structures of the system itself had to be recognized and readjusted. One of the resulting consequences of this intellectual configuration can be seen in the ways in which the practitioners of epistemological transmutation took seriously the varying scripts as a medium of representation to both construct and communicate knowledge.

Taisei honzō meiso and the Visuality of Scripts

Keisuke's 1829 *Nominal Differentiations in Western Materia Medica* was one of the first expositions on the basic principles of the Linnaean classificatory and binominal system in Japan.[16] The Shōhyaku-sha's effort to introduce the Linnaean system culminated in 1856, when Iinuma Yokusai, a retired physician, published the twenty-volume *Illustrated Explanations of Plants and Trees of Japan: Grasses* (*Sōmoku zusetsu, sō-bu*), the first publication to systematize the flora of Japan using the Linnaean system.[17]

The Linnaean taxonomy divides animals and plants into the hierarchical order of kingdom, genus, and species. The kingdom (animal, plant) is further divided into classes.[18] In the case of the plant kingdom, Linnaeus organized plants according to their reproductive organs (known as pistils and stamens), identifying twenty-four classes that derived from the number, length, and positions of stamens.[19] In each class, a further categorization of order, which relied solely on the number of pistils, was created. This standardizing schema then allowed Linnaeus to refine the system further with the even more specific categories of genus, species, and variety. Linnaeus's binominal scheme for plants, which he adapted from his classificatory system, used the categories of genus and species to compose a two-part name for each species of plant, thereby standardizing the naming process. Keisuke uses the example of a willow tree in his book to explain the structure of the binominal system. The varieties of willow trees constitute distinct species, but their grouping together under the genus of willow signaled their close affinity.[20] This nomination scheme stands in stark contrast to the Chinese and Japanese names the Shōhyaku-sha were accustomed to.

Neither the Chinese nor Japanese names indicated the abstract classificatory categories and the taxonomic relationship of different plant types. In contrast, the nine general categories consistently deployed in the previous publications related to *materia medica*—water, stone, grass, tree, insects, fish, shellfish, birds, animals—did not, and could not, reflect an intellectual hierarchy for organizing nature, because there was no hierarchy. In addition, the names of objects did not persistently reflect the categories to which the objects belonged within the system of *materia medica*.[21] Even without specific knowledge of a plant, a reader familiar with Linnaean nomenclature could recognize it as a plant and divine its basic appearance and characteristics by comparison with known plants in its genus. On one hand, the Chinese characters in *materia medica* were laden with culturally and historically embodied meanings that the scholars in Japan readily understood. On the other hand, they could not help dissolving the issues of names precisely because of, or despite, those attached meanings, which were absent in the Roman alphabet.

In fact, in the introduction to Keisuke's *Nominal Differentiations*, Ishiguro Sei'an, the domainal physician of Owari and a member of the Shōhyaku-sha, aptly explains how introduction of the Linnaean binominal scheme would allow the scholars of *materia medica* to move beyond the convoluted process of decoupling ascribed names and trying to find a sharable name:

> The descriptions [of the plants] given by the Chinese are clear, and their assigned Chinese characters are also clear [*akiraka*] and understandable. However, when it comes to their practical application in Japan, [using their texts] would be like falling into fissures and holes. I have always felt frustrated at this process of application, and have established a method [*hō*] in the past to differentiate and categorize plants. . . . Although the method introduced here differs from the method I concocted, as long as a method is based on a

> principle, its meanings and outcome come from the same trail. When studying, there must always be a principle (*ri*) and the consistent application of the principle. With this publication, all variations of names and their characteristics will be on a temporarily shared platform. In our country . . . we must know their names, understand their beneficial attributes, and bring welfare to the people.[22]

Sei'an prioritizes the welfare of the people as the primary goal, while acknowledging the insurmountable gap between the Chinese and Japanese contexts. He explicitly acknowledges that the problem does not arise at a translingual level; on the contrary, the letters and names in Chinese texts are clearly readable and understandable to him. Rather, Sei'an identifies the issue as the lack of what he terms a "principle." He envisions that a systematic principle would bring about a stable and sharable foundation on which issues of nomenclature could be resolved. Extending the traditional purpose of *materia medica* as a curative discourse, Sei'an securely positions creation of a uniformly sharable principle that assembles knowledge of plants to help the sick as the primary aim of *Nominal Differentiations*. By sharing the Linnaean binominal system, Sei'an hopes that scholars of *materia medica* could temporarily sidestep the convoluted issue of nomenclature and reorient the discourse toward efficient determination of the appropriate remedies for illnesses.

Keisuke's 1829 *Nominal Differentiations* draws largely on Carl Peter Thunberg's *Flora Japonica* (1784). He acquired the book from Philipp Franz von Siebold in 1828 as a farewell gift when Keisuke returned to Owari after a six-month stay in Dejima studying with Siebold.[23] Thunberg had studied under Carl von Linné at Uppsala University and came to Japan as a medical doctor working for the Dutch East India Company in 1775.[24] *Flora Japonica* outlines his findings from his fourteen-month stay in Japan.[25]

Upon his return to Owari in the third month of 1828, Keisuke immediately began to study and translate the book in order to present its contents for a Japanese audience, and he published his work in the tenth month of the next year. *Nominal Differentiations* also includes a title page in Dutch: *Naamlyst van gewassen* (Name list of plants) and a subtitle *door den beroemden natuuronderzoeker C. P. Thunberg, op Japan gevonden. Herzien en met Japansche en chineesche namen verrykt door Jtoo Keiske* (By the renowned natural researcher C. P. Thunberg in Japan. Revised into Japanese and Chinese names by Itō Keisuke). That Keisuke dispensed with the Latin title of Thunberg's work, instead inventing a different title in Dutch, points to the inadequacy of treating *Nominal Differentiations* simply as a word-for-word translation of *Flora Japonica*.

Indeed, Keisuke's decision to revise the title is only the most immediately noticeable change in the contents of Thunberg's book. These differences reveal that Keisuke conceptualized and regarded his *Nominal Differentiations* as a publication based on *Flora Japonica*, but not as an equivalent text distinguished only by language. In other words, Keisuke modified the contents of *Flora Japonica* for his *Nominal Differentiations* in order to cater to a specific audience. Thunberg's book, for instance, assumes that readers were already familiar with the structures of Linnaean taxonomy, but Keisuke's work presumes a need to instruct its readers in the primary organizing principles of the Linnaean system. Keisuke's gesture to differentiate *Flora Japonica* from *Nominal Differentiations* acknowledges the difference in the general state of knowledge of his audience from that of Thunberg, and it indicates his aim to introduce this new epistemological framework to his readers. Central to Keisuke's attempt to present this new classificatory schema to his Japanese readers is the concept of *shin*:

> For those of us who are studying Western science, it is urgent that we collect products from across Japan and discuss them in the context of Western theories in order to correct their names and investigate their real [*shin*] so that we can take advantage of the knowledge in curing patients. However, there are only a few people who can do this and I have not seen a book that articulates the necessary process to carry out our tasks. The reason is that those who are translators are not familiar with *materia medica* and those who are interested in *materia medica* are too preoccupied with arguing the proper Chinese names and are not interested in Western theories. Because these two types do not converge, *materia medica* is not elucidated.[26]

For Keisuke, knowledge of the real (*shin*) is the key component in pursuing more expeditious and judicious medical practice. Keisuke argues that *shin* is inherent in an individual plant—*shin* can be revealed, identified, and used by the scholars of *materia medica*—and, once properly known, can lead to the cure of an illness. Keisuke contends that applying the Linnaean binominal system facilitates assessment of a plant's *shin*. His approach merges three areas of knowledge: knowledge of the existence of particular plants in Japan, aided by the exhibitions and pictorial practices of the Shōhyaku-sha; identification of the plant's Western scientific name from texts; and information on the plant's medicinal efficacy gained through both texts and experience. Keisuke's efforts to transmit the Linnaean scheme through publication sought to close the incredible gap in knowledge between the experts of translation and those of *materia medica*, and navigate the disorienting array of names in Chinese and Japanese that confronted physicians.

The aims of *Nominal Differentiations* stand in stark contrast to a work published in 1822 by Keisuke's colleague in Edo, Udagawa Yōan (1798–1846). Yōan participated in the translation of "Encyclopedia" we explored earlier, and he was one of the most famous scholars of Western learning (*rangaku*). Yōan's *Botanical Sutra* (*Botanica kyō*) is a portable booklet that outlines the history and foundational principles of the Linnaean system in the form of a Buddhist sutra. Published seven years prior to Keisuke's book, Yōan's text summarizes the conceptual workings of the Linnaean hierarchical schema for plants and animals, but it does not explain the binominal system. Yōan's lack of interest in the issue of names and naming both demonstrates the general focus of scholars of Western learning in Edo on theoretical and abstract framings and in turn carves into relief the difference between the work of scholars in Edo and in Owari. Yōan aims to expound the truth (*shinri*) of the taxonomic theory in his book, much like, he imagines, Buddhist sutra. His focus on apprehending an abstract conceptual framework for classifying living things accounts for his disinterest in Linnaean names, since he is unconcerned with bringing the binominal taxonomy into practice. Yōan employs the traditional method of rhythmically punctuating his sentences to facilitate memorization and recitation, thus adopting a familiar pedagogy for instructing readers as a way to introduce a completely different conceptual system of understanding plant life. As the word *botanika* in the title—a transliteration of the Latin—suggests, Yōan aimed to move from the descriptive and cure-oriented *materia medica* to the structured and principled *science of botany*. The underlying frustration with naming in Keisuke's *Nominal Differentiations* is completely absent in Yōan's text, which instead strives to explain the structural principle unifying animals and plants. For Keisuke the real (*shin*) of *materia medica* resided in coordinating the name and efficacy of the plants; for Yōan it resided in the fundamental structural principles of the plant world. Yōan's book is an attempt to *reproduce* the theoretical erudition of Western botany in Japanese by assuming the universality of botany as an organized and principled way of studying plants. In this light, Yōan's translational practice in *Botanical Sutra* operated with a presumption similar to the one guiding the translators of *Compendium*, who also assumed the transparent authority in which the system of knowledge was embedded.

Yōan began the "sutra" with a short list of scholars of botany in Europe, including names of "great sages" and claims that these pioneering scholars had expounded the truth (*shinjitsu*: *shin* as in *shashin*, *jitsu* as in *jitsugaku*) and educated numerous students.[27] He then summarizes the content of these teachings, promoting a division of the world into two categories—animals and plants—the pivotal difference being the former's ability to walk. Yōan emphasizes the common structural principles that unite plants and animals by relying on a simile of mechanisms and functions between human anatomy and plants.[28] Whereas Keisuke presents the Linnaean system as a way to untangle the knot of names in order to cure patients more effectively, Yōan values Western botanical *theory* as a source for a unifying principle that grasps the world of plants. Put differently, Yōan's work attempts to map out the theoretical framework expressed in Western books by discrediting and eradicating the framework embedded in Chinese *materia medica*. In this sense, the innovation of Yōan's book is to introduce the theoretical structure of the Linnaean system without interpreting or modifying its constellation of knowledge. Similar to the translational practices of *Compendium*, Yōan assumes the transparency and universal applicability of the system. Keisuke's work, by contrast, is invested in concreteness. He sought to maintain fidelity in representational form of Linnaean names through the visuality of the Roman alphabet.

The main body of Keisuke's *Nominal Differentiations* consists of a complex list, markedly different from Yōan's *Botanical Sutra* in which Chinese characters are organized in the order of classical Chinese. Each page is divided into ten vertical rectangular compartments. Each compartment, in turn, is dedicated to identification of a single plant species, giving its Linnaean name in the Roman alphabet (Fig. 2.1, B) at the top, followed by its Japanese name in *katakana* below (C), with the Chinese name(s) in *kanji* (D) at the bottom. The circles below the Chinese names (E) indicate that the correlation between Japanese and Linnaean names was verified by Siebold, while the square frame (F) placed around certain Chinese names signifies that these are nominal relationships provided by Thunberg in *Flora Japonica*.[29] The numbers written in Chinese *kanji* characters above some of the compartments at the very top (Fig. 2.1, A) indicate the taxonomic category under which these species are grouped within the twenty-four classes of plants and their more specific orders within the Linnaean system.[30] The list is oriented in a way compatible with the direction of reading in Japanese: right to left, top to bottom. The Roman letters are positioned perpendicularly.

The plant species in the list are arranged alphabetically by their Latin names. This alphabetical arrangement departs significantly from *Flora Japonica,* which organized the lists of plant species by their taxonomic order. Thunberg's book divided the plants found in Japan into the twenty-four class categories. Take, for instance, the case of *Aconitum japonicum* in two books (Figs. 2.1 and 2.2). The plant is listed in *Flora Japonica* under Polyandria, the thirteenth category of genus that includes plants whose flowers

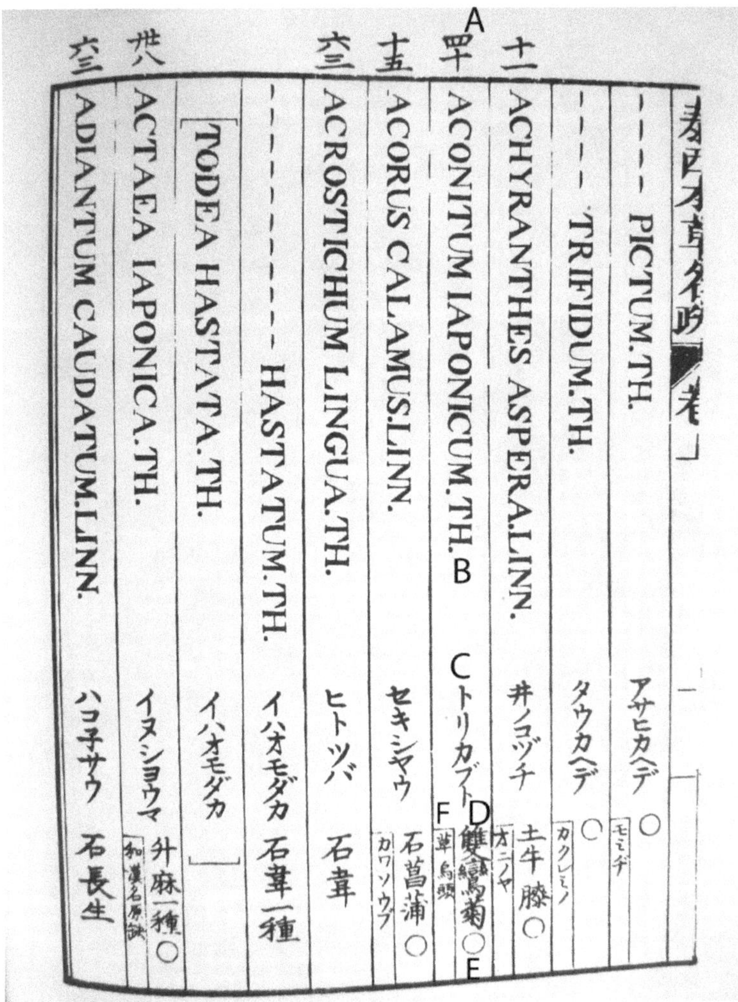

Figure 2.1 A page from Itō Keisuke, *Taisei honzō meiso,* 1829, woodblock print, Hōsa Library

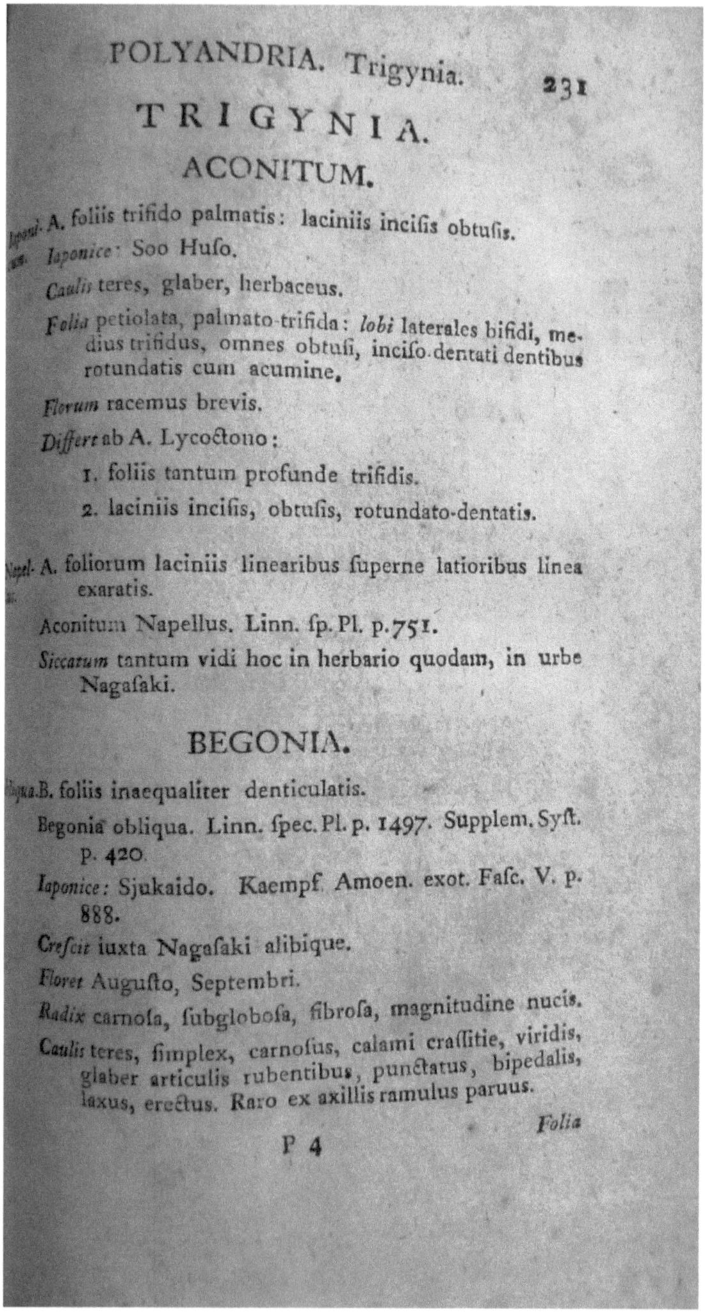

POLYANDRIA. Trigynia. 231

TRIGYNIA.

ACONITUM.

Japonicum. A. foliis trifido palmatis: laciniis inciſis obtuſis.

Iaponice: Soo Huſo.

Caulis teres, glaber, herbaceus.

Folia petiolata, palmato-trifida: *lobi* laterales bifidi, medius trifidus, omnes obtuſi, inciſo-dentati dentibus rotundatis cum acumine.

Florum racemus brevis.

Differt ab A. Lycoctono:

1. foliis tantum profunde trifidis.
2. laciniis inciſis, obtuſis, rotundato-dentatis.

Napellus. A. foliorum laciniis linearibus ſuperne latioribus linea exaratis.

Aconitum Napellus. Linn. ſp. Pl. p. 751.

Siccatum tantum vidi hoc in herbario quodam, in urbe Nagaſaki.

BEGONIA.

obliqua. B. foliis inaequaliter denticulatis.

Begonia obliqua. Linn. ſpec. Pl. p. 1497. Supplem. Syſt. p. 420.

Iaponice: Sjukaido. Kaempf. Amoen. exot. Faſc. V. p. 888.

Creſcit iuxta Nagaſaki alibique.

Floret Auguſto, Septembri.

Radix carnoſa, ſubgloboſa, fibroſa, magnitudine nucis.

Caulis teres, ſimplex, carnoſus, calami craſſitie, viridis, glaber articulis rubentibus, punctatus, bipedalis, laxus, erectus. Raro ex axillis ramulus paruus.

Folia

P 4

Figure 2.2 A page from Carl Peter Thunberg, *Flora Japonica*, 1784, University of Michigan Herbarium Library

contain multiple stamens (more than twenty) with receptacles. Thunberg's listing indicates the Polyandria category and identifies two distinct specimens found in Japan that belong to Aconitum: *Aconitum japonicum* and *Aconitum napellus*. In the entry for each plant, he then provides the Japanese name(s) ascribed to the plant and describes the plant itself, including the location where the particular specimen was found. In Keisuke's listing of the same plant, in comparison, he uses the number forty at the top of the page (written in Chinese characters, A) to indicate the "polyandria, trigynia" order, and then he provides its binominal name in Roman letters oriented toward the right (B). Omitting where the plant was found and the description of the stems and leaves, Keisuke's listing rearranges the information so as to focus solely on the names. This highlights the centrality of the issue of names in the pursuit of *materia medica* for the Shōhyaku-sha. Both in the configuration of the listing and the prominence of the large font used to render Linnaean names, Keisuke addresses the import given to matching and stabilizing the names of plant specimens. He then lists a Japanese common name in *katakana* script in its proper orientation, read from top to bottom (C). Its Chinese name follows in Chinese characters (D), which are also read from top to bottom, and the circle below indicates that Siebold verified the information (E). The smaller set of Chinese characters is the Japanese name of the plant according to Thunberg, *Aconitum japonicum* (F). Unlike Thunberg's listing, which relies on the categories of genus and species to structure the publication, Keisuke's book concentrates on the names ascribed to plants, thus adapting the Linnaean system to the specific challenges of naming and knowing plants that he and his colleagues encountered in Owari.

By prioritizing the alphabetical names, *Nominal Differentiations* makes the Roman alphabetical order more pronounced and visually asserts the divisions of class and order as prominent and decisive elements of the binomial system. As the Latin names in the Linnaean taxonomy are constructed from the categorizations for the orders of genus and species, Keisuke's rearrangement of *Flora Japonica* relies on the relationship between the organizing principles of the Linnaean scheme and their Latin names, maintaining fidelity on both representational and conceptual levels. For instance, a plant known as *tennansho* is listed alphabetically under "Arum dracunculus. Linn." followed by the plant's other names in Japanese *katakana* and Chinese characters. *Musashi abumi*, a plant that resembles *tennansho* and was sold as such, is listed as a separate species six columns over as "—Ringeus. Th.", followed by the *katakana* rendering, *musashi abumi*, and Chinese characters noting that it is "a kind of *tennansho*" (Fig. 2.3). What could not have been gauged previously, both intellectually and visually, however, is that the five plants between *tennansho* and *musashi abumi* belong to the same genus of *Arum*. Their grouping and proximity in this newly applied

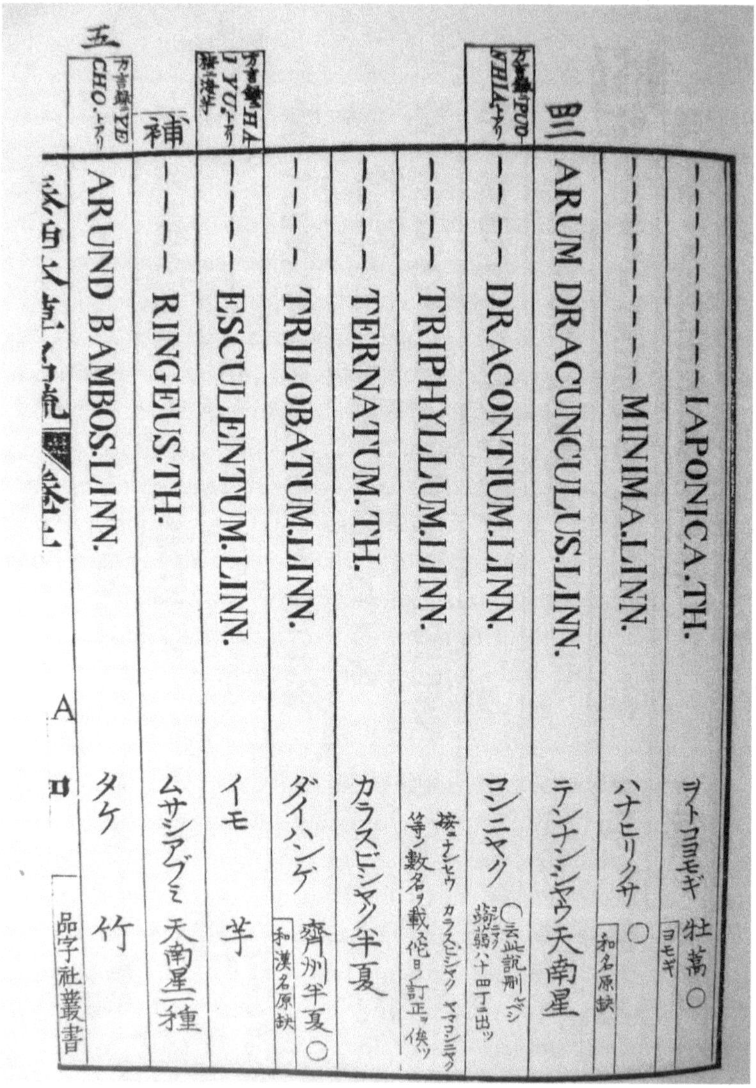

Figure 2.3 A page indicating *tennansho*, from Itō Keisuke, *Taisei honzō meiso*, 1829, woodblock print, Hōsa Library

order of listing allows the reader to imagine the relationships of lineage and similarity among the seven listed plants.

Comparing the entry for *tennansho* with the entry for the same plant in *Clarification on the Names of Things*, published twenty years prior by Keisuke's teacher Hōbun, underscores the significance of the visuality of the Roman alphabet in *Nominal Differentiations*.[31] In *Clarification*, which is organized according to the Japanese syllabic order of *iroha*, *tennansho* is listed under the category of objects whose names start with the "te" sound. Within this phonetic category, there are sixteen further divisions adopted directly from Li Shizhen's *Compendium*, which was arranged according to the phonetic order of the Chinese names and provided an influential precedent for the organization of Hōbun's book. Similarly, in *Clarification*, the plant *musashi abumi* is listed under the category of objects whose Japanese name starts with a "mu" sound. The Chinese character below indicates that it is a kind of *tennansho*. Unlike *Nominal Differentiations*, however, Hōbun's book did not—and could not—suggest visually through the organization of the text the taxonomic relationship among *tennansho*, *musashi abumi*, and other plants that are grouped together under the *Arum* genus in Keisuke's work. The visuality of Latin text consequently usurps the auditory character of Japanese as the principle of categorization in *Nominal Differentiations*. Hōbun's syllabic organization could not have solved the issue of naming because the variety of regional names given any particular plant species did not necessarily begin with the same syllable. By visually grouping his schematized listing of plants, Keisuke mobilized the Roman alphabet as a visual medium that pictorialized the differences and similarities among plant specimens. In reorganizing the order of Thunberg's *Flora Japonica*, Keisuke decided to arrange the order of entries by prioritizing the genus of the Linnaean names expressed in Roman alphabet. Keisuke rejected the phonetic arrangements of *Clarification* and *Compendium*, and he avoided organizing his book with transliterated Latin names in Japanese *katakana* scripts. Arranging the plants alphabetically using Roman letters by their Latin names allows him to uphold and visually foreground the importance of the genus category within the binominal taxonomy.

Keisuke's clever harnessing of Chinese characters to express seminal concepts of the Linnaean system circumvented the Roman alphabet, maximizing the size of the potential Japanese audience. He rendered pistils and stamens—respectively, the female and male reproductive organs of flowering plants—into Japanese (*meshibe* and *oshibe*) by devising new combinations of two Chinese characters for each word. The first part of the new translated terms used the character for either female or male, paired with the character for flower for the second part. Keisuke thus articulated the functional meaning of these plant organs through Chinese characters in a way that the abstruse words *pistil* and *stamen*, whether written in the Roman alphabet or transliterated, could not conjure for his readers. As signifiers, the Chinese characters of male and female denoted and communicated the abstract meanings of pistils and stamens much more effectively to his audience. In other words, Keisuke strategically took advantage of the semiotic functions of Chinese characters familiar to his audience, rather than transliterate these words, which would have captured only the sounds of the terms and not their meanings. The prioritized fidelity in transposing these Chinese characters to the concepts pistils and stamens thus resides in conveying the meanings of the concepts.

Keisuke's innovative combination of Chinese characters parallels his use of the Roman alphabet in the naming of plants. Although he could have transliterated the Latin names into *katakana* script—phonetically mimicking the pronunciation of the Latin names as was commonly practiced by the scholars of *materia medica*—he chose to record the Linnaean names using the Roman alphabet, while rendering concepts central to Linnaean nomenclature through creative combinations of Chinese characters. Much as Keisuke's newly coined Chinese terms succinctly convey their meaning, the Roman alphabet preserves the fidelity between Linnaean names and the plants to which they are ascribed in a way that other textual representations, such as transliteration and translation, could not. Both uses of text reveal Keisuke's sensitivity to the visuality of text in construction of knowledge. The project of *Nominal Differentiations* is not merely a translation of *Flora Japonica* for the sake of novelty, or the introduction of Linnaean binominal taxonomy as an allegedly superior system. Rather, it is a product that emerged from and responded to his concern for the local status of *materia medica*.

In the Linnaean binominal scheme, Keisuke saw the potential to clarify the confused nomenclature of *materia medica* that hindered understanding of real plants. As indicated by his comments cited above, he identified the core of the problem with translators' ignorance of the *materia medica* and scholars' preoccupation with matching varied plant names in Chinese and Japanese. By introducing an entirely new and different text—the Roman alphabet—to render the names, the issues of translation and local variations that thwarted clarification of names in *materia medica* is partially resolved.

The scholars of *materia medica* need not be concerned with rendering names in either Chinese characters (*kanji*) or Japanese *hiragana* or *katakana* scripts, and the translators need not translate the names into Japanese or render them into *katakana*. Knowledge of the Roman alphabetical order and pronunciation of Latin names, moreover, was not a prerequisite for profiting from *Nominal Differentiations*. Looking up the Latin name for *tennansho* (jack-in-the-pulpit), for instance, does not require the reader to know the proper pronunciation of the genus name *Arum*. The Roman alphabet did not so much serve here as a script from which one can extract meanings as a series of graphic symbols that expressed the Linnaean names visually without concern for pronunciation.

Writing the Linnaean names of plants in the Roman alphabet in *Nominal Differentiations* minimized the potential confusion caused by introducing a new taxonomy. Rather than guessing at how phonetically transcribed names in *katakana* would be rendered originally in the Roman alphabet, readers could focus on the relationships of names and plants. Keisuke's reservations about translating the Linnaean names of plants into Chinese characters or transliterating them using *katakana* emerge clearly in his introduction to the supplemental volume:

> In Western books, there are some that include Chinese names [in Chinese characters], similar to the method of *materia medica* scholars in Japan. These Chinese names are very helpful and useful as a reference. However, these names in Chinese characters sometimes phonetically transcribe the sound of Western names, or they are rendered wrongly. Therefore, it is difficult to recognize the object itself from the ascribed Chinese names alone [in the Western books]. There are also many names where it is difficult to clarify [the Chinese name] through discussions. It is because of these reasons that I included their names in Hamburg dialect in this appendix volume. While I was traveling [to Nagasaki], I had opportunities to examine several foreign books [on plants], some of which included direct translations of [Western names] into *kanji*.[32]

Keisuke thus observed problems with translating and transliterating plant names in Western books similar to those he encountered in Owari. In the West, too, attempts were made to translate foreign names into different languages, for instance, from Chinese to English, but they often resulted in further confusion. Observing the attempt to apply phonetically similar-sounding Chinese characters to English names or going through the process of matching names between two languages in Western books (precisely the same confusing process that the Shōhyaku-sha members and scholars of *materia medica* were grappling with), Keisuke decided to include an appendix with a list of names in English and Linnaean binominal taxonomy with varying Chinese names given in subscript.

The supplemental listings of translated Chinese names that appear in Western books further attest to Keisuke's awareness of how translation could exacerbate the disjunction between name and knowledge. Keisuke's listing for "Tigers Paw" in the appendix exemplifies his awareness of these problems (Fig. 2.4). According to Keisuke, "Tigers Paw" corresponded to the Chinese plant known in Japanese as *koshou* (Ch: *huzhang*, Keisuke Romanizes the Chinese pronunciation as "*hoo chang*"). This English name, moreover, is not a Linnaean name but comes from direct translation of the two characters that make up its Chinese name: "tiger" and "paw." Only those competent in both Chinese and English could discern this relationship between the different names for the plant and avoid mistaking the English name as one that conveyed knowledge about it. Keisuke, however, makes this translation visually apparent in the appendix by placing the Chinese characters for "tiger" and "paw" next to their English equivalents. The two-word translation "Tigers Paw" could mislead readers into thinking it is a Linnaean binominal name, but by placing the Chinese characters next to the letters Keisuke conveys that "Tigers Paw" is a transliteration of the Chinese name. Should a reader come across a Western botanical book that discusses "Tigers Paw," then, he would be able to refer to the appendix and identify the plant known to him through the two Chinese characters. By creating a visual correspondence between two scripts, Keisuke aimed to save further confusion by introducing the Linnaean binominal taxonomy.

In contrast, the awareness of confusion entailed in transcribing in a different script points precisely to what distinguishes the Linnaean classification system from others as a methodology. Because it is a system expressed only in the Roman alphabet common to European languages and was constituted in Latin among the scholarly circles of Europe, it serves as an anchoring device to communicate across language barriers. Certainly, Foucault's interest in elucidating the epistemological formulation of natural history was limited to cases in Europe. But here we witness the way in which, in hindsight, a stronger case can be made because the Roman alphabet and Latin served solely as coded visual representation. By implementing the Linnaean names, represented in the medium of Roman alphabet, the disarray of names that confronted the Shōhyaku-sha—the dialectic names, Chinese names, and

Japanese names—could be replaced by one sharable and stable name. By maintaining the Roman alphabet in the list of Linnaean binominal names in *Nominal Differentiations*, Keisuke omitted one layer of translingual mediation, keeping the fidelity of the Linnaean names and the named plant unbroken. Coordination of Chinese, Japanese, and Latin names in *Nominal*

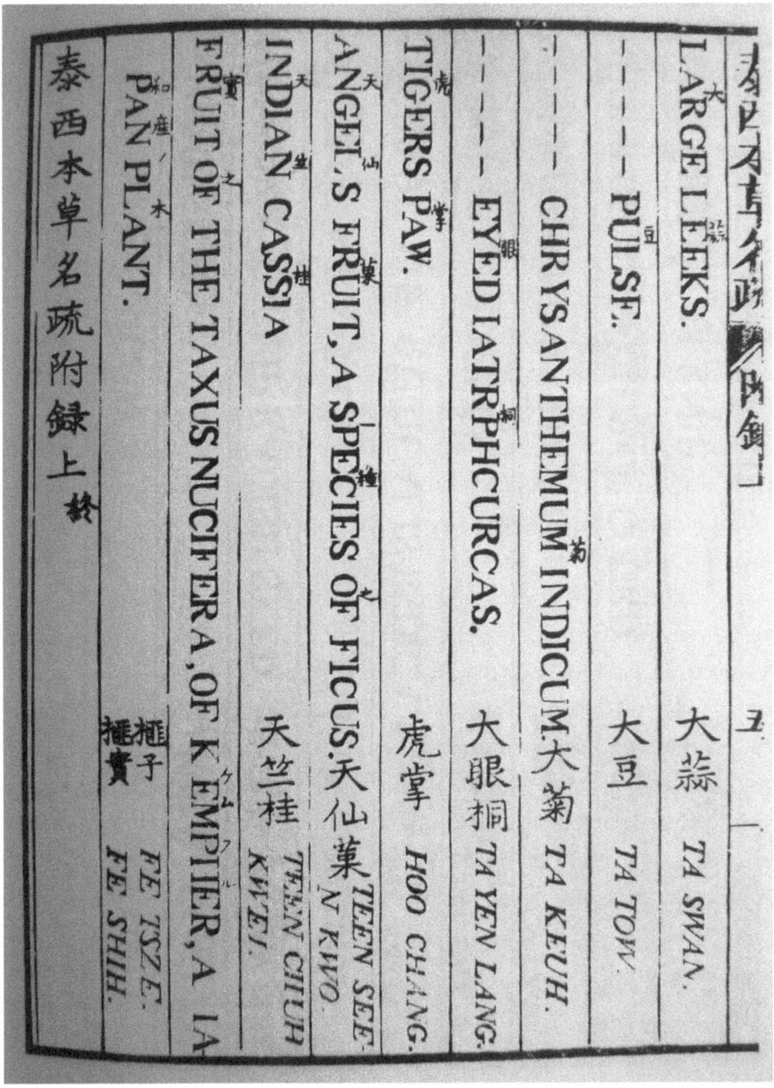

Figure 2.4 A page showing "Tigers Paw," from Itō Keisuke, *Taisei honzō meiso furoku*, 1829, woodblock print, Hōsa Library

Differentiations was effective because Keisuke kept the Linnaean names in their original language, thus preserving the clarity of the binominal system through which he hoped to reorganize botanical knowledge.

Along with the works by his teacher Mizutani Hōbun, such as *Clarification* and *Compendium of Materia Medica, Heard and Seen*, Keisuke's *Nominal Differentiations* was a direct response to the immediate and urgent issues of names and naming that confronted the Shōhyaku-sha. Keisuke's decision to maintain the Roman alphabet in *Nominal Differentiations* assisted in streamlining the process of untangling the mess of variant plant names and organizing the knowledge of *materia medica*. In this book, alphabetic letters functioned as a code that presented names as visual representations. At the same time, the creative combination of Chinese characters to translate Linnaean vocabulary facilitated readers' apprehension of the Western taxonomic system by extending the already available meanings of Chinese characters.

As we explore in the next chapter, the group regularly organized exhibitions in order to directly engage with the actual objects studied in *materia medica*, as a way to problematize and concretize the relationship of things, their names, and their known efficacy through observation. Keisuke's documentation of these exhibitions reveals that the Linnaean names for the plants in the displays were rendered in transliteration using *katakana*. Keisuke had clearly differentiated the purpose of *Nominal Differentiations* from the recording of the exhibition: in *Nominal Differentiations* he aimed to introduce and disseminate the Linnaean binominal scheme to the scholars of *materia medica* in Japan, while he published accounts of exhibited items in order to demonstrate to a wide audience that the very problem of matching names to reality persisted in the study of *materia medica*. Prioritization of the representational medium in *Nominal Differentiations* served to eliminate another layer of translingual mediation, while documentation of exhibitions foregrounded the problems of naming sharpened by direct observation and study of the plants themselves.

THREE

Modes of Observation and the Real

Exhibition Practices of the Shōhyaku-sha

> People used to say that those who sell medicine use two eyes, those who provide it use one eye, while those who consume it are blind. Today, however, it is not just the consumer who is blind: even the medicine seller and providers are all but blind.
>
> Hiraga Gennai, *Record of Bunkai Gathering* (*Kyokuzan sensei bunkairoku*, 1760)

The discourse of *materia medica* at the center of the Shōhyaku-sha's activities involved coordinating various sources in order to reconfigure their knowledge. The act of looking was fundamental to this process of synthesizing Chinese *materia medica*, Western botanical taxonomy, and the study of living plants in their environment. Hinged on the concept of the real (*shin*) as in the essential qualities of every plant that can be revealed and known, the introduction of the Linnaean scheme directed and shaped how the group examined plant specimens. The Western system of nomenclature helped stabilize the relationship between specimens and the names ascribed to them, but taxonomy did not—and could not—solve all the epistemological challenges that faced the Shōhyaku-sha: the medicinal efficacy of plants, the local availability of plants, and ways of discriminating between similar-looking plants required more than implementation of Linnaean binominal names. In fact, the group never disavowed the epistemological framework of *materia medica* articulated in *Compendium of Materia Medica* (*Bencao gangmu*) or stopped using names derived from the text. A practice that allowed them to address these issues was collective exhibition (*honzō kai*). In the process of constructing knowledge through

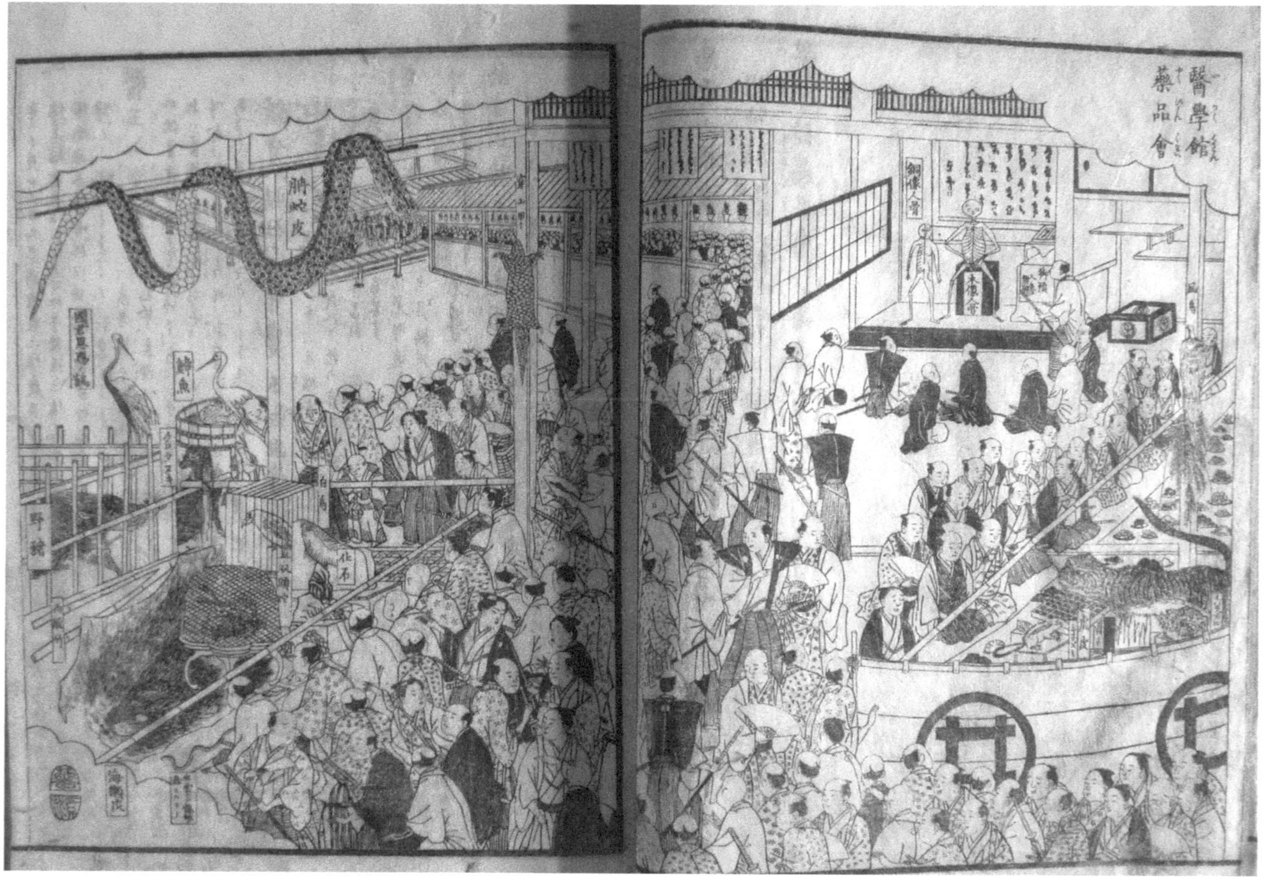

Figure 3.1 Odagiri Shunkō, *Igakukan*, from *Owari meisho zue*, 1844, woodblock print, University of California Library

the triangulated interchange among texts, illustrations, and actual objects, *honzō kai* events served to showcase actual objects as the public and collective site of inquiry. The various dialogues that occur in observing the objects themselves necessitated assessment of information previously attained in text, illustrations, or experiences, and yet the primary significance of the *honzō kai* as part of the epistemological formulation resides in the fact that it offered a place for contemplating and examining living plants.

Although not all catalogues of *honzō kai* survive, those that do indicate that the Shōhyaku-sha mounted these exhibitions monthly from as early as 1827 until 1874. The responsibilities of hosting would rotate among the group, and a theme was selected for each gathering.[1] A woodblock print from *Notable Places of Owari* (*Owari meisho zue*), published in 1844 by Odagiri Shunkō, allows us to imagine what these *honzō kai* were like (Fig. 3.1). The image depicts a room designated for "miscellaneous" at a *honzō kai* hosted by Owari's domainal medical academy, *Igakukan*, where several Shōhyaku-sha members studied and taught.[2] Like the *honzō kai* hosted by the Shōhyaku-sha, this annual fair was open to the public and featured a comparably large number of exhibited items organized and displayed in similar ways.[3] A diverse crowd, including samurai, merchants, and Buddhist monks, as well as a woman and a child, fill the large room, while people near the bottom right wait to enter. Those wearing black kimono coats (*haori*) and swords are members of the samurai class, possibly domainal physicians appointed by the Owari domainal lord and teachers at *Igakukan*. Some visitors seat themselves in front of the arranged objects, while others seem to be rushing elsewhere. The crowd clusters particularly around the live animals in the bottom left corner, and the figures wear expressions of curiosity and excitement at catching a glimpse of the animals— a wild boar, a black raccoon, a crane, a two-headed snake, and a swan, while a snakeskin and pangolin skin hang above. Far fewer figures gather around a wooden skeleton in the alcove (*tokonoma*) on the upper right, a reception area used to display precious objects. Objects included in the symbolically significant *tokonoma* area include, in addition to a wooden skeleton, a human figure made of copper used for teaching acupuncture, *ninjin* (Korean ginseng), and *yūtan* (Ch: *xiongdan*, dried bear bile), which was used in *materia medica* to dissipate heat from the body. The accompanying labels indicate that the *ginseng* and *yūtan* were borrowed especially for this occasion from Lord Tokugawa of Owari.[4] Similarly, to the right of the alcove the participants gaze at the minerals placed directly on the *tatami* mats, allowing the viewers to observe the items closely by lowering themselves to the floor. In the distant background to the left, the hallway connecting the two buildings hosts displays of potted plants, neatly located just above audience eye level so as to prevent any accidental breaking of pots.[5] The *honzō kai* hosted by the Shōhyaku-sha would have employed similar techniques to display the plants. The potted plants as well as snapped branches would have been positioned in such a way as to observe them directly and closely, while also maintaining a physical distance to protect the plants.

These display practices became significant for the group as a way to address collectively the issues of the naming, availability, and efficacy of plant specimens. As was the case in Linnaean taxonomy and *Nominal Differentiations in Western Materia Medica* (*Taisei honzō meiso*), the concept of the real (*shin*) was deployed to articulate the intended purpose of the exhibitions. In the context of exhibitions, the notion of the real operated dialectically with its antithesis of the false (*gi*): the former was negotiated and apprehended through the possibility of the latter. Observation of objects gathered at the exhibitions were thus embedded in a dialectics of knowledge, as the members of Shōhyaku-sha collectively sought to grasp the real through seeing and discussing the exhibited items.[6] The concept of the real pursued in these exhibitions, then, resembles an ideal of authenticity in the sense that the exhibition aimed to delineate the characteristic and attributed name(s) of the displayed items by connecting them to existing knowledge. In these collective efforts to identify and recognize the real, the known "facts" from publications conveyed as much weight as personal opinions and expertise. In other words, these exhibitions called for participants to suspend temporarily the authority and confidence bestowed on one's own knowledge and instead to approach an object with the possibility that modification of one's knowledge of its name and curative characteristics might well be necessary. These occasions of *honzō kai* served as a temporary laboratory in which knowledge was tested against the real through observation and discussion in hopes that by the time the exhibition closed (usually within a day), some of the relationships of the actual plants and their names, efficacy, and availability would be fully ascertained. This system of appraisal and contestation thus created a complex recursive process.

This chapter posits these exhibitions (*honzō kai*) as a hybrid medium of communication and site of negotiation that aimed to strengthen fidelity among existing objects, their names, and useful knowledge about them.

Anchoring the group's exhibition practices was the presumed power of observation to reveal, authenticate, or challenge the name and characteristics ascribed to the exhibited items. Protracting the triangulated relationship among the text, illustration, and the actual entity, these occasions served as a site of reassessing knowledge via attentive observation of displayed specimens and conversation with others. In the practice of translation, Itō Keisuke faced the challenges of mediation by types of script and chose to maintain the Roman alphabet in representing Linnaean names in order to uphold the fidelity between those names and their expression in the publication. Examination of actual objects at *honzō kai*, by contrast, was not hindered by the same issues of mediation. The objects were assumed to be bare entities to which names, medicinal efficacy, and uses could be attached. In short, the issue of representation was not pivotal or conditional in organizing these occasions. However, it would be misleading to posit the occasions as transparent and free of representational mediation. The primary purpose of these events was to examine objects with the naked eye; the *honzō kai* were always mediated by the very epistemological conditions—tags attached to the objects and conversations that took place within the space—that would filter and shape observation of displayed objects. In the triangulated process of decoupling and resynthesizing the discourse of *materia medica*, *honzō kai* offered actual objects as the central focus of inquiry while framing them by imputing previously known facts and names. The epistemological negotiations that took place in these events both assisted and hindered the quest for the real that stood at the center of the Shōhyaku-sha's enterprise.

The analysis that follows foregrounds the three interrelated ways in which knowledge and the observable were negotiated within the *honzō kai*: (1) the persistent need to locate the real (*shin*) in the name of particular specimens, thereby distinguishing them from misidentified medicinal plants; (2) conflicting and elastic notions of the real that conflated the real embodied in any single specimen with that of the broader species category exemplified by the specimen; and (3) the weakened hierarchy in the sources of information and knowledge used to identify the real. What underlies these practices of observation, and what our discussion brings forward, is how the real became imbued within the physically present specimen itself even as the conceptual import of the real and strategies to identify it through observation permeated these exhibitions. Although *honzō kai* served as the site of ongoing epistemological processes to identify the real by stimulating the exchange and synthesis of knowledge, the outcome of these events augmented the notion that *shin* was ultimately located within material objects. The voluminous knowledge and experience associated with the exhibited objects, therefore, was rendered relational, rather than constitutive to the real.

Honzō kai as a Medium

To be sure, the Shōhyaku-sha did not "invent" the *honzō kai* gathering. It had been organized ever since the late eighteenth century.[7] The strong personal interest of the eighth shogun, Yoshimune (r. 1715–1745), in the field of *materia medica* led him to allocate resources to sponsoring and expanding the pursuit of *materia medica*, which arguably led to increased numbers of collective groups and gatherings such as *honzō kai*.[8] Historian Isono Naohide counted more than 250 *honzō kai* between 1757 and 1867 from extant records, including 72 in the city of Edo, 88 in Kyoto, 21 in Osaka, and 58 in the Owari domain, where the Shōhyaku-sha was based.[9] The frequency and the popularity of the *honzō kai* format echoed and extended the seminal shift in *materia medica* discourse toward a closer and stronger relationship between ocular observation and knowledge. As a medium that framed the exhibited specimen within a discourse of *materia medica*, while also encouraging active renegotiation of knowledge through intellectual exchange and observation, the *honzō kai* constituted a new interactive configuration that both expanded and deepened the discourse of *materia medica*.[10] The assumption that the *honzō kai* events would and should facilitate the process of reevaluating and repairing the field of *materia medica* is found in the record of the *honzō kai* gathering in Osaka in 1760.

In the fourth month of 1760, Toda Kyokuzan, a physician and scholar of *materia medica* in Osaka, hosted a *honzō kai* gathering at a local Joan-ji temple. In the record of this gathering, published as *Record of Bunkai Gathering* (*Kyokuzan sensei bunkairoku*), the famous innovator and satirist Hiraga Gennai couples perception and *honzō kai* as powerful tools to combat the ignorance that permeates the field of *materia medica*:

> People used to say that those who sell medicine use two eyes, those who provide it use one eye, while those who consume it are blind. Today, however, it is not just the consumer who is blind: even the medicine seller and providers are all but blind. The seller knowingly identifies *honhakusai* [*Potentilla fragarioides*] as *saiko* [Chinese thorowax root], and *omoto* [*Rohdea japonica*] as *riru* [*Veraturum nigrum*, black hellebore Ch: *lilu*], and cases like these are countless. Our teacher Kyokuzan had been deeply concerned about such situations.

> Early this summer, he brought together those who are interested in *materia medica*, each submitting a few samples of medicinal items in order to discuss and distinguish their realness and forgery, and if the sample was not well known, he asked the artisan to make a picture of it. Our teacher Kyokuzan collected them and published [the record of the gathering] as a book here. If anyone wants to pursue this path, implementing [the knowledge provided in] this book, he would approximate avoiding the sorrow of being blind.[11]

Certainly, Gennai was both familiar and dissatisfied with the contemporary state of the *materia medica* field as he himself hosted *honzō kai*, known as *Tōto yakuhine*, in Edo in 1759 and 1763.[12] Whether intentional or unintentional, the confusion of identified names of the plant remedies serves here as a concrete and pivotal concern that motivated Kyokuzan to organize the exhibition. This introduction makes explicit not only the centrality of the medical function of *materia medica* but also the abusive practices of pharmacologists and physicians. Gennai sets forth ways in which the current uncritical and neglected state of *materia medica* directly affects the quality of care provided by physicians. The metaphor of blindness, as a metaphor for lack of knowledge, effectively underscores the pretense of ignorance performed by both seller and prescribers who acted uncaringly, the antithesis of what they were believed to practice. Gennai implies that the blindness of the sick to the remedies could not be helped, whereas misidentification of plants by medical professionals could.

In Gennai's mind, the synthesis of knowledge and observation promoted by *honzō kai* becomes not only the "cure," so to speak, of current misguided practices but also a means to eliminate future ignorance and ensure the accuracy of prescriptions. Consequently the *honzō kai* is rendered as the pivotal medium through which the real would be separated from false knowledge of medicinal plants through the power of perception and collaborative discussion in forging a necessary consensus. In this process of coordinating and reassembling knowledge and experiences, the correct name of the item moored the entire occasion. Recognizing the plant specimen as a medicinal plant was one step, but what was of paramount importance was identifying the plant with a name that withstood the questioning of the exhibition's participants. Calls circulating for submissions included in *Record* specified the aims of the gathering:

> As noted earlier, the purpose of this assembly is to inquire about and verify the real (*shin*) and the false (*gi*) objects through discussion. We will eliminate any suspicions about specific items and correct information related to

> the objects. If you have doubts about any object displayed, please do not hesitate to raise a question. Participants will discuss the item together and determine [the answer to your question]. Similarly, if we the organizers find the names of your submitted item improbable, we will not hesitate to raise questions regarding its authenticity. Please be patient and wait for the conclusion of the discussion. Of course, in regard to those [specimens] brought in by residents living far away, if there are any doubts cast on them, we will clarify the matter immediately. If the outcome of our inquiry does not correspond to your opinion, please accept our evaluation ungrudgingly.[13]

The primary goal of these *honzō kai* gatherings rests in identifying the real uncontaminated by misinformation and misidentification. The confidence expressed here that collective inquiry during the exhibition led to a definitive conclusion, however, would become less assertive six decades later at the Shōhyaku-sha's exhibitions.

It is helpful here to situate this effort to find existing specimens in their environment and match correct names to them in relation to the epistemological shift instigated by Kaibara Ekken. Ekken redirected *materia medica* by pointing to the marred state of the field, recognizing the disparity between the canonical texts and his own environment, thereby laying the groundwork for reassigning the names of plants in the Japanese environment. He encouraged scholars of *materia medica* to pair the results of direct observation of natural specimens with the accepted knowledge of *materia medica* by exploring their own regions. He then urged them to contrast their observations with the knowledge derived from Chinese texts. In doing this they would be acknowledging and promoting a symbiotic relationship between observation and knowledge in which firsthand examination is inseparable from the configuration of *materia medica*. Whereas local plants were accessible and thus observable, the plants described in Chinese texts were physically out of reach, rendering the knowledge offered by the texts unverifiable. By contrasting the specimens presented in the Chinese texts with information obtained firsthand, Ekken posited observation as a source of knowledge on equal footing with received tradition. This emphasis upon observation in Ekken's *Materia Medica in Japan* (*Yamato honzō*) proved as significant in redirecting the study of *material medica* as his bifurcation of the translational approach to Chinese texts discussed in the previous chapter.

Ekken's emphasis on exploring one's local environs and observation brought two seminal epistemological shifts to the field of *materia medica*: (1) the emphasis encouraged testing knowledge gathered in books to determine its compatibility with the Japanese environment; and (2) it also facilitated detection of subtle differences in shape, color, size, and smell, which in turn could result in accumulation of new information and knowledge.[14] Ekken envisioned the study of *materia medica*—involving excursions to nearby mountains; locating and identifying plants, minerals, fish, and animals in their own environment; as well as sharing and contesting information with others—as requiring skepticism, synthesis of knowledge and experience, and verification of knowledge and names through one's own bodily senses.

The proliferating *honzō kai* held in the latter half of the eighteenth century furnished ideal venues for engaging in the study of *materia medica* as envisioned by Ekken. Within the exhibition space, one was able to examine objects from various regions of the country and discuss discrepancies in knowledge and experiences associated with those objects and their use. These collective displays thus functioned as a new hybrid medium. They mediated the content of the inquiry, nested it in tandem with discursive questions, and reconnected the items through various types of representation, including regional and real names. The presiding concerns with discerning, identifying, and verifying the real necessarily fostered a dialectic of seeing that focused on comparative and contrastive viewing. At *honzō kai* events, then, with the material objects displayed before the eyes, observation engaged in a recursive process of turning the visible into the knowable and vice versa. The critical gestures of this negotiation were to point to visible characteristics that helped determine what one was looking at (from visible to knowable), and to delineate how one's knowledge about the plant was visibly manifested (from knowable to visible).[15]

It is important to emphasize that in these exhibitions the body of knowledge presented in *Compendium* per se was not questioned or refuted; rather, its applicability in the Japanese context was the purpose of the display. *Compendium* thus served as a basic reference that was reconfigured in accordance with local knowledge of *materia medica*. Indeed, the Shōhyaku-sha's two *honzō kai* in 1827 included a visual display of selected categories directly taken from *Compendium*: at the *honzō kai* hosted by Keisuke, the *Compendium* section on minerals and stones was displayed; and at the one held by Ōkōchi Zonshin the section on plants growing on stones formed the basis for part of the exhibit.[16] Although it is unclear if the corresponding pages of *Compendium* itself were displayed along with the items, the order of the objects in the exhibition record corresponds to the order in which items were listed in the text. The inclusion of *Compendium* indicates the centrality of this publication to the pursuit of *materia medica*, while simultaneously highlighting the unstable state of names one hundred years after Ekken's *Materia Medica in Japan*. In the displays structured upon these specific categories from *Compendium*, the presumed name given in the text was written on a tag, along with a note that indicated the variety ("one of its kind"), the name of the submitter, and the names of the town and mountains where the item originated—literal manifestation of the Shōhyaku-sha's desire to situate specimens in their locality. These *honzō kai* practices thus decoupled and reconnected objects, textual information provided on the tag, and one's perception with the goal of more firmly grasping the real.

This practice of exhibiting living plants, which is already mentioned in the Chinese canon, offers insight into the issue of translation explored in the previous chapter and helps elaborate the various ways mediation worked in *honzō kai*. Ekken's call in *Materia Medica in Japan* (1709) to approach *Compendium* and other Chinese canonical texts with skepticism gave rise to a group of scholars who sought to reformat and reshape the epistemology of *materia medica* in the eighteenth and nineteenth centuries. These scholars abandoned the idea of transparent and uniform equivalence between the flora surrounding them and the plants described in Chinese texts, and instead they assumed heterogeneity between their environment and the body of knowledge presented in the canons of *materia medica*. In this process of rematching their local flora with the textually rendered entries in Chinese canons, the variety of dialectic names complicated any attempted synthesis. The introduction of Linnaean taxonomy was but one of many attempts made to rearrange the names ascribed to plants into a sharable and stable format in order to facilitate calibration between the local flora and names given to plants in texts. Yet implementation of the Linnaean binominal scheme alone did not fully resolve all the epistemological issues. The regional names still continued to be used in *materia medica*, and the efficacy of medical plants needed to be rearticulated. Displaying the existing plants from their region and attaching a tag that noted the location where the plants grew, these *honzō kai* practices attempt to enact the study of physical objects called for by Ekken a century and a half earlier. Practicing these exhibitions served to answer epistemological issues that the effort to implement a Linnaean taxonomy could not fully solve. Just as the Linnaean binominal system could not solve the issue of names once and for all, the *honzō kai* gathering would not have settled the issue either. Rather, the *honzō kai* functioned as a means to ascertain pieces of information and verifications incrementally.

The *honzō kai* served to mediate the study of *materia medica* in a way that printed texts could not, by bringing together in a physical location both actual examples of plants—from species listed in *Compendium* as well as others absent from the text—and knowledge associated with them.[17] As a medium that facilitated the study of *materia medica*, the interpersonal dimension of such occasions was crucial, as the 1760 flyer for *Record* indicated, for the exchange enabled individuals to extrapolate, modify, or challenge the prescribed information at an accelerated pace.

Here, Ian Hacking's reminder of the need to recognize the distinction between "construction-as-process" and "construction-as-product" is particularly useful to further frame the epistemological stakes of *honzō kai* gatherings.[18] Addressing the discomfort among both scientists and social constructionists regarding the relationship between scientific facts and the historical and cultural contingencies at stake in the discovery of such facts, Hacking emphasizes the distinction between the two as the key recognition that allows us to tease out the nature of claims made in natural sciences. Although construction-as-process and construction-as-product are intimately related, Hacking claims that sociologists are interested in the activity in the laboratory and in other social activities in which scientists partake. The focus of the study thus entails the element of time. On the other hand, Hacking argues that construction-as-product, a category that scientists are interested in, is the assemblage of truths, a search for universal and autonomous truth and the law of nature that exists outside of temporal spectrum. This distinction "between an activity and an assemblage of truths" enables us to historicize the material while deemphasizing the "scientific correctness" of its epistemological content. This perspective highlights the processual nature of inquiry in the discourse of *materia medica* and the *honzō kai* by seeing both as ongoing interactions with materials and ideas that presented solutions, questions, and new ideas to pave the way for the pursuit of knowledge. Understanding the *honzō kai* events as construction-as-process, rather than as reflections of construction-as-product, allows us to study the Shōhyaku-sha's continued practice of *honzō kai* as an open-ended historical process while avoiding ahistorical judgments of *materia medica* as scientifically unsound. Instead of comparing their understanding against current knowledge of medicinal plants, this approach opens the more productive potential of seeing in their activities the very process of "assembling truths."

Reassessing Knowledge and Shin

The confusion by the scholars of *materia medica*—intentional or unintentional—of the real (*shin*) and the false (*gi*) that concerned Gennai in 1760 continued to recur in *materia medica*, as the Shōhyaku-sha highlighted the identical issue in their call for submissions for their 1830 *honzō kai*. The first entry on the list of types of items they are seeking reads:

> 1. Medicinal items. These include items for which descriptions can be provided—whose realness (*shin*) and qualities can be debated, or ones originating in Japan whose Chinese name is unknown, or has a Chinese name but cannot be found in Japan, or originates in Japan but whose Chinese and Japanese name is uncertain. Also included are those that are inaccessible and whose Chinese names should be discussed.[19]

This passage reveals that for the Shōhyaku-sha, too, the notion of the real and the false hinged on names. The existing variation of names, taken from Chinese sources or local dialects, and the epistemological tension that Keisuke's *Nominal Differentiations* sought to ease, continued to perturb the group. And indeed, despite Keisuke's publication the year before, the investigations pursued in *honzō kai* remained concerned primarily with grasping the real through identifying proper Chinese and Japanese names. Placing medical items at the head of the list, moreover, demonstrates that the Shōhyaku-sha's persistent concern with medical practice also directed their *honzō kai*. Let us explore through a particular case how the question of locating the real name unfolded in events.

The case of a plant known as *baimo* in *materia medica* exemplifies how participants in an 1827 *honzō kai* sought to locate the real as a basis for knowledge. This involved a complex circular process of negotiating the real by constant comparison and correlation to the already available textual rendition of the plant. This method of back-and-forth between what one sees with one's own eyes and recalled known facts about the plant demanded that the participants not only be familiar with what had been published previously but also correlate observations with accumulated knowledge as a way to assemble a new set of checklists to verify the real name and efficacy of a plant such as *baimo*.

In classical Chinese medicine, *baimo* contained curative properties released upon consuming the dried bulb. At the 1827 *honzō kai* hosted by Keisuke, Mizutani Hōbun submitted a blossoming *baimo*, which was then known in Japanese as *haruyuri*.[20] Since the eighteenth century, scholars and physicians had puzzled over exactly how and where this popular remedy for a cold could be found in Japan. Claims of imported and domesticated *baimo* known under various regional names spread, along with assertions that a particular kind of *baimo* was more potent than others. Ekken's *Materia Medica in Japan* (1709) described the plant as follows:

> There are *haruyuri* indigenous to our country. According to *Honzō yakugen* [an imported text from China], *haruyuri* relieves discomfort in the chest area, alleviates uncomfortable fever, and enriches the heart and lungs. Thus, when taken orally, it eliminates phlegm. The same book also notes that when the bulb has no center, or consists of one part, not two, it is not appropriate for medicinal purposes and should not be mixed into the medicine.[21]

In Owari the identifications and properties of *haruyuri* became even more confused. In Matsudaira Kunzan's *Accuracy and Flaws of Materia Medica* (*Honzō seigo*, 1776), he listed *baimo* under the entry of a plant known as *engosaku* (*Corydalis yanhusuo*; Ch: *yan hu suo*) and wrote that "*baimo*, both Chinese and Japanese, are useable." Two years later, in *Response to Accuracy and Flaws of Materia Medica* (*Honzō seiseika*, 1778), Kyoan wrote, "*Baimo*: *hatsuyuri*, sprouts in winter, flowers at the time of the New Year. About one *shaku* [33 cm] in height at most, the leaves and flower belong to those of *himeyuri* save for the pale yellowish flower with purple dots." After listing a few associated names and claims about this plant, Kyoan ends the entry by noting, "the misinformation and mistakes concerning this plant are for no good reason extreme. There are about a hundred thousand false claims about this plant, and one should know them thoroughly"—signifying that despite efforts to correct and ascertain the efficacy of *baimo* in imported and domesticated specimens, the confusion between *haruyuri* and *baimo* persisted.[22] The medical efficacy of bulbs from the plant had been widely acknowledged, but opinions diverged sharply over whether the plant grew locally.

Ono Ranzan, a physician in Kyoto and the teacher of Hōbun, responded to Ekken's call for correction approximately one hundred years later by pointing out errors in *Materia Medica in Japan*. He integrated new information on this plant in his "Criticisms and Corrections of *Yamato honzō* ("*Yamato honzō hisei*").[23] Ranzan contends:

> *Baimo*; *hahakuri*: ancient name, *haruyuri*: found in Hanado area; *baimo* is a more commonly used name. Regarding Ekken's sentence beginning with "there are," there are those which are produced in our country today, but were

> unavailable in the past. After being imported, the details of the real thing (*shin mono*) became widely known in Japan. Regarding Ekken's sentence on "the bulb consisted of one part," the bulb of the real thing (*shin mono*) is always composed of two parts, and contains a central piece inside. According to *Compendium*, a bulb consisting of one part is known as *tanryūsei*, which was imported into our country during the Han dynasty. The appearance of *tanryūsei* is exactly the same as the real thing (*shin mono*) [of *baimo*], except that their flowers are purple. *Baimo* imported long ago bear white flowers that are few in number. They are known as *kawa-baimo* and these are of better quality for medicinal use. Newly imported *baimo* are darker and also turn yellow. They are known as *Shozan baimo* and are of lesser quality.[24]

The term "real thing," which appears three times here, plays a central role in this entry and throughout Ranzan's rearticulation of the knowledge of *materia medica*. In each case, the concept is rendered through the real (*shin*) and *butsu*, meaning "things." What I translate as "real thing" thus designates a proven physical entity possessing physical attributes that, even though not always sufficient for distinguishing one plant species from another, have a distinct, "real" efficacy. Ranzan, indeed, ascribes the uses given in *Compendium* to four plants that do not share the name *haruyuri* as Ekken claimed, but he nevertheless categorizes all four as being the real thing (*shin butsu*) of *haruyuri*. These continuous shifts in textual descriptions of what is real and false remained a pivotal and contested arena in knowledge of *materia medica*. In turn, the physical object itself continued to exist, as agreed upon, irrefutable, unequivocal material evidence in the construction of knowledge. This example of *baimo* illustrates how the notion of *shin* functioned as an axiomatic and elusive aspect in configuring *materia medica*. It created ongoing tension between the objects held to be real and those resembling it: plants known as *hahakuri*, *haruyuri*, and *kawa-baimo* are deemed to be a "real thing" of *baimo* in Ranzan's account, whereas *tanryūsei* resembles *baimo* but is understood to be not a "real thing." *Shozan baimo*, finally, is a "real thing," but its medical efficacy is less potent. As seen in Ranzan's account, the presumption that the "real thing" contains distinct and unmistakable physical properties was perpetually inscribed and foregrounded in the writings left by the Shōhyaku-sha. It is in this process that the concept of *shin* pointed to the irreversibly ontological object itself and proved imperative in the study of *materia medica*.

Hōbun's *Clarification on the Names of Things* (*Buppin shikimei*, 1809) also includes this plant: "*Haruyuri*, also known as *Amigasa yuri*: Chinese name *bei mu*: hollow stem, has medicinal benefits, rare object," while Shibata

Masabumi writes in *Medicinal Items for Everyday Use* (*Nichiyo yakuhin kō,* 1810) on *baimo* that "the root of *haruyuri,* those sold at medicine shops are all imported and are of good quality. There are some that sell domestic ones as well, and those are of bad quality."[25] Masabumi reaffirms Ranzan's observation that imported *baimo* is more efficacious than local varieties. In other words, exactly what constituted the real thing (*shin mono*) of *baimo* remained to be determined.

The record of the 1827 *honzō kai* notes that the flower of *haruyuri* brought by Hōbun was in bloom during the occasion, but it fails to mention the color of the flower; purple would indicate it is *tanryūsei,* not the "real thing" of *baimo* in Ranzan's account. Yet, the contrast of the differing accounts of *haruyuri* by Ekken, Ranzan, and Hōbun indicates the motivation and significance of Hōbun's submission of *haruyuri* to the exhibition: weighing observations of the plant itself ("a rare object," according to Hōbun) against Ranzan's opinion would enable identification of the presented plant with the older or newly imported variety. In addition, since the information attached to all of the displayed items in this *honzō kai* included the name of the person who submitted the specimen, any participant could have linked this plant to Hōbun, thus making it possible for a viewer with doubts to take them up directly with Hōbun himself. Participants were, in this sense, capable of clarifying with Hōbun the identification he ascribed to the plant. The fidelity between the name provided on the tag and the actual specimen was exposed as a vulnerable link that could be strengthened or weakened through collective inquiry.

In this sense, the real of *haruyuri* at this gathering had to be sifted from layers of previously known facts and descriptions in tandem with the visible characteristics of the exhibited plant. The process of inquiry at the *honzō kai* thus would entail recalling the knowledge accumulated before one sees the item at the exhibition. Subsequently, at the site itself it was hoped that fidelity between the concept of the real and the displayed item would be established by matching the observable and observed physical attributes with one's own prior knowledge of it. As such, knowledge about a particular plant remained in a state of constant readjustment and modification.

The case of *haruyuri* is a compelling example of how the types of evidence to qualify the real were in flux at these occasions, and the hierarchy of evidence (textual renditions, one's experience and knowledge, and the actual object on site) had to be negotiated. Through this process, participants in the *honzō kai* solidified, undermined, or even obliterated the perceived relationship between the real and the material object. Drawing a visual correspondence between the Chinese text and the Japanese context, these sections of *honzō kai* demonstrate the authority bestowed upon observation, and one's ability to coordinate information extracted from observation with opposing descriptions furnished in texts. In this respect, these *honzō kai* events participate in the epistemological transmutation explored in the previous chapter. The presumption of difference between the items in the Chinese canon and their environment persisted in the organizing intent of the exhibition, and the impulse to verify the state of the incongruity through displaying physical objects allowed the participants at the exhibition to modify and adjust their knowledge of *materia medica* collectively.

Heterogeneity, Shin, *and* Materia Medica

Although comparative evaluation of textual descriptions and names served as a key approach for locating the real at *honzō kai*, this was not the ultimate recourse. The dialectic of appraising the real at the exhibitions, in other words, did not rest entirely on known and published facts. The gathering aimed to amass contributions of hitherto unknown information from any participant, thus opening the discursive field to potentially unexpected possibilities. A flyer distributed in advance of the Shōhyaku-sha's 1830 exhibition explicitly states how they envisioned the gathering as an open forum in which every person in attendance might possess and offer knowledge of the real.

> We have a taste for *materia medica*, but have been unable to research notable objects in the realm between heaven and earth [which would require] traveling difficult routes and widely collecting various objects scattered in various domains. That our knowledge is skewed and shallow had been, in our thoughts, a major and regrettable obstacle. With this flyer, we hope to expand the interactions and communication with those who are familiar, however arbitrarily, with a variety of items within the confines of our oceans. We beg that you will allow for our intention to come to fruition and let us display the objects listed here. And later, let us provide you with a record of this display, which will include a wide range of objects. We also hope to provide a meritorious explanation that will invite more queries for correct information. It will also be helpful if you could send items from the Owari domain and nearby regions. We will not act arrogantly on this occasion.[26]

The Shōhyaku-sha's desire to correct, expand, and deepen their knowledge that underpinned the *honzō kai* gatherings stemmed from this acknowledgment of their uneven understanding. The deliberate framing of *honzō kai* as an event concerned with the discourse of *materia medica* allowed the exhibited items to be constituted as an object of epistemological inquiry in a public format. What stands out in the group's stated objective for hosting the *honzō kai* is the recognition that "interaction and communication" with any person familiar ("however arbitrarily") with the specimen could deepen their understanding of *materia medica*.

If we return to the woodblock depiction of the *honzō kai* organized by the domainal medical academy, this heterogeneity of the crowd is brought into relief: the groups of *samurai* appear as the largest figures, noticeable by their swords and topknots, but there are also merchants, who lack swords. The inclusiveness of the open-submission policy and the varied backgrounds of the submitters thus extended to an effort to invite a broad range of visitors and participants to the exhibition. In some *honzō kai* records, including those of the Shōhyaku-sha, farmers and merchants submitted items and participated.[27] This fact is provocative, considering the relentless effort of the Tokugawa *bakufu* to implement and control a social structure divided into four distinctive and hierarchical strata—samurai (who received a stipend from their lords), farmers, craftsmen, and merchants—in the hopes of realizing an orderly society governed by Neo-Confucian ideals. Establishing regulations and penalties to secure and maintain these boundaries and extend privileges to the samurai class permeated the 260-year reign of the Tokugawa *bakufu*. Recent scholarship, however, indicates that professional and amateur groups such as the Shōhyaku-sha frequently ignored these artificial divides.[28]

Although this policy of open admission to the *honzō kai* was not unique to the Shōhyaku-sha, their inclusiveness is still significant and telling of their intended goal. Another circle of scholars of *materia medica*, by contrast, known as the Shabenkai in Edo, was headed by domainal lords who spent alternate years in Edo as the result of an attendance system (*sankin kōtai*), and their activities tended to be more economically costly and socially exclusive.[29] Indeed, the reassurance that "we will not act arrogantly on this occasion" from the flyer quoted above indicates that even though the Shōhyaku-sha acknowledged class differences, explicit and implicit, among the organizers and participants, they sought to maintain the openness of the *honzō kai*. The Shōhyaku-sha strove to prevent social differences from interfering with the study of *materia medica*. This inclusiveness indicates the value they placed on the possibility that someone uncommitted to *materia medica* or unconnected to the medical profession could offer critical knowledge or experience to help discriminate the real of a specimen. Because *honzō kai* were intended to be open and the validity of any person's knowledge, regardless of his status, was assumed, the process of negotiation that one would go through at the site entailed both insinuating and contradicting pieces of knowledge, including one's own, about the displayed item. Compared to the 1760 invitation from Osaka, in which the organizers acerbically claimed that the participants will "eliminate any suspicions about specific items and correct information related to the objects," the Shōhyakusha avoid positioning themselves as authorities or insisting upon certitude. Instead they downplay their own expertise by indicating inability to gather information and specimens from all corners of Japan and characterizing their knowledge as "skewed and shallow."

This socially inclusive and equalizing approach to identifying the real does, however, create a conflicting tension vis-à-vis the concept of the real. As we saw in the previous chapter, in the process of translating the Linnaean binominal scheme Keisuke expounded that the real can be revealed, identified, and shared to provide better cures for patients. To Keisuke, the realness of the plant consisted of its existence in his region, its Linnaean name, and its medical efficacy. In the context of *honzō kai*, the first of the three elements of Keisuke's rendition of the real became obviously verifiable by perception. The physical object, regardless of whether its Linnaean name and medical property were known, was displayed at the site. But what about the latter two aspects? Especially if the event was to advance the study of *materia medica* through its inclusiveness, how did the group envisage a collective or an individual coming to a satisfying point of equilibrium where the real is completely and thoroughly identified while also encompassing the variety of one's particular knowledge and perception? How could acceptance of heterogeneity be reconciled with the notion of the real?

Keisuke's concept of the real, Ranzan's repeated call to identify the "real thing" (*shin butsu*) through physical attributes of a plant specimen, and the participatory format of the *honzō kai* are intimately bound up with the conflation of the real on two levels: as an idealized state where all three aspects of the plant's existence, its name, and efficacy are verified, and as a physical and actual entity of a plant available in the natural environment. This conflation, moreover, points to the intellectual weight that the concept of the real carried within their study of *materia medica* as a discursive notion that pertained to the construction and rearticulation of knowledge.

The Shōhyaku-sha's strategies for discerning the real (*shin*), however conflated the concept was, is best understood by contrasting the flyer for the exhibition and the inventory of items exhibited at these *honzō kai* that we examined earlier. In the flyer inviting submissions for the 1830 *honzō kai*, the Shōhyaku-sha listed these desiderata:

> Plants, trees, minerals, stones, insects, fish, animals.
>
> Note its place of origin, which includes the names of the domains, counties, villages, or the mountains and rivers where the item is found. Also provide the regional names, and descriptions of its shape, or methods of cultivating them. This will include medicinal items as well as edible ones, which will help during periods of famine, and those that can be utilized to make tools.
>
> —Items that have been known in regions as beneficial or poisonous; in case of the latter, provide descriptions of the antidote.
>
> —Rare and odd items that belong to families.
>
> —The studies of *materia medica* and names of things (*meibutsu*) should consist of a variety of knowledge and by assembling them. However, many products, due to the soil and temperature of their place of origin, appear differently [although they are the same product], thus producing varied degrees of quality.
>
> —If it is at all possible to bring the actual items to the exhibit, that will be most delightful. These we would like to see in actuality: plants and trees (from medicinal items to vegetables, grains, and others that are worth evaluating). In addition, even if the item is small and appears to be a weed, if it is rare, please send us the plant including its roots. If you live far away and cannot send the actual living plants, please send us pressed leaves (preferably with its fruits) or fruits, at its ripest point, indicate when the seed was planted.
>
> —Minerals and stones: from fossils to soil and sand.
>
> —Insects and fish: insects should be pinned at the center of its back with a needle, and fish should be dried.
>
> —Shellfish.
>
> —Animals: There is a method to take out the guts and skin the meat, then mix old leaves, lime with hemp or cotton and sew the skin closed.[30]

The knowable, the attributed name(s), places of origin, and method(s) of use became integral aspects of the submitted items. Furthermore, categories of such as "potted plants" and "snapped branches" demonstrate the Shōhyaku-sha's recognition that different forms of knowledge may be obtained from observing the stages of a plant's life cycle. Here the real seems to waiver

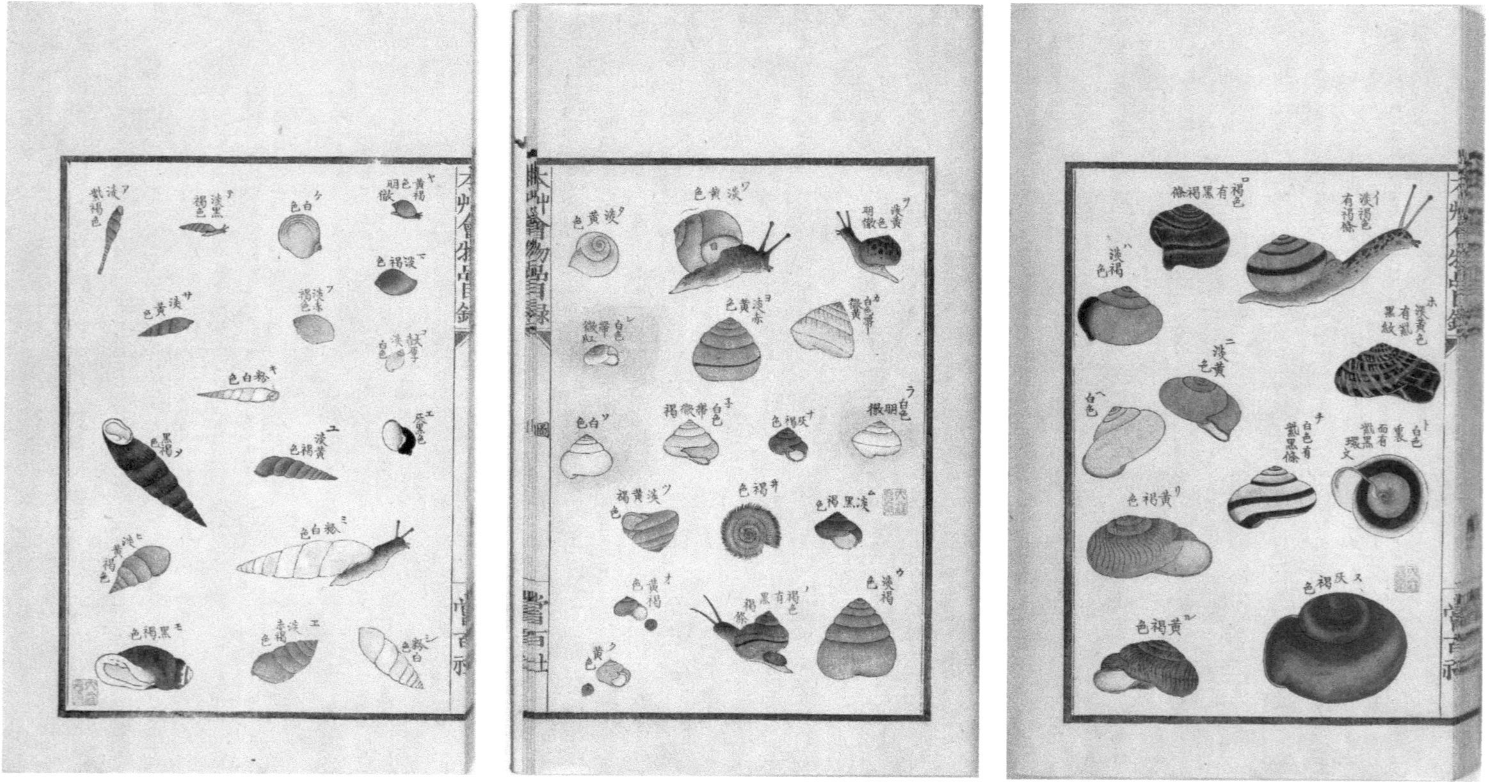

Figure 3.2 Ōkubo Masa'aki, *Snail shells*, from *Honzō-kai buppin mokuroku*, 1835, woodblock print, Hōsa Library

between verifiable and reliable information about an item and the actual item itself. Whereas the unknown or unfamiliar grass is being accepted as the entity that requires careful consideration of its realness at *honzō kai*, the verdict of its name or efficacy could either confirm or dispute the attributed knowledge and names. The real is negotiated between the perceptible attributes of the individual item and the outcome of the collective inquiry; it could be revealed by observation or refuted through an account of one's experience and knowledge. Correct and reliable identification of a plant was understood, on the one hand, to overturn false identification and information, but the real was also imagined in a much more elastic way.

A submission of forty-five snails by the member Ōkubo Masa'aki to the 1835 *honzō kai* exemplifies this point (Fig. 3.2 and color plate 2). This three-page woodblock representation is included in the record of the *honzō kai*, and each page bears the stamp of Masa'aki, indicating that he himself had depicted the snails. Each snail in the print is identified with a *katakana* script, which corresponds to the list of exhibited items in the same record. Next to the identifying *katakana* letter are descriptions of the colors and patterns on snail shells rendered in Chinese characters. By matching the *katakana* script placed next to the represented snail with the listing of the exhibit, one is able to decipher the location from which each snail shell came. For instance, one can learn from the list that snail shells marked in the *iroha* syllabary of *ro*, *ha*, *wa*, *re*, and *ke* are all from the city of Nagoya. Other shells are identified as being from regional mountains and villages near Owari, and there are two specifically noted to be a variety originally found in China and Korea. This list does not reveal the logic of order in which the shells are listed, but their pictorial representations indicate that the shells are listed according to their size and shape, with the larger shells listed first. In other words, the visible quality (size and shape) was prioritized over the imperceptible quality (origin) of the shells in ordering the forty-five snails. Some snails were live; others were just shells.

The comparative mode of displays, by necessitating comparison of objects placed under one classificatory name, visually positioned an accentuating spotlight on individual items placed within the nominal category, while the broader classificatory class asserted a seemingly equalizing epistemological weight to the grouped objects themselves. The type of observation encouraged to identify and differentiate these snails rests on two propositions: (1) there are unifying and identifiable physical attributes in the specimens that correspond to the generic name "snail" (*katatsuburi*), a category of classification used in the listing itself and taken directly from *Compendium*, in which snails are noted to be particularly effective for healing centipede bites; and (2) there are also diverging, distinctive, and observable differences among them. The tension between the systematic and taxonomic on one hand, and the particular and individuated on the other, becomes intensified in this mode of display. The traditions and accepted knowledge in *materia medica* render the category of "snails" as a valid and verified classificatory entity, and the differing shapes, sizes, and patterns on the snails simultaneously manifest the realness of each one exhibited at this *honzō kai*. If we return to Keisuke's three-pronged idea of the name, the medicinal efficacy, and the existence of a plant in his region, this comparative display of the snails addresses two: the efficacy of the exhibited mollusks was accepted and the samples were taken from nearby regions. What this collection of snail shells seems to challenge is the nominal designation of "snail" itself by demonstrating the divergence and heterogeneous types of snails from their region. By imparting the category of knowledge "snails" with the same degree of reality that showcased the divergent size, colors, and patterns on the shells, Masa'aki's collection seems to challenge the overcompensating and overextended nominal class "snails." The effort to locate the real in this comparative display is suspended between the known medical efficacy and the category of "snails" embedded in the discourse of *materia medica*, and the divergent samples of the specimen on display from their local environment. The idealized real—that is, the state of knowledge where the name, efficacy, and existence of particular objects are verified and shared—and the actual real, visibly heterogeneous snail shells, had to be negotiated and accounted for.

These arrangements grouped for comparison underscore the irreducibility of each object contained within the category of "snails." By staging these comparative displays at the *honzō kai*, the Shōhyaku-sha brought to the fore the philosophical disposition of the discourse of *materia medica*. As if to undermine the simplicity of the category "snails," the record of the *honzō kai* notes that each shell was accompanied with a tag that specified where it was found along with a description of color, size, and visible patterns of the shell. These conflicting positions of the real as ascribed to the individual specimen as well as the broader category to which it belongs resonate with contemporary scholar Aramata Hiroshi's observation that scholars of *materia medica* sought to recognize and understand the outside world with an assumption that there will always be novel, uncategorized, and unique objects.[31] Subtle visible differences carried as much weight in their reformulation of *materia medica* as the category itself. If we recall the group's flyer from 1830, they noted that "many products, due to the soil and temperature of their place of origin, appear differently [although they are the same product], thus producing varied degrees of quality," acknowledging the morphological variety in appearances as well as medicinal efficacy that exist in nature. The case of snails from 1835 echoes and manifests their interest in articulating the cases of divergence and heterogeneity found in nature. These instances of comparative display point to the participants' reliance on heterogeneity at these exhibitions. As much as the Shōhyaku-sha assumed that heterogeneity would be observed in the exhibited items, they also understood that heterogeneity marked the sources with which the real could be established.

Unnamed Objects and the Real

Another feature of the *honzō kai* that influenced modes of observation and transformed the concept of the real were the displays of "unnamed" objects. In the previous cases, various conjectures concerning the names and characteristics of a plant already existed in the realm of public knowledge. Conversely, objects gathered under the category of "unnamed" stood devoid of prescribed descriptions and opinions. In other words, identification of the real and the false involved a completely different set of tools for looking at unnamed objects. How then does the notion of *shin* operate in contemplating these objects? What comes to light through analysis of the "unnamed" is the ways in which the concept of the real was not necessarily predicated on formally provided explanation. Indeed, the real had to be acknowledged on the level that these objects were existent and observable. Without the sources of knowledge used to identify the real, the existing object itself came to be positioned as the most trusted aspect of the real in these instances. In other words, contrary to the comparative display in which the observable and the knowable were constantly negotiated, the observable was the only source of the knowable in this type of display.

Exhibiting "unnamed" objects indicates that the Shōhyaku-sha were unable to identify the items, and that the group had hoped some participant would contribute information about the specimens. The display of "unnamed" objects suggests confidence in the process of inquiry, that observations and discussions could lead to insights into the general characteristics of the object. These insights could, in turn, produce a consensus about the unnamed, ultimately perhaps even revealing its name. The notion of the real, in these cases, cannot be derived from prior knowledge but has to rely on the presence of unnamed objects at the *honzō kai*.

The particular case of "grass with unspecified name" illuminates the process in which the existence of an object served as an unyielding assertion of the real, although extrapolation of the identity of the plant did not come to fruition. Masa'aki submitted "grass with unspecified name" at the 1835 exhibition, noting that it originated from Hakusan Mountain, about 150 kilometers north of Nagoya.[32] Apparently, no one at the exhibition could identify the grass, and it remained nameless in the published catalogue. In other cases of "unknown fruit," "bearing on the barley branch," or various imported plants and objects from Europe, the mere existence of the items remained largely unknown until *honzō kai* gatherings, and thus the display served as a revelation of their existence to most of the participants. Placing quotation marks around a given name, as in the case of objects of European origin, or ending the names with a phrase "it is called" (*to iu mono*), the Shōhyaku-sha members expressed their uncertainty about the uses, properties, and given names of these unfamiliar and novel objects. The ontologically verifiable entity that stood in front of their eyes had to be acknowledged, while its relation to *materia medica* or names remained unknown. Aramata's assertion that the assumptions of scholars of *materia medica* was to expect the novel, uncategorized, and unique specimen in the world is helpful in understanding the motivations to include unnamed objects. Even if the names and efficacy were unknown, the undeniable existence of the plant constituted and offered one aspect of the real for the participants of *honzō kai*. The tension in observation arose from an ontologically verifiable entity and epistemological uncertainty over its uses, name, and even the category of things to which it belonged.

The collective desire to identify the real was at the crux of *materia medica* inquiries. As a hybrid medium, the *honzō kai* organized by the Shōhyaku-sha thoughtfully and carefully coordinated knowledge and perception in ways that textual translation could not. Connecting textual information, verbal exchange, and observable characteristics, these occasions mediated the process of confirming, challenging, and refuting the real of specific items. The two presumptions that inform the *honzō kai* events may seem conflicting: (1) the exhibited items could be named or identified wrongly, and (2) the unnamed objects could be identified through intensive observation and discussions. The underlying desire to showcase these items, however, lay in Shōhyaku-sha's hope to strengthen their understanding and knowledge of *materia medica*, and in this sense these two presumptions do not conflict. The lingering tensions created by these suppositions stem from the Shōhyaku-sha's desire to see and evaluate the real of things.

In the triangulated process of constructing their knowledge of *materia medica*, the occasions of *honzō kai* were a unique and integral opportunity in their pursuit. Observation of the specimens permitted collective calibration of the names and efficacies attached to the item exhibited, while the unknown objects would remind the participants of the perpetual existence of the novel and unique brought from their regions and beyond. Since it was hoped that textual translations of the Linnaean scheme would bring a sharable and stable system of nomination to the disorganized state of names in *materia medica*, the *honzō kai* occasions were opportunities to concretize the answers to questions derived from other textual sources or to challenge accepted knowledge. The case of *baimo* shed light on the process of negotiating among various kinds of knowledge associated with the plant. The conflation of the idealized "real" and the physical "real" entity, as attested in the case of snails, accelerated the opacity of the idealized real on one hand, while on the other hand, each specimen displayed under the category directly addressed the realness of its own existence. The heterogeneity of the natural world (forty-five specimens) and the epistemological nomination derived from *materia medica* ("snails") were couched on the notion of the real. Inclusion of "unnamed objects" in the ongoing process of navigating *materia medica* gestures to both the Shōhyaku-sha's acceptance of the incompleteness of their knowledge of *materia medica* and their readiness to test the item, as it were, to extract and deduce names and uses through observation and interpersonal exchange. The fundamental intellectual drive embedded in the strategies of observations and display at play in the context of *honzō kai* points to an unrelenting effort to identify and solidify the "real thing" of specimens included in *materia medica*. Understanding the discursive centrality of the concept of *shin* uncovered in the historical records of *honzō kai* lends further significance to the concept of *shashin* (transposition of the real) that permeated characterization of pictorial representations used in the study of *materia medica*. What types of logic can we discern in the illustrations named as transcriptions of the real? How did the three-tiered concept of the real come to be ascertained in the pictures of plant specimen? The following chapter explores these questions through examples of illustrations made by or used by the Shōhyaku-sha.

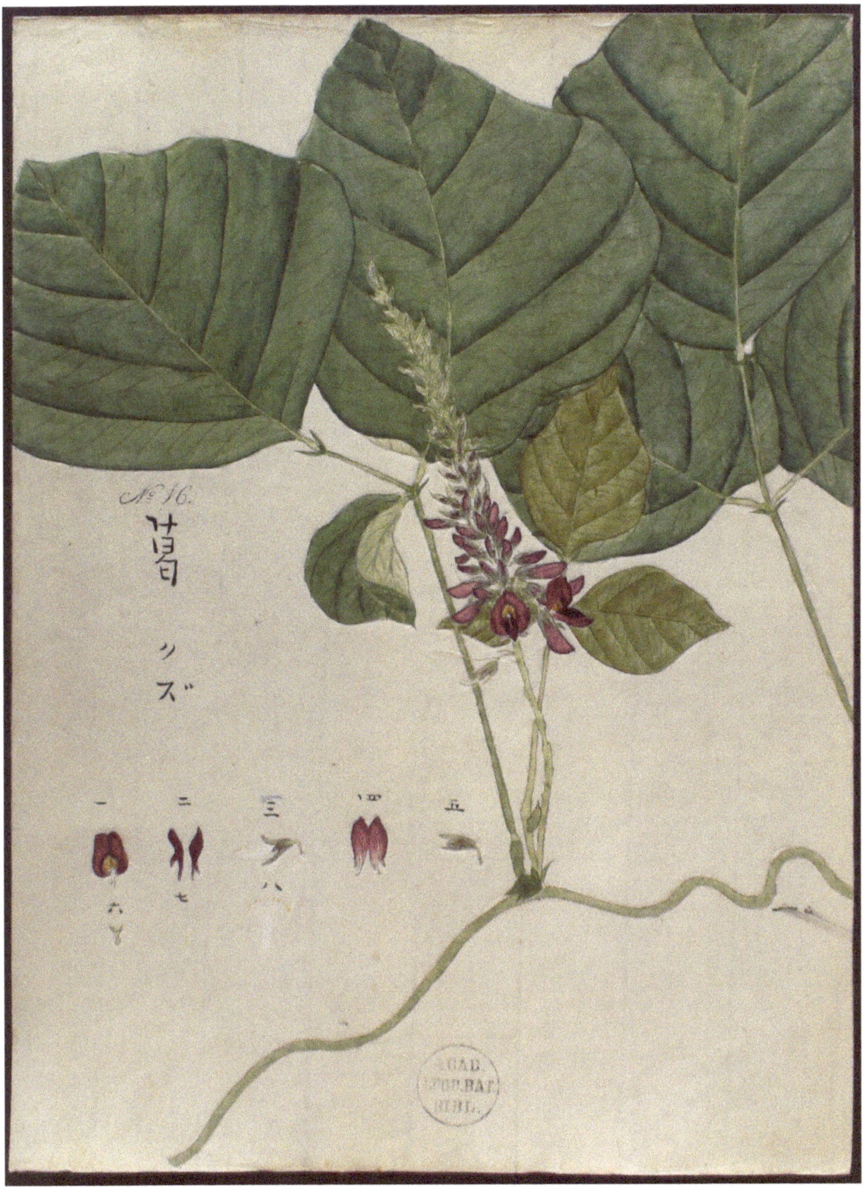

Plate 1 *Kuzu*, from "*Honzō shashin*," before 1826, ink on paper, Leiden University Library

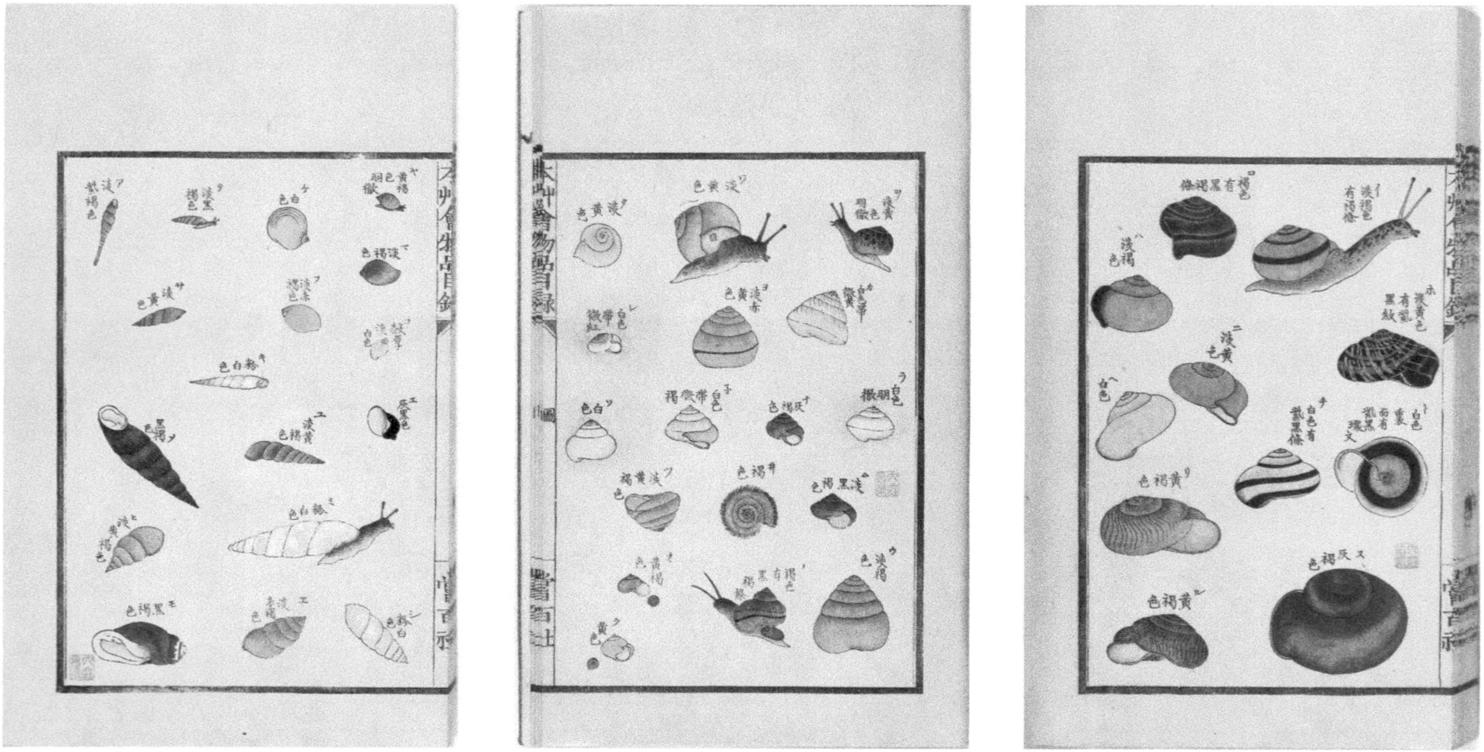

Plare 2 Ōkubo Masa'aki, *Snail shells*, from *Honzō kai buppin mokuroku*, 1835, woodblock print, Hōsa Library

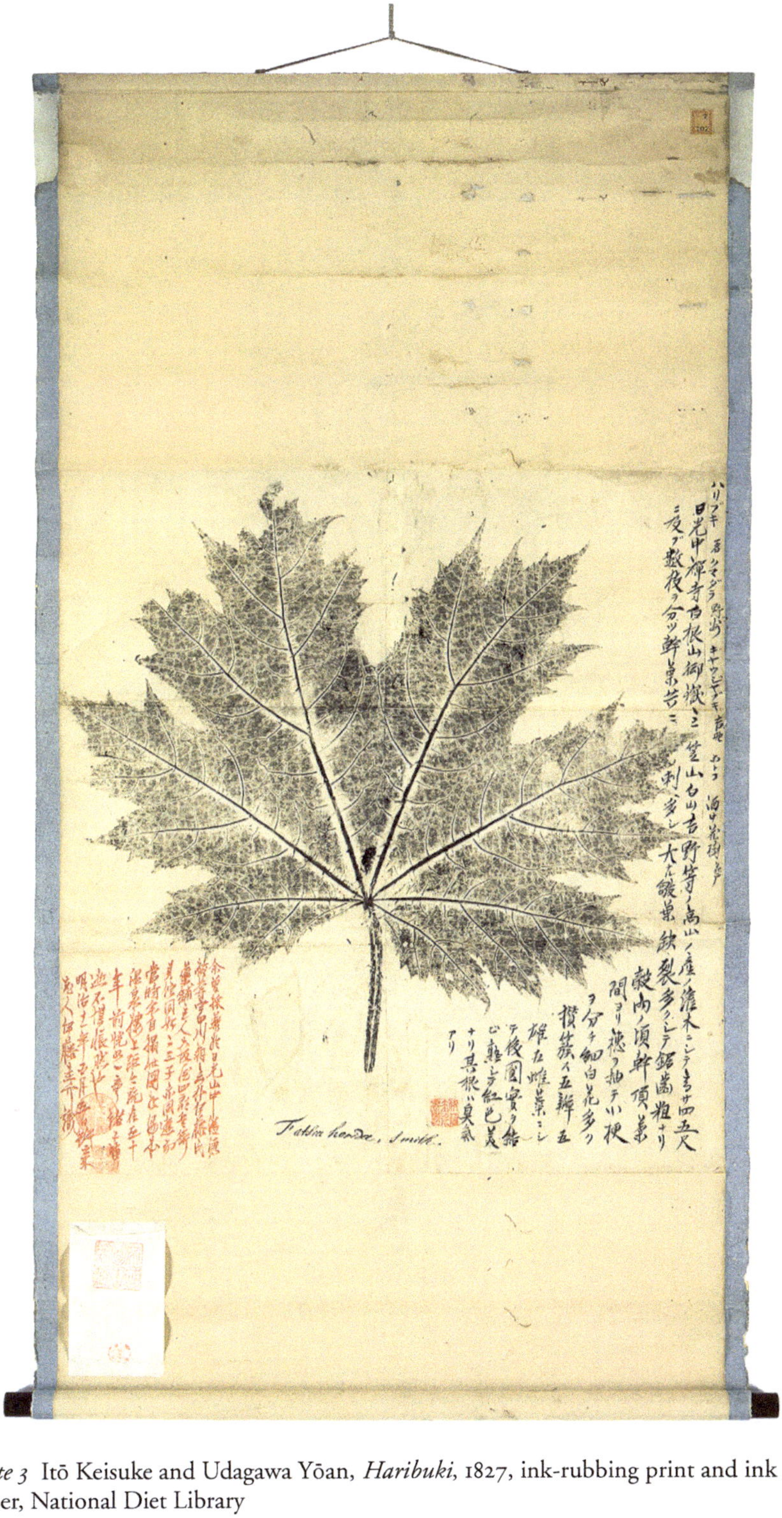

Plate 3 Itō Keisuke and Udagawa Yōan, *Haribuki*, 1827, ink-rubbing print and ink on paper, National Diet Library

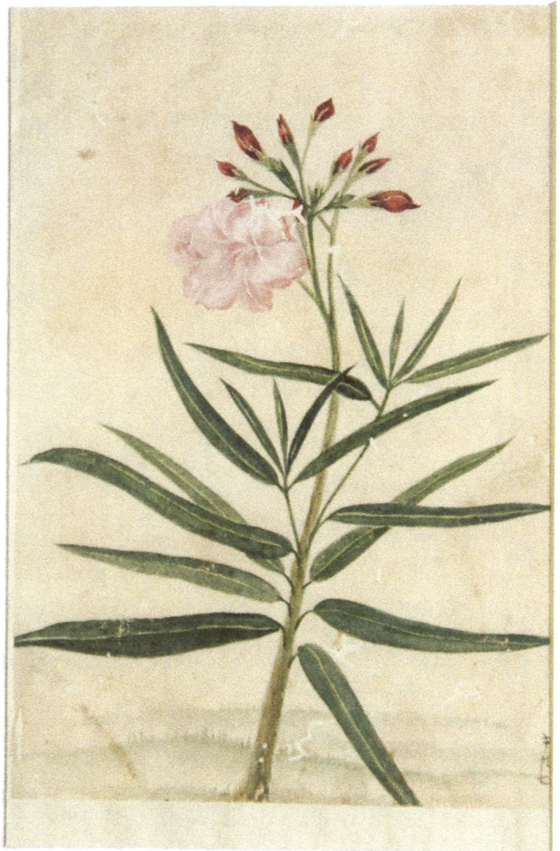

Plate 4 *Oleander*, from Itō Keisuke, "*Kyōchikutō*," c. 1860, ink on paper, Itō Keisuke Collection, Higashiyama Botanical Garden

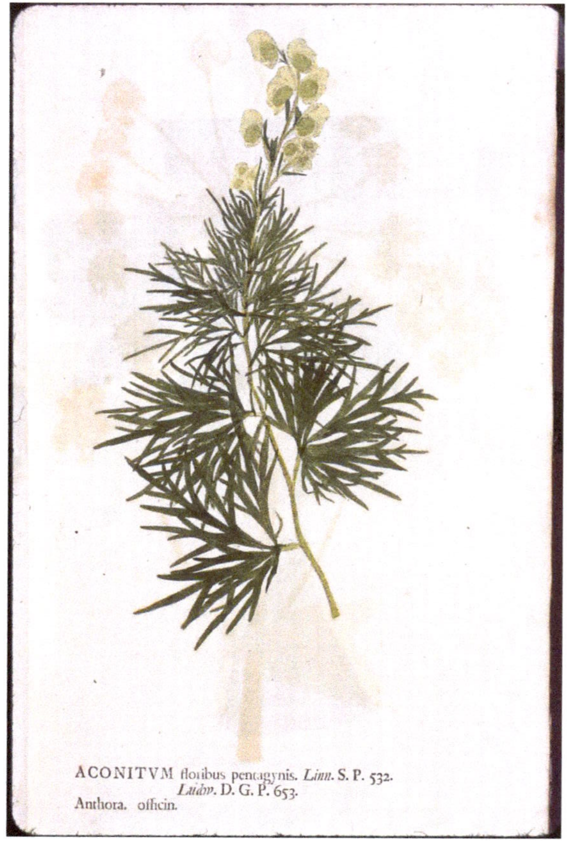

Plate 5 Johann Hieronymus Kniphof, *Aconite*, from *Botanica in Originali . . .*, 1765, nature print, University of Chicago Library, Special Collections Research Center

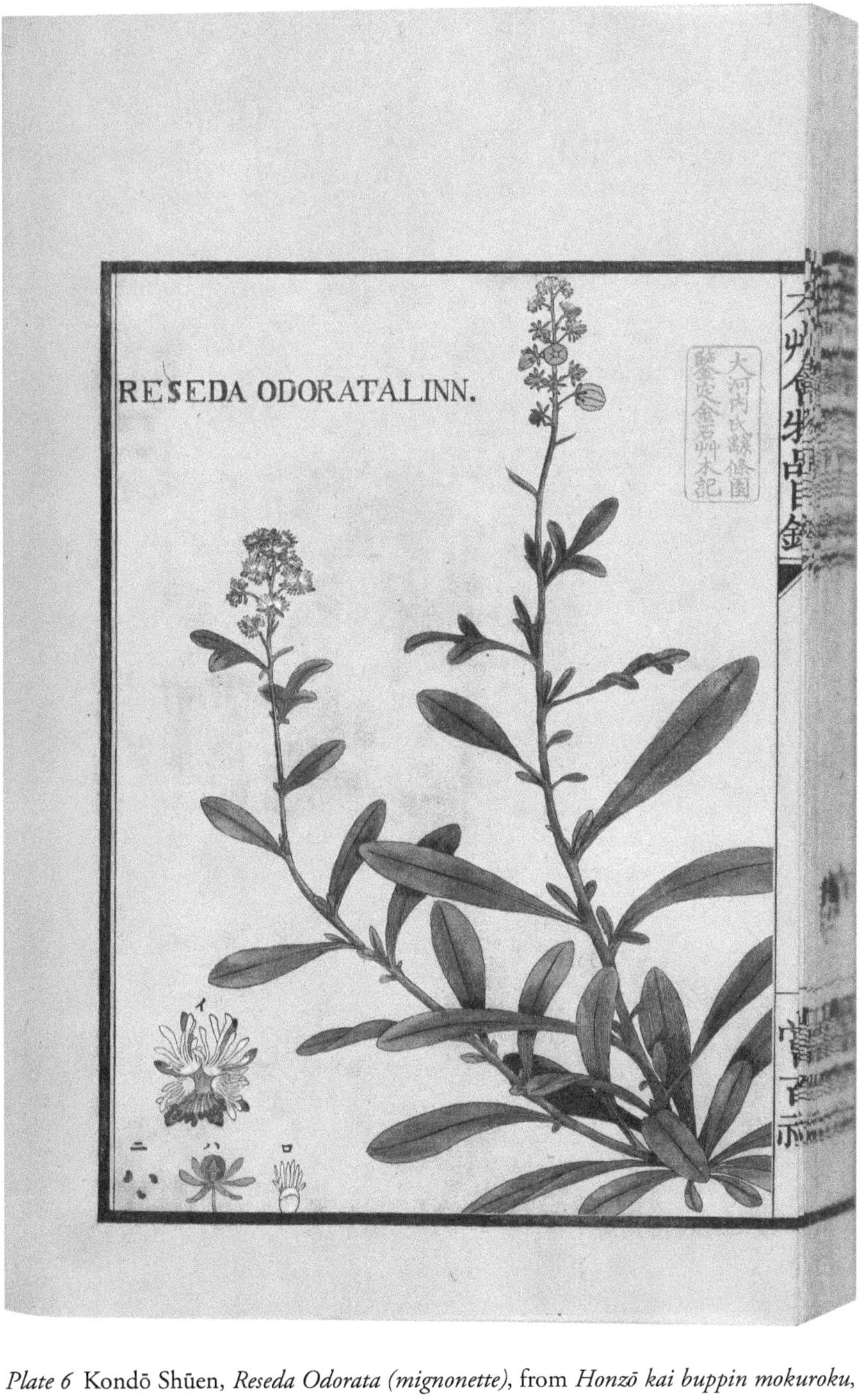

Plate 6 Kondō Shūen, *Reseda Odorata (mignonette)*, from *Honzō kai buppin mokuroku*, 1835, copperplate on paper, Hōsa Library

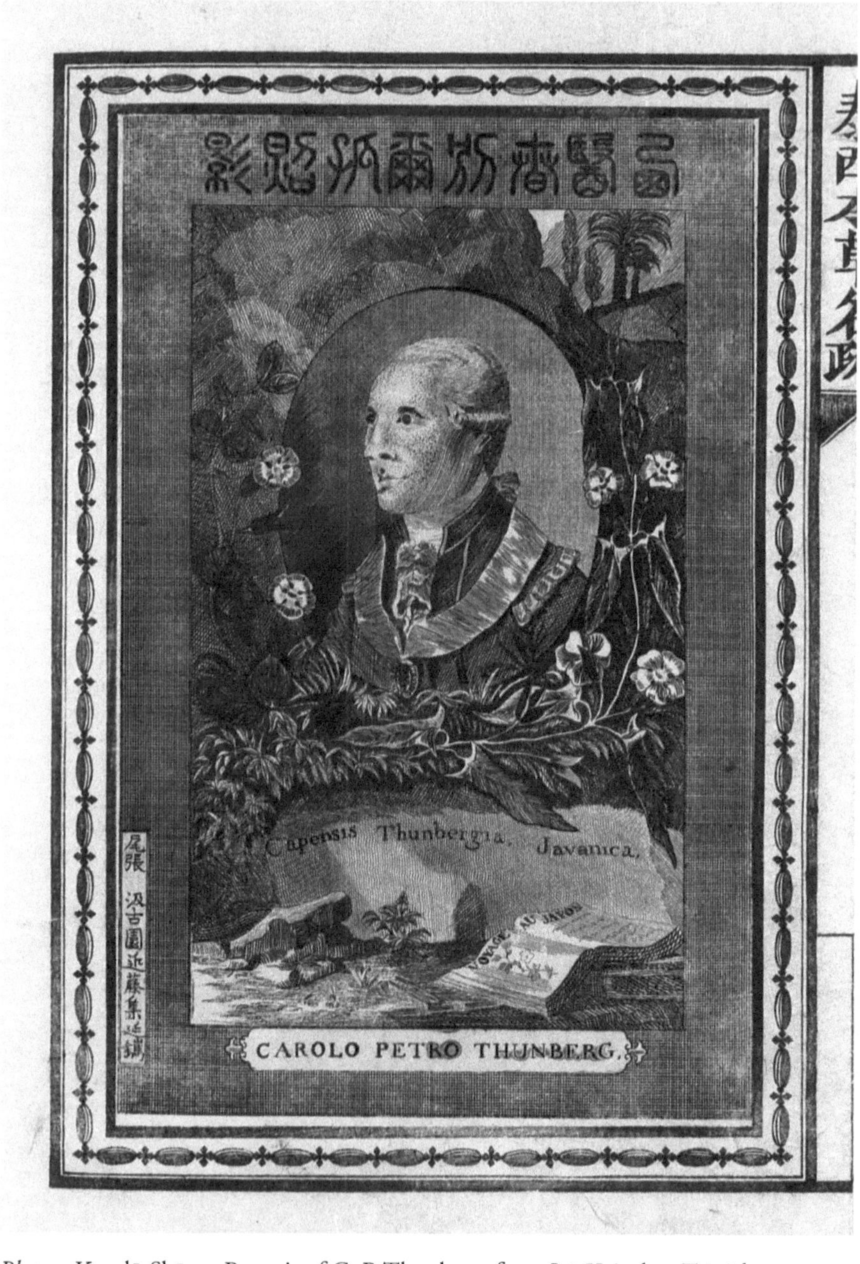

Plate 7 Kondō Shūen, Portrait of C. P. Thunberg, from Itō Keisuke, *Taisei honzō meiso*, 1829, copperplate on paper, Hōsa Library

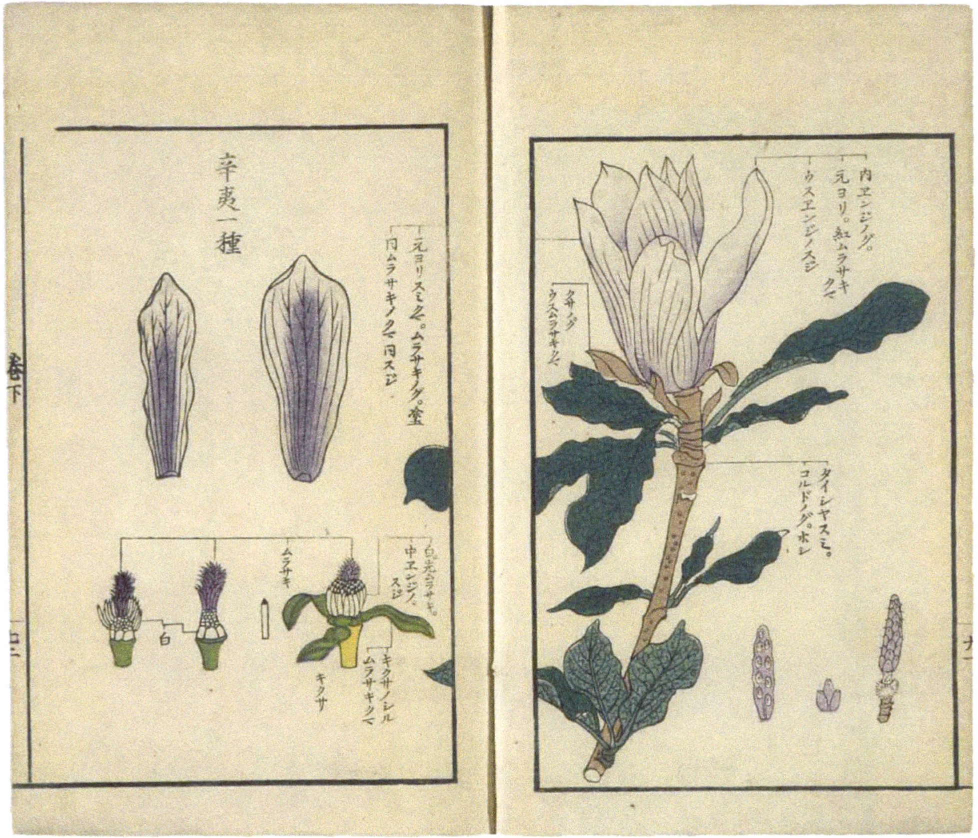

Plate 8 Kawahara Keiga, *Magnolia*, based on *Keiga shashin sō*, republished with colors in *Sōmoku kajitsu shashin zufu*, 1868, woodblock print, Waseda University Library

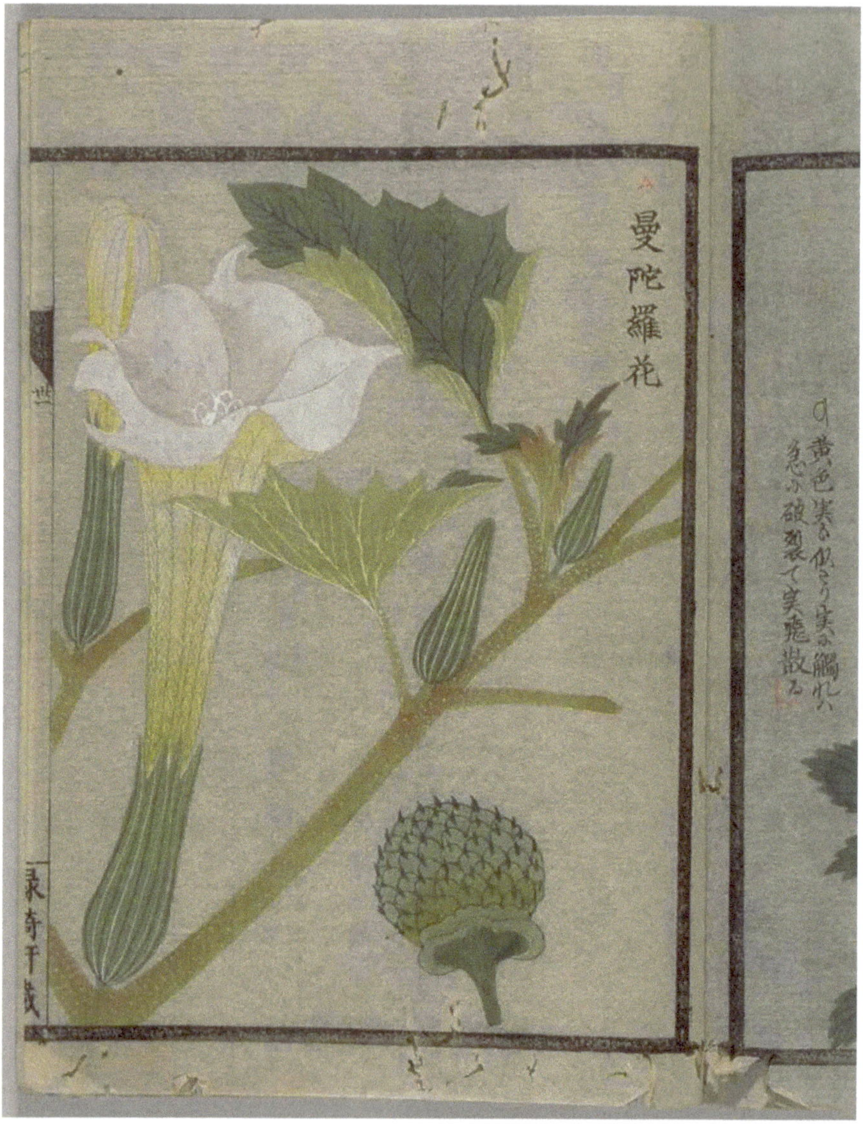

Plate 9 Iwasaki Kan'en, *Chōsen asagao*, from *Honzō zufu*, before 1830, woodblock print, Tokyo National Museum

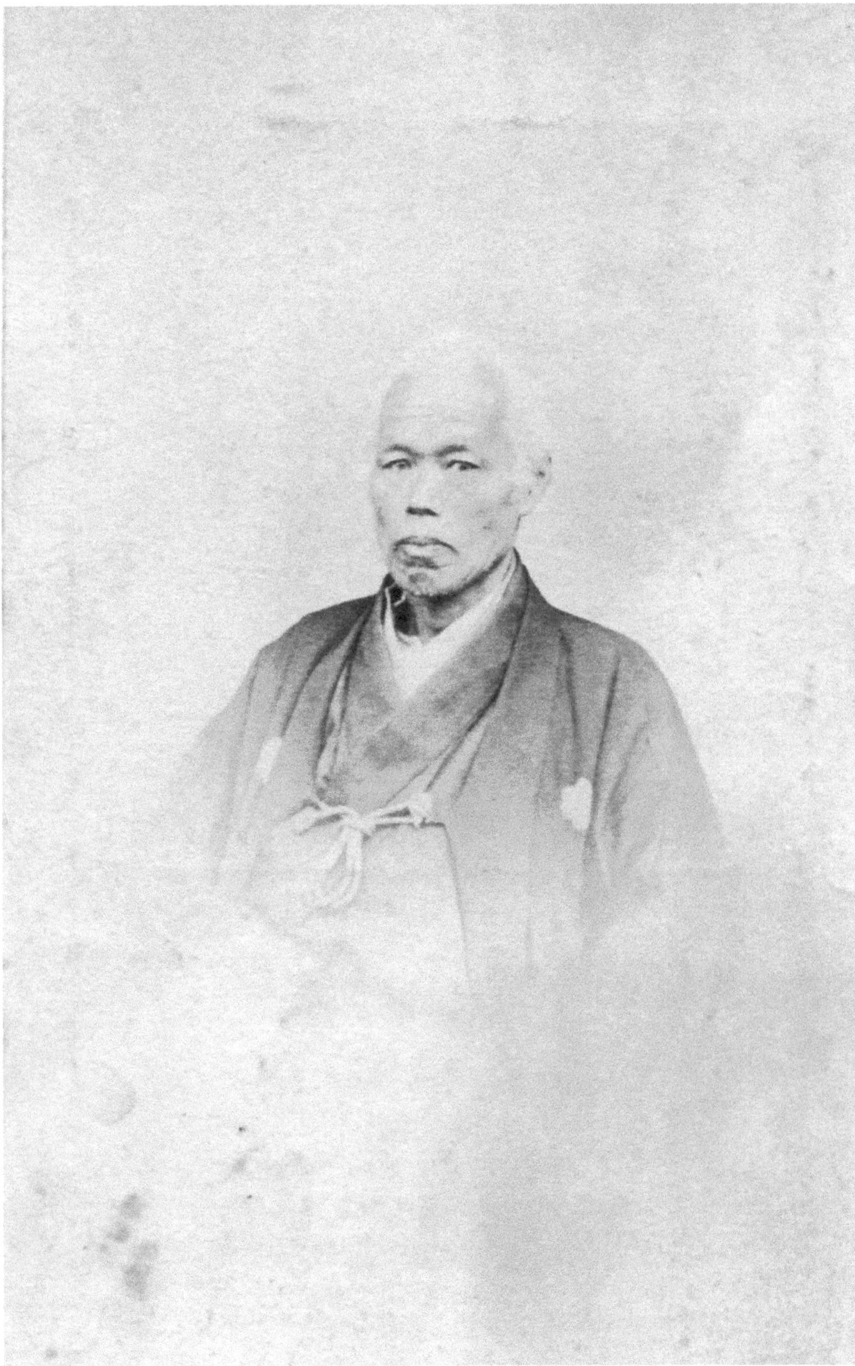

Plate 10 Uchida Kuichi, Portrait of Itō Keisuke, 1871, collodion process on paper, Itō Keisuke Collection, Higashiyama Botanical Garden

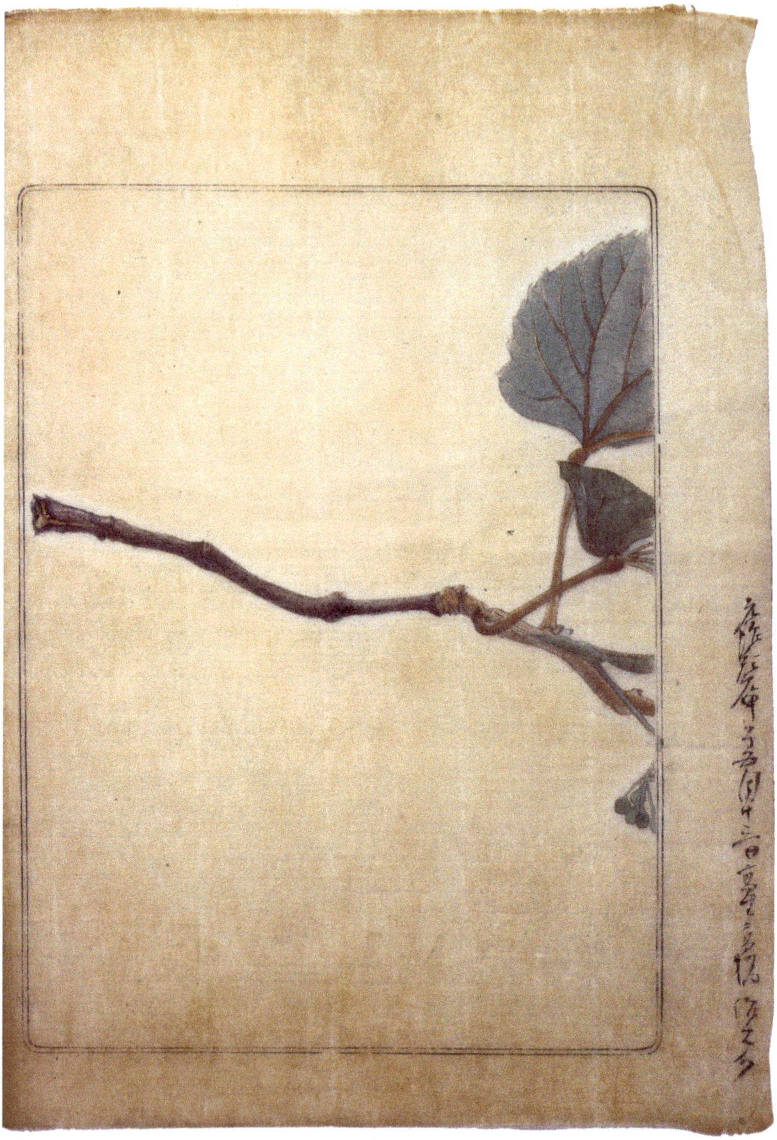

Plate 11 Takahashi Yuichi, *Bodaiju*, from "Kaiunrō zufu zassan," 1863–1865, ink on paper, Special Books, Tokyo Metropolitan Library

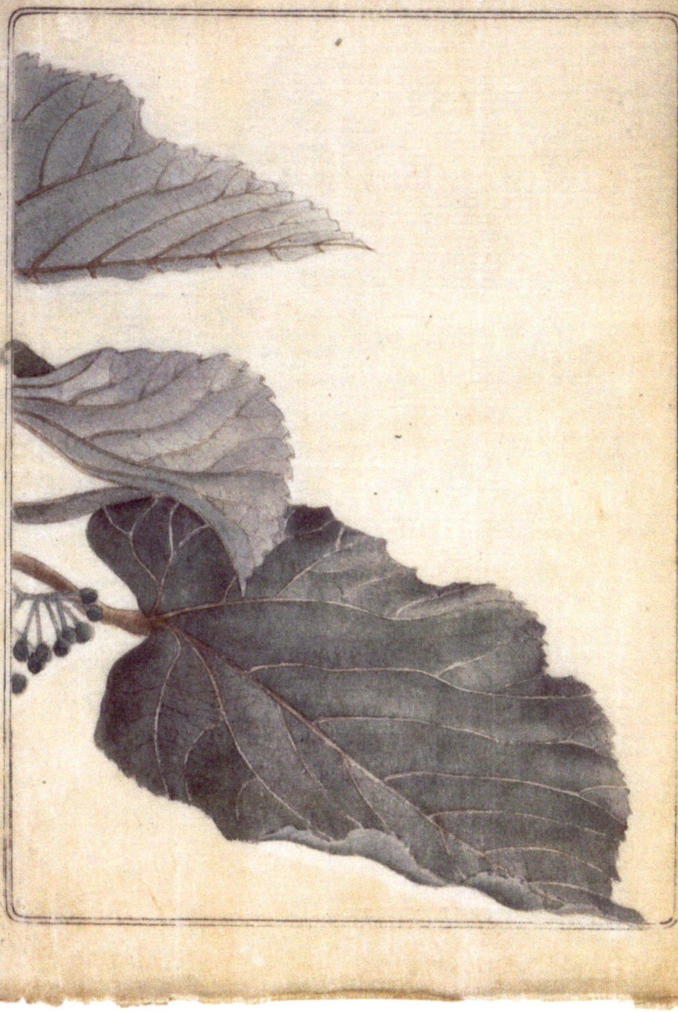

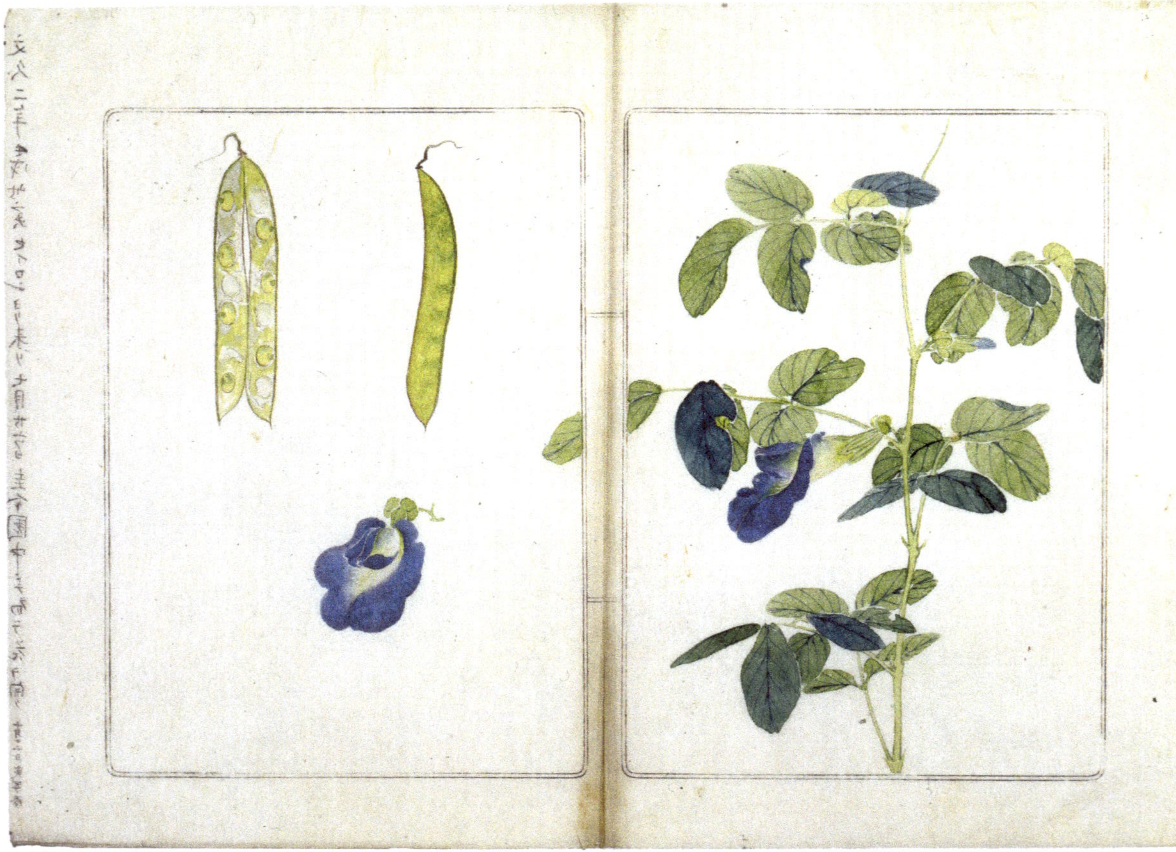

Plate 12 Itō Keisuke, *Snow pea*, from "Kaiunrō zufu zassan," 1863–1865, ink on paper, Special Books, Tokyo Metropolitan Library

FOUR

Picturing the Real

Questions of Fidelity and Processes of Pictorial Representation

> However, a picture produced by a craftsman is, for the majority, overtly colored and focused too much on *tenkei* [embellishing persons and animals added in landscape paintings].
>
> Udagawa Yōan, "Treatise on Botany" ("*Shokugaku dokugo*", circa 1829)

For their investigations of *materia medica* the Shōhyaku-sha deployed various techniques and senses—observation, translation, and exchange of textual information with others—in search of the unvarying reliability of the name(s) and efficacy of plants. The *honzō kai* gatherings served as just one of the ways to identify and solidify the relationship of a plant, its name, and its use through collective participation and interrogation. Central to these activities was the concept of the real, and the meaning of the concept continuously shifted and developed. It was used as a limiting tool (in relation to the false, *gi*) as well as an expansive concept (in regard to the classificatory category and individual specimen). The comparisons staged at *honzō kai* events supported complex negotiations of the real (*shin*) of the plant through a constant feedback loop that employed textual and personal sources. This process of desynthesizing and resynthesizing a body of knowledge both shaped the outcome of their discussions and anchored their investigations. For unnamed plants in the exhibitions, the Shōhyaku-sha would not rely on knowledge contained in texts but could instead depend upon collective associations and conjectures while assuming the real (*shin*) of the plant to be inherent in the plant itself. The negotiation that took place in the discussions about objects and knowledge at the *honzō kai* events was based on the actual specimens themselves.

The manuscript "*Honzō shashin*" from the 1826 meeting with Siebold discussed previously reveals the significance of illustrations in the group's effort to verify and solidify their knowledge of *materia medica*. The multimedia manuscript was as concerned with the exchange between Siebold and the Shōhyaku-sha as it was with the actual specimens that the group brought to the meeting. Indeed, the group examined and produced their own illustrations as a means to accelerate their epistemological pursuit of *materia medica*. How did the illustrations allow the group to assess particular knowledge? In what way did the status of illustrations in constructing knowledge compare to that of actual specimens? And how are we to understand Mizutani Hōbun's deployment of *shashin* in the manuscript as an embedded aspect of the group's constant effort to recalibrate their knowledge vis-à-vis the reality of flora around them?

This chapter addresses these questions through historical and discursive interpretations of illustrations characterized as *shashin*. As Hōbun's use of the term in "*Honzō shashin*" attested, the word circulated among the Shōhyaku-sha without a definitive singular association with any particular medium. In the process of evaluating the relationship between illustration and the names and efficacy of plants in *materia medica*, the concept of *shin* was repeatedly evoked. The overlapping and multiple uses of the term *shashin* in relation to illustrations provides a starting point for exploring what this term encompassed in their investigations of *materia medica*. Careful scrutiny of how the Shōhyaku-sha applied the term reveals that its discursive utility lay in the ability to designate faithful copies of *already* represented pictures and the existing visible forms of the specimens. Contextualizing the meanings attached to *shashin*, moreover, throws two imbricated meanings of *shashin* into relief: (1) the emergence of distinct meanings that were different from the application of the term to particular kinds of sketches; and (2) the efficiency of *shashin* in attesting to the existence of a particular specimen and the direct observational experience of the specimen. These overlapping meanings will be examined in three interrelated contexts: in relation to ink-rubbing prints, which the group called *shin'ei* (literally, "the shadow of the real"); in relation to the uses and production of copper-etching prints, which the group referred to as *shashin*; and comparison to the uses of *shashin* among the Shōhyaku-sha's contemporaries.

Shin'ei Prints and Fidelity

For the scholars of *materia medica*, particularly the Shōhyaku-sha, the business of ascertaining and verifying the existence of a particular plant dominated their activities. Unlike other collectives studying *materia medica*, the Shōhyaku-sha assiduously produced literally dozens of manuscripts using a method of picture making called ink rubbing. In fact, the Shōhyaku-sha were distinguished from other collective groups pursuing *materia medica* in their continued use of this method of printmaking, known in Japanese as *inyō-zu-hō* (literally, "method of picture making through impressing leaves"). How did this particular representational technique support negotiating the triangulated relationship among text, illustration, and the plant in question? The following analysis of ink-rubbing prints will illuminate the ways in which the group's concerns with representation and with finding fidelity in texts, illustrations, and real specimens shaped the investigation of *materia medica* as a pictorialized testament of both the existence of the represented object and the image maker's direct encounter with it.

This printing technique involves rubbing ink on botanical specimens and taking an ink transfer from them through a direct or indirect method of impression. Both methods required flattening the specimen prior to rubbing.[1] In the direct method ink is applied directly to the object and a piece of paper is then pressed delicately over it, producing a bilaterally symmetrical mirror image of the object. Indirect impressions are made by first placing paper over the object and then applying ink onto the paper, thus leaving the object clean of ink. The indirect method produces a properly oriented image, an important consideration when reproducing texts, but it cannot capture the details of the object as effectively as the direct method. The simplicity of this print-making process and its ability to accurately capture the size of the specimen made it an especially attractive technique (Fig. 4.1). Although the term *inyō-zu-hō*, coined at a later date, misleadingly suggests that this method was used only for representations of leaves, there are numerous images depicting roots, flowers, and fruits of plants; in one case, the method was actually used to document a *honzō kai* event.

The popularity of this method among the members of the Shōhyaku-sha carries particular significance because these images, too, were characterized as and understood through the concept of the real. Put differently, the concept and representation of *shin* was not tied to a particular technology, but was part of a larger discourse of representation that ran across a variety of media.

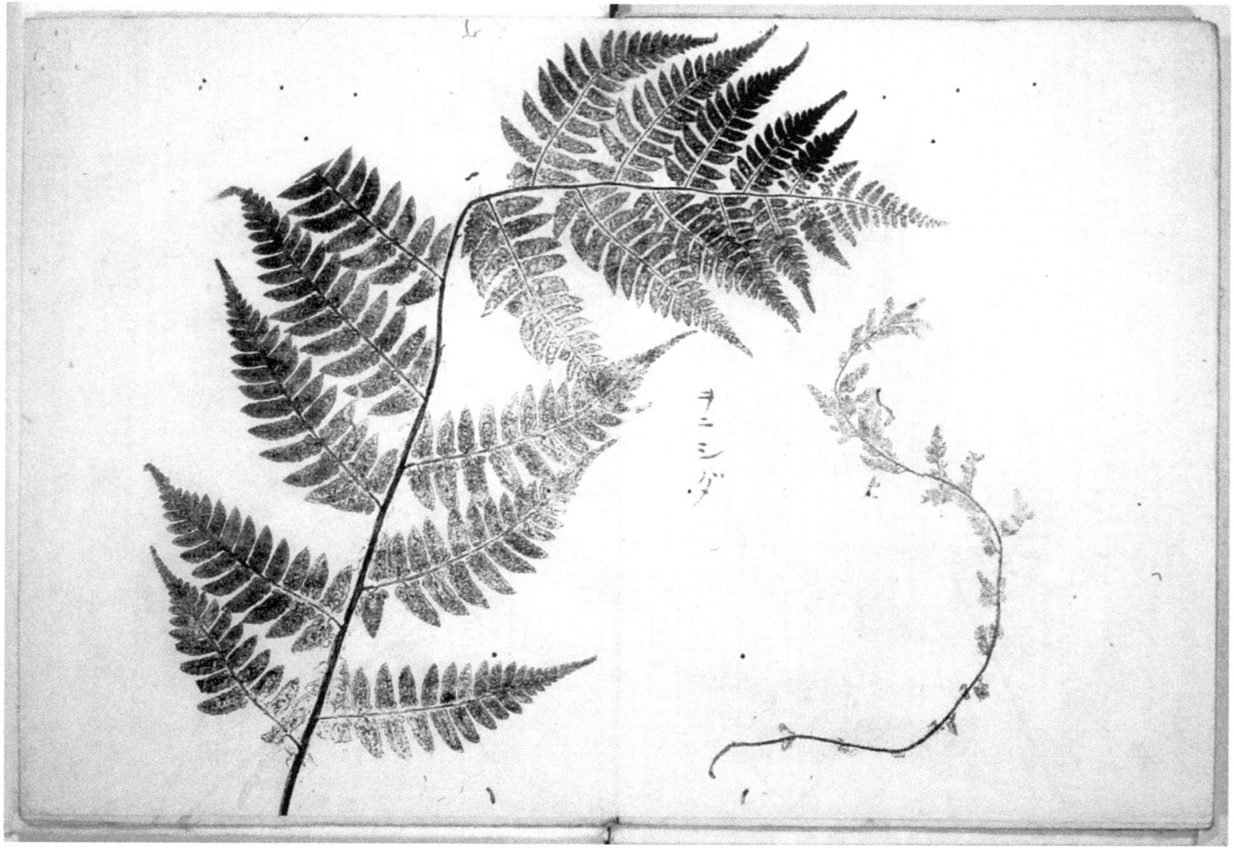

Figure 4.1 Tanba Shūji, *Onishida*, from "*Honzō shin'ei*," before 1861, ink-rubbing print, Kyōu Library

The specific uses of these prints reveal to us that the Shōhyaku-sha used the process of ink rubbing as a way to anchor the plant in the world. The overwhelming majority of the Shōhyaku-sha referred to the ink-rubbing prints as *shin'ei* (shadow of the real) in titles for albums into which the prints were gathered.[2] The pertinent incorporation of the concept of *shin* raises two significant questions: What kind of hermeneutics was deployed to understand these images as explicitly related to the concept of the real? And how does *shin* in this case relate to other uses of the term?

Here it is helpful to recall earlier discussion of the concepts of *shashin* and *shin* in "Encyclopedia for Household Use" in which Udagawa Zonshin and Yōan foregrounded, through these concepts, their interest in the process of producing representations. Their encyclopedia entry on camera obscura was particularly notable for avoiding conventional use of the term *shashin* as employed by the scholars of Western learning (*rangaku*), who added the word lens or mirror (*kyō*) to describe the apparatus in Japanese. Instead of simply adapting the previous rendition of camera obscura in Japanese, *shashin kyō*, Yōan referred to the instrument in a title, calling it "a device that makes the *shin'ei* of scenes and objects appear in a darkened room." In the translated entry for camera obscura, the concept of *shin* is evoked twice more, once in translating *waar gedaante* as "*shin* form" and then in translating *zijgen* as a verb, "*shashin*." Yōan, moreover, insinuates *shin* as a quality found only in already existing, observable physical entities rather than those whose existence is uncertain or imaginary.

Yōan's characterization of the ink-rubbing technique in an unpublished manuscript, "Treatise on Botany" ("*Shokugaku dokugo,*" circa 1829), speaks to this.

> On the technique that can preserve the *shin* form of plants other than a herbarium
>
> Although a herbarium sufficiently maintains the real form [*shin kei*] of plants by pressing the entire plant, it cannot be preserved for long. To maintain the real form and also to preserve and transmit it forever, there is nothing better than colored sketches [*shasei*].... The saying "a glance is worthier than listening to hundreds of accounts" holds true. However, a picture produced by a craftsman is, by and large, overtly colored and focused too much on embellishing persons and animals added in landscape paintings [*tenkei*]. As a result, pictures become nothing but pleasing trifles for women and children.... Recently, someone from our country has concocted an excellent method. In its ability to maintain *shin*, it surpasses a herbarium, and it can

> also be preserved for a long time. This method begins by placing a leaf or an entire plant in a herbarium book, and allowing the specimen to be pressed for a day or half a day. After pressing, the specimen is taken out of the book and black ink is applied on one side of the specimen using a brush. An inked specimen is placed on a wooden board, and a clean sheet of paper is placed over the specimen. Rubbing the paper with the palm of the hand, the shape of a leaf is clearly transferred to the paper. In addition, by using this method, the faint vein and the folded areas of the leaves, which are impossible for a man to depict, can be depicted. In this sense, it is no different from copper-etching prints. I hear that this method also exists in the West and that what is known as Kimura Kenkadō's herbarium from Holland is made using this method.[3]

This passage underscores the ability of ink-rubbing prints to preserve *shin* as their most significant feature as a representational medium. Yōan points to the immediacy of ink-rubbing prints as bypassing the mediation of observation by the image maker, and then he characterizes the representational quality of ink rubbing as on equal footing with herbariums. Yōan thus creates a hierarchy of visual media based on how the media *perceivably* preserve the "*shin* forms" of plants, with the ink rubbing and herbarium the highest, followed by sketches, and text as the lowest. The "*shin* forms," in this sense, are judged by the represented details that, according to him, would have been missed had they been overlooked by an artist in making a sketch. In comparing the properties of hand sketches and copperplate etchings, Yōan posits ink rubbing as equivalent to the latter in the extent and quality of attainable details, which surpass any sketch. Yōan implies here that he viewed copperplate prints to be devoid of the image maker's intentional ornamentation. He heralds the ability of ink-rubbing prints to represent detailed information as one of the characteristic and advantageous aspects of this printing method; this serves precisely as the basis for his claims that the "*shin* forms" are retained in ink rubbing.

Regarding the durability of the *shin* form, Yōan sees ink-rubbing prints as superior to herbariums. In his mind, the two are intimately related. In translating the entry on herbariums in "Encyclopedia," Yōan grasped the collection of herbarium as *shashin* pictures. This observation indicates two aspects of ink rubbing that he held significant: reproducibility and temporal durability. Given the fact that multiple prints could be taken from one inked specimen, thus reproducing the *shin* form of a herbarium, the priority Yōan places on durability suggests his awareness of the reproducibility of ink-rubbing prints. Yōan sees no difference between indexical ink-rubbing prints and a herbarium. Bearing the physical traces of the plants from which the ink is transferred, Yōan evaluates these indexical marks on paper as the unmediated specimen itself, offering a direct experience of the plants equal to handling the herbarium. In Yōan's view, these ink-rubbing prints permit the same direct access to the specimen as a herbarium and are superior to it in the pursuit of knowledge in that the prints can physically outlast fragile dried specimens. Yōan's view that ink-rubbing prints offer the equivalent experience of direct observation characteristic of herbariums—that they equally express *shin* form—is expanded in the case of the Shōhyaku-sha.

In addition, when Yōan draws a contrast with "pictures produced by craftsmen" and prioritizes the direct experience of seeing, touching, and handling the actual plant from which the illustration is made, he is asserting the importance of producing pictures by one's own hands in the study of botany. In this respect, the ink-rubbing prints came to be seen by the Shōhyaku-sha as embodying an uncontestable physical link—however invisible—of the encounter between the image maker and the object from which the representation is made.

The ink rubbing that Itō Keisuke and Yōan made together in 1827 during their excursion to Nikko Mountain exemplifies how the Shōhyaku-sha would expand the uses and significance of ink rubbing, as elaborated by Yōan, to interpret it as a visualized testament of an encounter with a particular plant. The ink rubbing in Figure 4.2 (see also color plate 3) is of a plant known as *haribuki* in common Japanese. The texts in Japanese that surround the ink rubbing are written by Keisuke. The black text describes the plant in detail and supplies other relevant information: the three regional names given to the plant, the three locations where the same plant was found, the approximate height, the patterns of bifurcation of the branch, the texture of the plant's leaves, the color of its flower, the number of pistils and stamen, a description of its fruit, and the smell of its root. In red ink, Keisuke informs us that the rubbing was made during an excursion with Yōan more than fifty years ago (and he inserts the date of his writing in red as May 1878). In other words, Keisuke revisited this print half a century after he helped make it and added the text in red. The Roman letters under the rubbing, perhaps by Keisuke although possibly by another hand, give the Linnaean name for the plant, "Fatsia horrida, Smith." This particular use of ink rubbing to provide immediate access to the experience of handling the plant and to transmit knowledge is predicated on Yōan's understanding, recorded in his encyclopedia entries from around 1826, that these prints embody the "*shin*

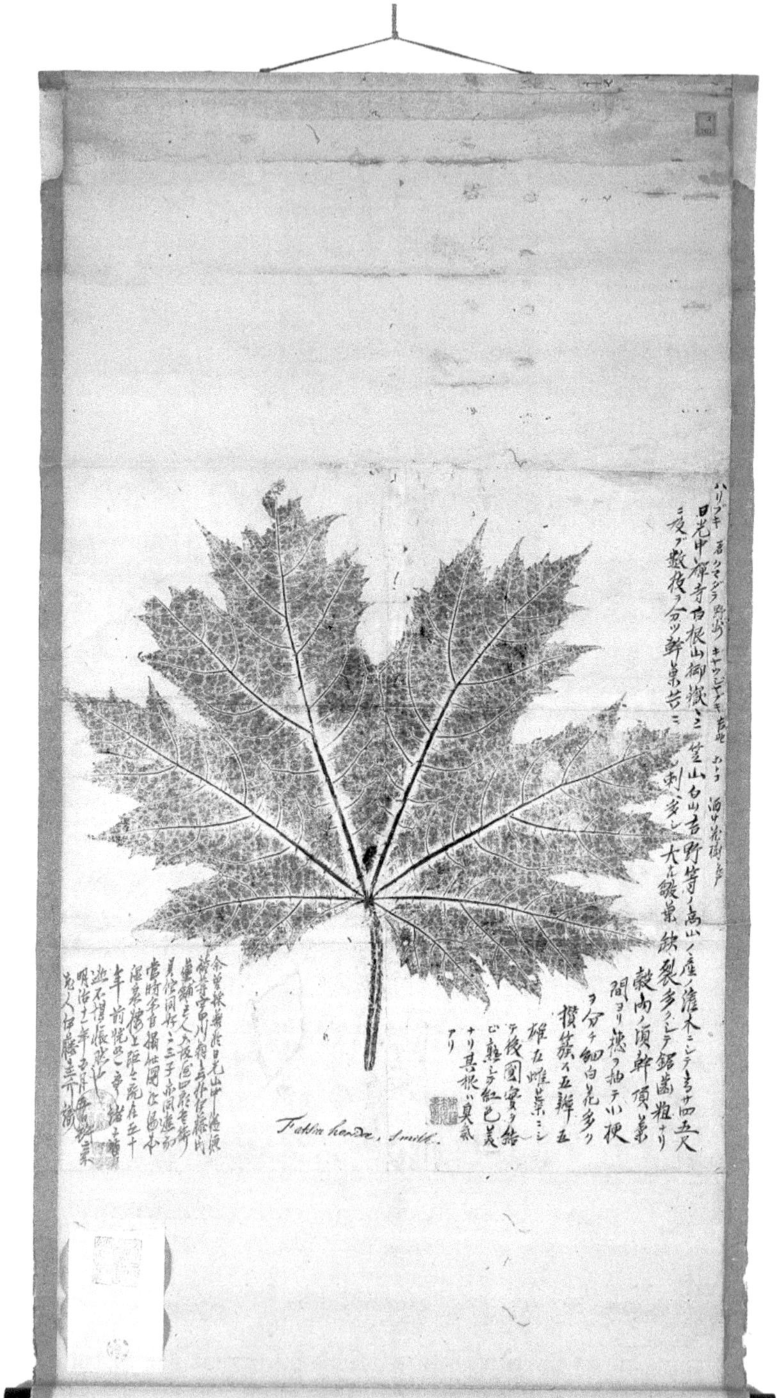

Figure 4.2 Itō Keisuke and Udagawa Yōan, *Haribuki*, 1827, ink-rubbing print and ink on paper, National Diet Library

form" of the plants, the real form that is not tempered by artistic embellishment. Ink-rubbing prints by Keisuke and Yōan suggest the ways in which the veracity of these prints, nested in the indexical relationship of object and representation, resulted from two separate but interrelated facets of fidelity: from the visualized physical contact that took place between the specimen and the impression itself, and from the invisible yet implied physical contact between the image maker and the represented specimen. But as we see below, the Shōhyaku-sha would place more emphasis on the second implied fidelity, and thus the meaning of *shin'ei* prints, to highlight the significance of a physical encounter with and experience of particular plants represented in ink rubbing.

Keisuke's undated personal pamphlet "Oleander" ("*Kyōchikutō*") is helpful in discerning how the Shōhyaku-sha envisaged the particular epistemological use of *shin'ei* prints (Figs. 4.3, 4.4, 4.5, and 4.6; see also color plate 4).[4] Keisuke pasted into "Oleander" assorted texts and illustrations related to oleander plants, creating a multimedia and multilayered database. The pages on oleander ripped from Chinese texts, his own notes of corrections, and Dutch and Latin names in Roman letters are assembled in this manuscript, creating an expansive and diverse collection of pieces of knowledge regarding the oleander plant. Among this textual and pictorial miscellany, Keisuke also inserted three *shin'ei* prints. Although none of the illustrations in the pamphlet are signed, it is likely that either Keisuke himself or other members of the Shōhyaku-sha produced these prints; they are the only group that incorporated ink rubbing in their study of *materia medica*.

For the Shōhyaku-sha, the Linnaean system of nomenclature served as a key epistemological framework for reformulating their knowledge about medicinal plants, but Keisuke also sought after other pictorial and textual information not necessarily confined to and construed in accordance with this taxonomic system. This voracious, wide-ranging interest is exhibited in open-ended expansion of the heterogeneous pieces of oleander-related knowledge in the pamphlet, which appears boundless. Construction of a normative standard of knowledge about the oleander plant thus did not remain restricted to one prioritized epistemological frame; rather, Keisuke evidently presumed the existence of an inexhaustible supply of information on oleander that needed to be amassed and culled. "Oleander" documents this process, indicating that one of the primary purposes of such collections was to delineate misinformation in order to deter further confusion.

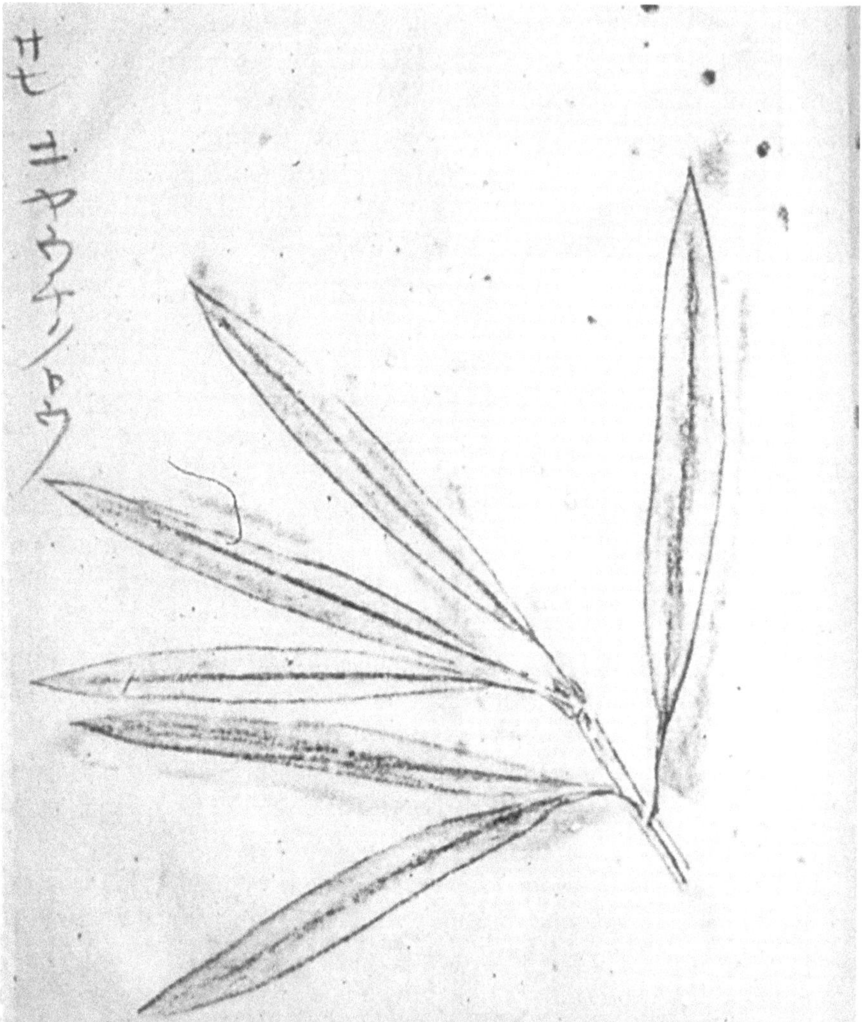

Figure 4.3 Oleander, from Itō Keisuke, "*Kyōchikutō*," c. 1860, ink-rubbing print, Itō Keisuke Collection, Higashiyama Botanical Garden

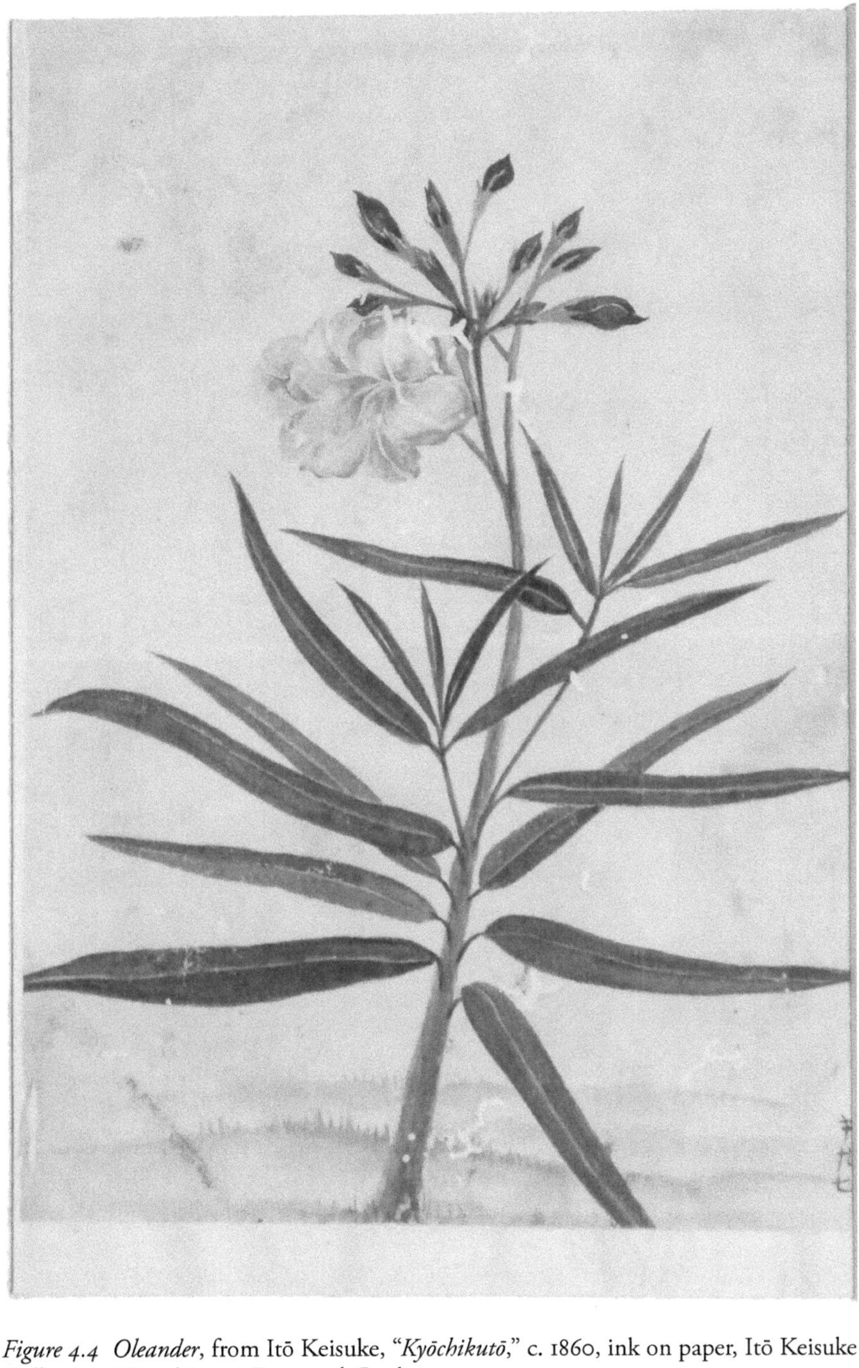

Figure 4.4 *Oleander*, from Itō Keisuke, "*Kyōchikutō*," c. 1860, ink on paper, Itō Keisuke Collection, Higashiyama Botanical Garden

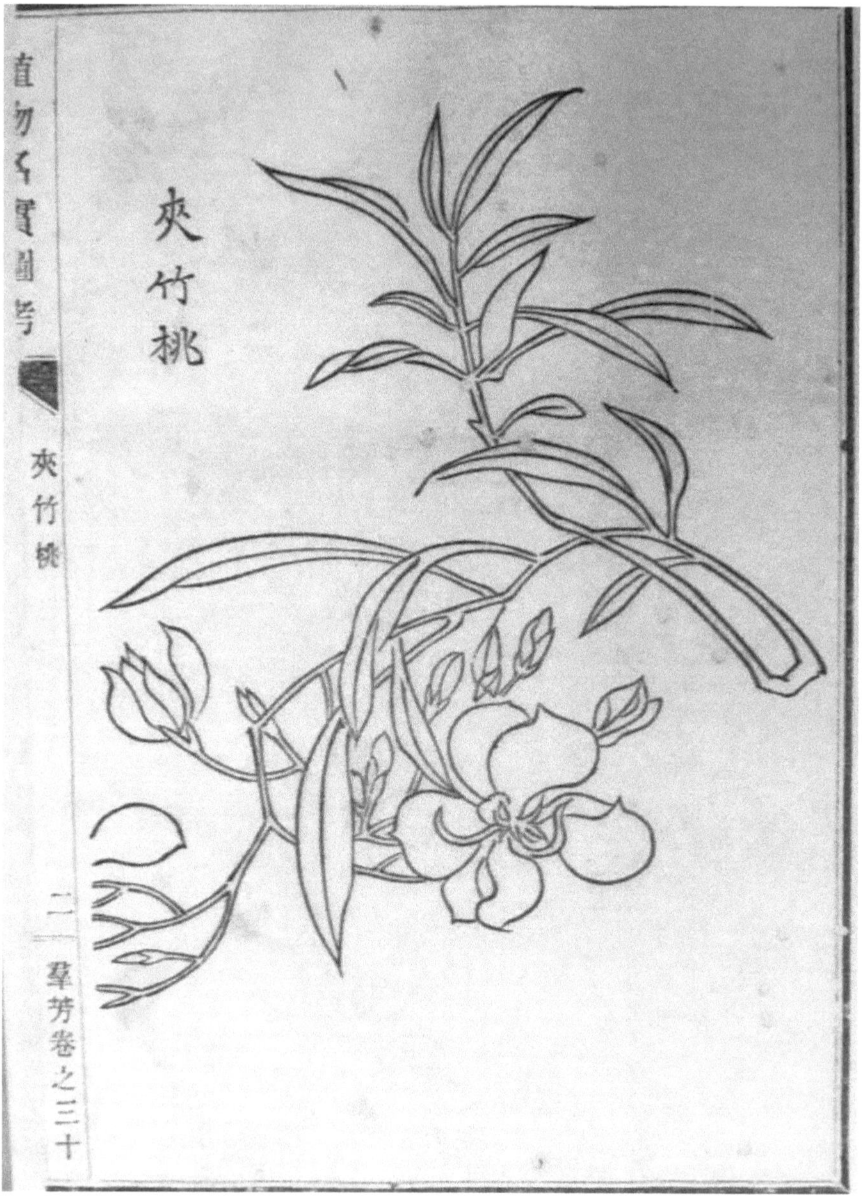

Figure 4.5 Oleander, originally from Wu Qijun, *Zhíwù míngshí túkao*, 1848, woodblock print, pasted in Itō Keisuke, "*Kyōchikutō*," Itō Keisuke Collection, Higashiyama Botanical Garden

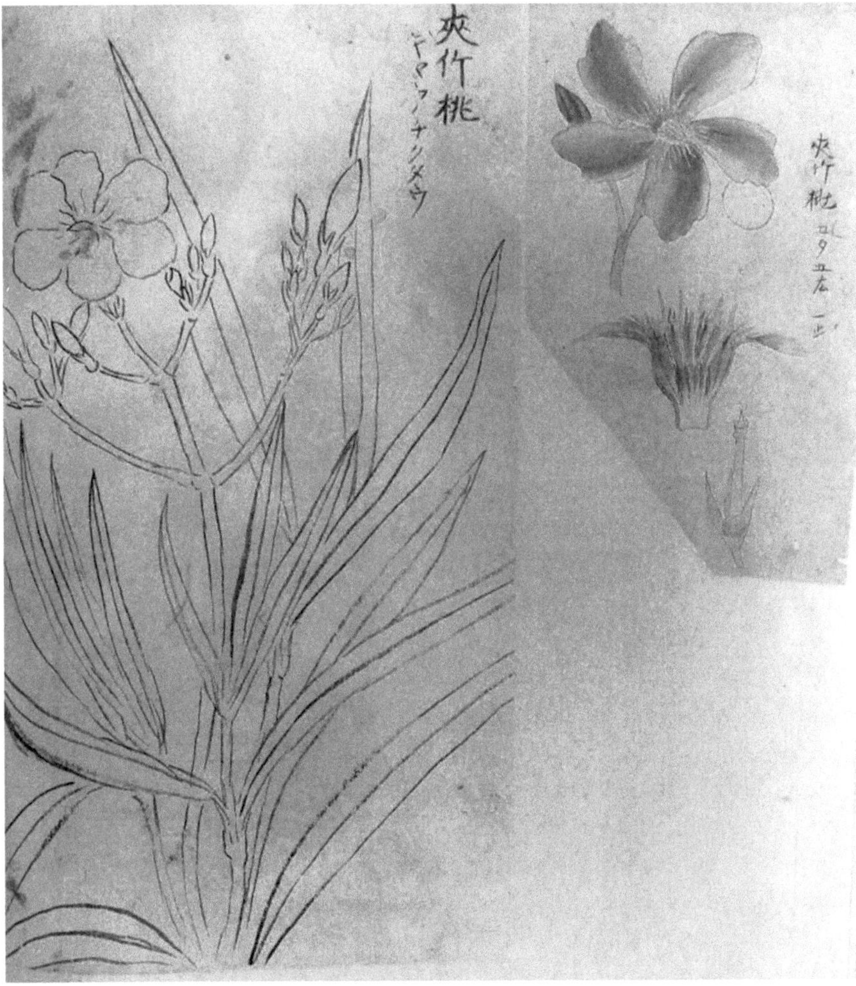

Figure 4.6 Oleander, from Itō Keisuke, "*Kyōchikutō*," c. 1860, ink on paper, Itō Keisuke Collection, Higashiyama Botanical Garden

In this manuscript, Keisuke calls attention to the fact that garden balsam and oleander are often confused in both Chinese and Japanese books; he insists that the two are completely different plants. The issue of correct names discussed in a previous chapter emerges again here as a pivotal epistemological concern for the Shōhyaku-sha. The inconsistency of names given to particular plant species left knowledge in disarray, and misinformation continued to circulate. Keisuke refers to the *Sino-Japanese Universal Illustrated Compendium* (*Wakan sansai zue*, by Terashima Ryōan, 1713) and *Primer on Eight Varieties of Paintings* (*Hasshu gafu*, imported from China between 1661 and 1667, and published in Japan in 1672) as publications perpetuating this confusion. Indeed, the ability to correctly match plants with names was vital for practicing physicians: in the case of garden balsam and oleander, both were used as herbal remedies, but careless consumption of oleander could be fatal. Proper identification of each plant, therefore, was critical.

Keisuke describes the general physical characteristics of the oleander, noting that it "grows three leaves per branch, and the branches grow opposite of each other on the stem." The illustrations and ink-rubbing prints in "Oleander" contribute visual evidence to resolving the problem of identification (Figs. 4.3, 4.4, and 4.7, and color plate 4).[5] Figures 4.3 and 4.7 are ink-rubbing prints, Figure 4.3 is produced using the indirect method, and Figure 4.7 is in direct method. Both prints focus on the tip of the Oleander branch. The script in the upper left corner in Figure 4.3 identifies the plant as *Kyōchikutō*, written in *katakana* script. In Figure 4.7, the writing in black reads "*serichia neriiho*" in *katakana*, and *Kyōchikutō ka* (class of oleander) in Chinese characters. Sketched in color ink by hand, Figure 4.4 (see also color plate 4) depicts a branch of the plant and includes the flowers as well as buds. Comparing the three figures, especially around the areas where branching begins, we begin to see one of the advantages of using the ink-rubbing technique. The image maker (Keisuke himself or someone from the Shōhyaku-sha) seems uncertain about how oleander branches grow: around the central branch, the bifurcation appears alternately, but the younger branches are depicted growing opposite one another. In ink-rubbing prints (Figs. 4.3 and 4.7) this point on the pattern of branching is consistently clear: they grow opposite each other.

Keisuke criticizes the images in *Universal Illustrated* and *Primer*, for incorrectly labeling as oleander what is actually garden balsam. He mentions these publications in writing but does not include the illustrations. However, he does include a page from *Zhíwù míngshí túkao* by Wu Qijun,

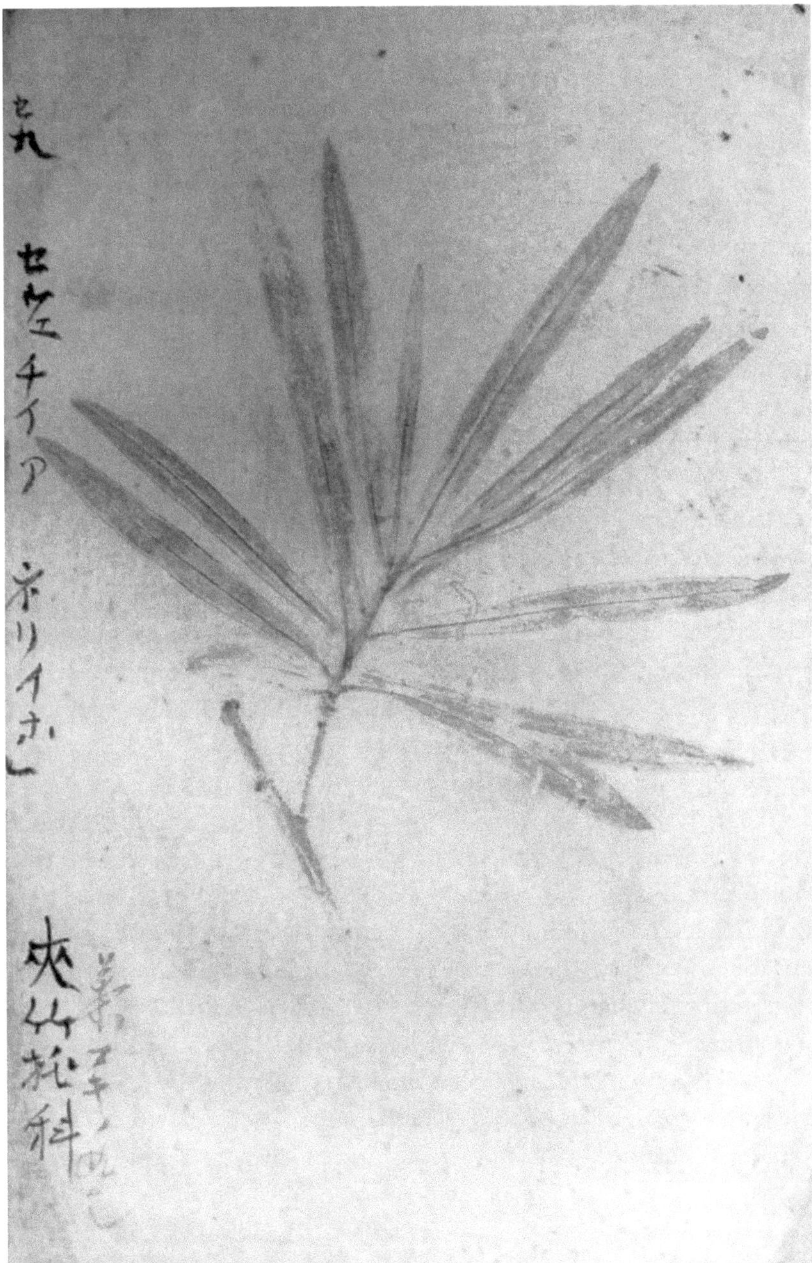

Figure 4.7 Oleander, from Itō Keisuke, "*Kyōchikutō*," c. 1860, ink-rubbing print, Itō Keisuke Collection, Higashiyama Botanical Garden

published in 1848 (Fig. 4.5). This illustration also reveals uncertainty about the arrangement of the branches in this species. In the branch that fills the bottom half of the image, the leaves seem to grow alternately, whereas on the branch depicted growing upward the leaves grow opposite from each other. In Figures 4.4 (and color plate 4) and 4.5, the bifurcation of branches from the stem contradicts Keisuke's characterization; Figure 4.6 represents an oleander plant that corroborates the textual information. Contextualized thus, the inserted ink-rubbing prints (Figs. 4.3 and 4.7) demonstrate the leaves growing in a whorl. With these differences in representation in mind, we can infer why Keisuke included these prints along with illustrations gathered from various sources. The *shin'ei* prints verify the particulars of how the branches form on the oleander plant. They corroborate branch formation patterns more reliably than illustrations in other media do. Already, as the illustrations from *Zhíwù* (Fig. 4.5), *Universal Illustrated*, and *Primer* attest, Keisuke was fully aware that published illustrations do not necessarily represent the physical appearance of the actual plants truthfully, and instead they may perpetuate confusion. The *shin'ei* print here serves as visualized testimony of both the textual information and the access Keisuke had to the plant, thereby transforming the perceptible into an intelligible and reliable account of the real oleander plant that Keisuke (or other members of the group, if they made the images) had encountered. Keisuke seized on ink rubbing as the technique of representation that could produce correct knowledge of oleander in a way that woodblock prints and hand-sketched illustrations could not. Keisuke writes on one page that the largest oleander tree in his region is in Mino; he jots down the name of the town and the approximate circumference of the trunk. Perhaps because this was his personal notebook, intended as a repository of accumulated information rather than a manuscript for sharing with others, he does not explicitly note the locations of the oleander trees from which the ink rubbings were made.

These ink-rubbing prints in Yōan's and Keisuke's work, moreover, did not result in the abandonment of all other representational practices. Keisuke's pamphlet "Oleander" concretely demonstrates varying types of representation that served the study of *materia medica* in an effort to identify trustworthy relationship among the texts, illustrations, and the existing plant itself. For instance, two colored illustrations from "Oleander" exemplify the type of images and information Keisuke was after: one is of a single-petal oleander and the other of a double-petal oleander (Figs. 4.4 and 4.6, and color plate 4). Later inscriptions in red ink that appear throughout this pamphlet, as in the ink-rubbing he made with Yōan, testify that Keisuke returned to these pages to make corrections to the previously gathered information. Echoing Kaibara Ekken's mistrust of textual knowledge and prioritization of empirical experience, this small pamphlet reveals the importance of observation and direct knowledge of plants in the unrelenting reworking and reformulation of the discourse of *materia medica*. What the ink-rubbing prints demonstrated as *shin* in this context was its indexical relationship to the real plant, attained through undeniable physical contact between the existing specimen and the representation. The role of ink-rubbing prints within the context of this personal pamphlet indicates the specific role and expectation that Keisuke ascribed to these prints: to record the existing oleander plant as a pictorial eyewitness to his own experience of observing and handling the plant.

The striking visual simplicity of the representations made in ink rubbing owes much to the stark background and the comparatively bare appearance of the images themselves. In fact, except for the few colored images, most of the ink-rubbing prints contain nothing but the rubbing of the represented plant and a few handwritten names. The reversed and pressed contact image remains free of manipulation or other interference, and this indexicality allows Yōan to claim that the herbarium and ink-rubbing are equal in their ability to maintain the "*shin* form." Keisuke and members of the Shōhyakusha extended the advantage of this medium as a way to not only record the specimen but also to use the representation as a visualized testament to the experience of handling the represented specimen. On the contrary, the notion of embellishment (*tenkei*) that Yōan evoked in his description of ink rubbing requires evaluation by an image maker (and later a viewer)—a judgment on the imperfection, deficiency, and immaturity of an image. Yōan seizes on certain types of detail as proof of the fidelity of ink rubbings, but the ability of these images to represent the size of a plant and the arrangement and scale of its parts also contributed to the perception that ink rubbings effectively preserved *shin*. Because the prints were made directly from the plant itself, they maintained the actual size of the specimen in the printed form, offering access to verifiable information about the relative size and shape of plants in a comparable format (see Figs. 4.2 and 4.8, and color plate 3). These comparative compositions of *shin'ei* prints use the edges of the page as X and Y axes with which to measure the height of each plant. Rather than envisioning the size of various plants through textual information, the comparisons made in *shin'ei* prints address questions of scale almost instantaneously through visual perception. In a page from Tanba Shūji's compilation of ink rubbing titled "Shin'ei honzō" (1856), Shūji calls visual attention to the difference in size, and thus the scale of growth, in the *onishida* plant by placing the sprout and the full-grown plant next to each other (Fig. 4.1). Similarly, Ishiguro Sei'an, another Shōhyaku-sha member, juxtaposes three plants—known as *sekishō*, *shanohige*, and *eyanukakusa*—so as to demonstrate similarities and differences in the shape and size of leaves of the species (Fig. 4.8). The simplicity of these pages, in which the only text is the plant names written with red in *katakana*, forces the eye to find corresponding similarities and locate their subtle difference within the found similarities.

However, it is critical to recognize that the Shōhyaku-sha did not use ink-rubbing prints as substitutes fully capable of replicating the experiences of seeing and touching the represented plants. The Shōhyaku-sha did not believe that ink rubbings could reproduce the experiences of encountering

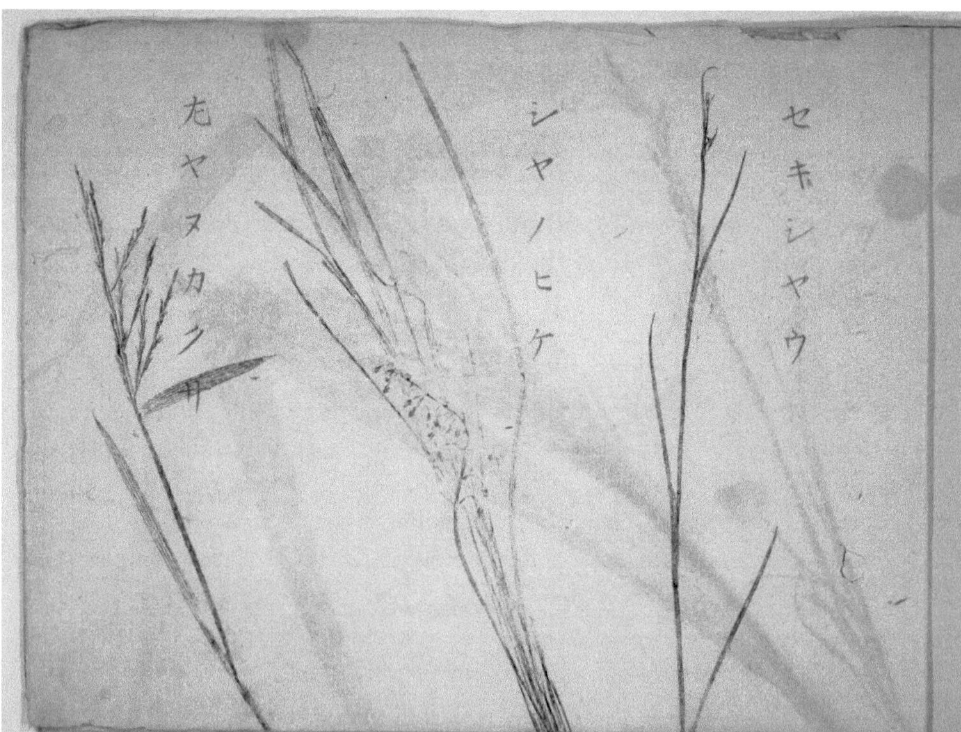

Figure 4.8 Ishiguro Sei'an, *Sekishō, shanohike, eyanukakusa*, from "Shin'ei honzō," compiled in 1857, prints made before 1836, ink-rubbing print, Kyōu Library

the represented plants. Rather, they understood the prints to be the visual record of a specific experience with the plant, and they were fully aware of the representational limits of these prints. As Ōkubo Masa'aki, one of the most avid practitioners of this method, rightly points out, these rubbed imprints do not represent the plants "as they are" or "the way they look."[6]

Masa'aki demonstrates a keen awareness of the limitations and advantages of representations that the Shōhyaku-sha shared. And it is here that the Shōhyaku-sha's approach to ink rubbing departs from that of Udagawa Yōan. The sculptural aspects of the plant specimens are flattened in these representations, and the colors of the specimens are translated into varying degrees of gray. In this regard, there is a strong similarity between the use of ink rubbing on plants and a practice known as *takuhon* (Ch: *tuoben*), a technique invented in China before woodblock printing. Yoshida Jakusōan's 1864 catalog of a *honzō kai* in ink rubbing includes numerous objects: a woodblock used to make woodblock prints, an archeological finding of an old piece of wood, dolls, musical instruments, a Buddhist statue, old roofing tiles, numerous mirrors from China, old coins, spearheads, swords, a flute, fossils.[7] Of the 126 objects displayed at this *honzō kai* gathering, Jakusōan produced thirty-five rubbing prints. In addition to failing to capture the three-dimensionality of these objects, the images lack accurate details, since they were created by the indirect method of rubbing, which is less adept at capturing fine detail than the direct method. These representations of the displayed objects are an unfailing record of an encounter with objects within this specific *honzō kai*.

Taking advantage of the same conditional aspect of this printmaking method, several of the Shōhyaku-sha members used ink rubbing as the means of recording a particular occasion. For instance, in "Catalogue of Pressed Leaves from Ezo" ("*Ezo sakuyō mokuroku*"), Keisuke relies solely on ink rubbing to represent a herbarium. Keisuke notes on the cover of the pamphlet that these prints were made when he had access to the herbarium of plants of Ezo (present-day Hokkaidō) in Edo. Here, he uses the prints as reliable evidence of his experience of witnessing, that is, seeing and handling the herbarium. Executed in the indirect method so as not to ink the actual herbarium directly, the ink-rubbing prints in this context remain faithful in respect to the existence of the herbarium and thus are evidence of the opportunity Keisuke had to handle and examine the herbarium. The portability and simplicity of the ink-rubbing process positions it as a method capable of producing irrefutably trustworthy evidence for one's encounter with the plant. The goal of these applications of ink rubbing was not "accurate" representation of a specimen. The visual accuracy of the objects was not the primary concern. The goal was to preserve and pictorialize one's encounter with the specific plant in ways that cannot be contested.

Examining Keisuke's small pamphlet of his collection of old roofing tiles and Jakusōan's record of a *honzō kai* he organized in 1864, both of which use ink rubbing to record the encounter with a particular object, one wonders whether to designate the technique used in these books, differentiated in Japanese as *inyō-zu-hō* or *takuhon*. The questioning becomes rather difficult—and problematic—because the historical accounts of the arrival of these techniques emerge out of a particular teleological view of adaptation and absorption. In fact, several historians of *materia medica* point to Johann Kniphof's *Botanica in Originali*, which uses a technique known as "nature printing," as the sole inspiration behind the group's use of botanical specimens as objects for ink-rubbing prints (Fig. 4.9 and color plate 5).[8] The surviving four hand-copied facsimiles of Kniphof's book attest to its warm reception and popularity among the Shōhyaku-sha members: Ema Ransai made one copy, Ono Ranzan and Mizutani Hōbun authenticated another copy owned by Tanba Shūji, Iinuma Yokusai had a copy in his possession, and Yoshio Jōsan made a copy for himself.[9] These hand-copied facsimiles were created by placing thin rice paper over the original Kniphof ink transfers and tracing the shape of the original seen through the paper by hand in black calligraphic ink.

The direct causal relationship asserted by this narrative of influence implies that the members of the Shōhyaku-sha absorbed and incorporated the "Western" example unreflectively, with no regard for their own epistemological concerns. Certainly, the Shōhyaku-sha shared enthusiasm for the novelty of Western ideas and goods, but the reasons for adapting and experimenting with Western ideas and methods had nothing to do with the exotic nature of their origin. In fact, as we read in Udagawa Yōan's appraisal of ink rubbing in his "Treatise on Botany," Yōan implies that the ink-rubbing method originated in Japan. According to him, it was a mere coincidence that a similar method was concocted in the West. This sense of serendipity is expressed in the last sentence of his description of ink rubbing, when he states: "I hear that this method also exists in the West and that what is known as Kimura Kenkadō's herbarium from Holland is made using this method." Ueno Masuzō, one of the most prominent historians of natural history in Japan, discredits Yōan completely in his *History of Natural*

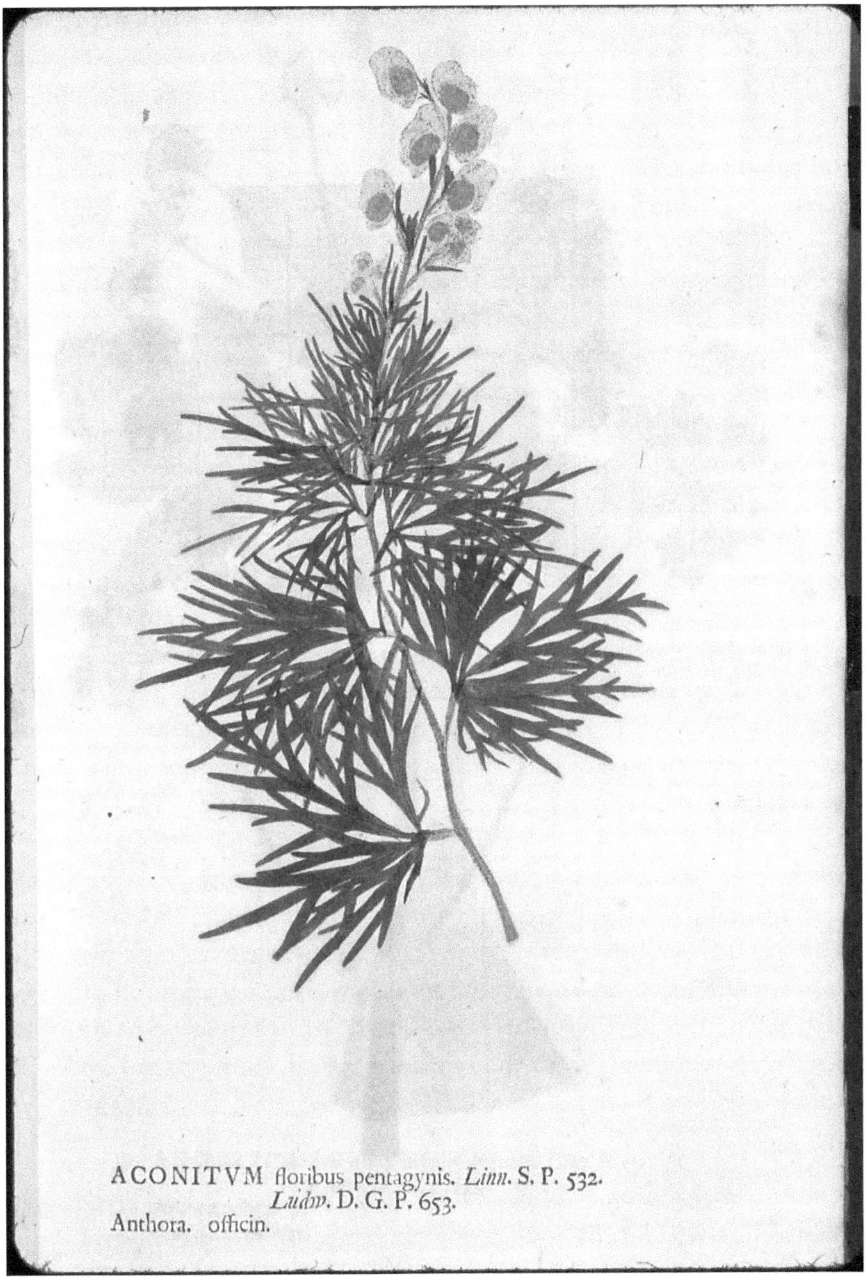

Figure 4.9 Johann Hieronymus Kniphof, *Aconite*, from *Botanica in Originali . . .* , 1765, nature print, University of Chicago Library, Special Collections Research Center

History in Japan (*Nihon hakubutsugaku shi*, 1988). Ueno argues that "it is probably the case that this ink-rubbing method was crafted after Kniphof's book when a copy of it came to Japan and someone from the Shōhyaku-sha laid a hand on the copy... [Yōan] writes as if someone in Japan invented [ink rubbing], but it is actually a technique that came from the West," thereby implying that Kniphof's book was the only source from which the Shōhyaku-sha would be inspired to make ink-rubbing prints.[10]

I am interested in situating the use of ink rubbing historically. That is to say, my interest is not in locating the originary moment of ink rubbing, nor in deciding which one of the terms associated with this method is more apt—*inyō-zu-hō*, *takuhon*, or a deviation of "nature printing." As we will explore more fully in Chapter Five, this historiographical gesture to single out the "Western" origin of a representational technique as the overriding cause for its incorporation in Japan is also found in the conventional narrative arc of photographic history in Japan. This correlation in the historiography of the two media offers a critical nexus for reflecting on the cultural conditions in which these techniques were employed, and the process through which it became desirable to incorporate them into local study of *materia medica*. For the Shōhyaku-sha, driven by their ongoing attempt to calibrate their knowledge of *materia medica*, examining prints of *Botanica in Originali* would have struck a particularly strong chord: the "nature printings" are the physical traces of the dried specimen.

First published in Germany in 1747, *Botanica in Originali* contained 1,186 images produced using the method of "nature printing."[11] The second edition was published eighteen years later in 1765 in Halle, and the number of images increased to 1,254. The pivotal difference between the two is the organizing framework: the first edition followed the classification proposed by Joseph Pitton de Tournefort (1656–1708), while the second version ascribed names in accord with the Linnaean classification system. In *Nature Printing and Its Origin and Objects* (1856), Henry Bradbury explained "nature printing":

> [Nature printing] adopted to produce such results at this period (1650) consisted in laying out flat and drying the plants. By holding them over the smoke of a candle, or an oil lamp, [the entire specimen] became blackened in an equal manner all over; and by being placed between two soft leaves of paper, and being rubbed down with a smoothing bone, the soot was imparted to the paper, and the impression of the veins and fibres was transferred.[12]

Wilfrid Blunt adds that Kniphof's nature printing substituted "printer's ink for lampblack and flat pressure for the smoothing bone." According to Blunt, Kniphof also introduced "coloring the impressions by hand according to Nature, a proceeding which though certainly contributing to the beauty and fidelity of the effect, yet had the disadvantage of frequently rendering indistinct and even of some times totally obliterating the tender structure and finer veins and fibres."[13] It is still unclear, when, how, and among whom Kniphof's book circulated in Japan. Keisuke owned the second edition of *Botanica in Originali*, which was passed down to his grandson Itō Tokutarō: this copy has not been located after the bookstore Maruzen acquired it.[14]

Similar techniques of visualization existed in Japan well before the arrival of *Botanica in Originali.* In the technique known as *takuhon*, the epitaphs of commemorative stones and graves were preserved by transferring the inscriptions to a piece of rice paper. This technique of rubbing (*takuhon* literally means "imprinted original") uses the indirect method in making the print by applying a cotton sponge that has absorbed black calligraphy ink over the paper covering the object. The characters on these stones were carved into the piece of stone; thus the applied ink creates an image on the paper according to the depression of the characters, creating a negative image in which the letters remain the color of the paper. As a result, the *takuhon* prints made a faithful reproduction of the writing on the stone with the correct orientation. The children's game of placing a piece of paper on a coin and rubbing a pencil over the paper is technically the same as the indirect method save for the use of sponge and ink. This technique had been known in Japan since the Kamakura period (A.D. 1192–1333).

What, then, can account for the fact that the popularity of ink-rubbing prints was limited to the Shōhyaku-sha?[15] Kniphof's "nature printing" and ink rubbing are semiotically and materially analogous: they both produce pictorial representations of a plant specimen through physical contact, an index, between the plant and the piece of paper, and they both use ink on paper, although their ink and paper are quite different. But the status in which we find the prints made from the two pictorial methods differs greatly: Kniphof's prints are bound in leather and published, whereas the Shōhyaku-sha's ink-rubbing prints remain mostly unbound and never published. How the ink-rubbing technique was actually incorporated into the Shōhyaku-sha's study of *materia medica*—as a way to provide a visual testament to the existence of a particular specimen or object, and also as a record of one's encounter with an object—echoes the ongoing epistemological concern that the group shared. From these patterns in how the group incorporated this technique into their research, we can infer that the Shōhyaku-sha recognized the advantage in using prints to provide uncontestable access to and witness of the real, precisely the premise of Kniphof's publication.

What differentiates the two media—and this is the key to understanding the popularity of ink rubbing among the Shōhyaku-sha—is the technical printing process. To do nature printing, one would have needed a herbarium sample that was already dried, a source of fire (candle or oil lamp), two pieces of paper, and bone to smooth over the paper. The minimum requirements for making ink rubbing, on the other hand, were an object (whether fresh or dried) to make the impression of, a piece of paper, and ink (and a sponge for the indirect method). The material status of ink-rubbing prints left by the Shōhyaku-sha—incorporated into personal pamphlets (Itō Keisuke's oleander, Figs. 4.3 and 4.7), in manuscripts collecting ink rubbings (Tanba Shūji and Ishiguro Sei'an, Figs. 4.1 and 4.8), as a testament to the handling of a herbarium (Itō Keisuke's "Pressed Leaves"), or in the record of excursions they took (Itō Keisuke and Udagawa Yōan, Fig. 4.2 and color plate 3)—signals their desire to use the prints for much more personal purposes. In this respect, it is important to recall that, similar to the Shōhyaku-sha's use of Linnaean taxonomy, incorporation of ink rubbing among their investigative tools did not lead them to abandon or repudiate other representational means altogether. The technique of ink rubbing offered advantages to the Shōhyaku-sha that other means of representation could not. The Shōhyaku-sha continued to use a variety of picture-making methods in the rearticulation of discourse of *materia medica*.

The potentials of the visualizing technique, therefore, were not entirely predetermined by the origin of the technique in the West or China; rather, new potential for the medium emerged with the development of a new epistemological orientation and its challenges to established modes of representation. What needs emphasis here is the role historically rooted desire for medicinal knowledge played in propelling the adaptation of the ink-rubbing technique for botanical representation. *Shin'ei* prints were not a technical "invention" but an innovative incorporation of the existing technique of ink rubbing into the epistemological project of the Shōhyaku-sha. This indispensable synchronization of a particular technology with the desire to apply it will resurface in the next chapter in discussing photographic technology and its use by the Shōhyaku-sha members.

This new application of ink-rubbing techniques further sharpens the contrast between the Shōhyaku-sha, as they sought to verify the existence of plants, and the scholars of Western learning (*rangaku-sha*). As Imahashi Riko elegantly demonstrates in her study, the act of "borrowing/copying" (*tensha*) was popularized among the Western pictures (*ranga*) artists in the northern region of Akita in the late eighteenth century. Imahashi argues that at the time, *tensha* illustrations related to *materia medica* and nature were viewed with a surprising amount of "trust," to the extent that the representations were conflated with their object. For instance, Imahashi argues that when Satake Shozan copied the illustrations of butterflies by Hosokawa Shigetaka in his "sketchbook of insects," he treated Shigetaka's illustrations as if they were the actual living butterflies.[16] She adds that viewers also adhered to this understanding of pictorial representations as real objects. In other words, insofar as illustrations related to *materia medica* were concerned, the difference between borrowing or copying an illustration and looking at the actual object of representation in life—a critical difference in current perceptions—did not play a significant part in the pictorial practices of the Western pictures (*ranga*) artists in Akita.

Imahashi's interpretation allows us to highlight the difference between the Akita *ranga* school in the late eighteenth century and the Shōhyaku-sha in Owari in the nineteenth century. For the latter, access to the object pictured in the representation was pivotal for their accumulation of knowledge; only after verifying its existence in the natural world could the Shōhyaku-sha trust the illustration. In fact, this concern with the image maker's access to the depicted object runs through the archival records and the vast volume of ink-rubbing prints of the Shōhyaku-sha as an obsessive quest.[17]

Copper-Etching Prints and the Concept of the Real

We can discern the association made between the real (*shin*) and pictorial representation in the Shōhyaku-sha's approach to copperplate prints. Members of the group also characterized illustrations in this medium as *shashin* and understood the pictures as tangible proof of the existence of represented plants. In 1831, for example, Keisuke sent a copperplate print of a nonindigenous herb, mignonette, to Kaku Sukeyuki, a scholar of *materia medica* who lived in distant Shimabara domain. In the accompanying letter, Keisuke wrote, "Mignonette is a particularly rare kind, and I am including a *shashin* picture of that as well for you."[18]

The "*shashin* picture" to which he refers is a copperplate print of mignonette produced by Kondō Shūen, a local Owari *samurai* (Fig. 4.10; see also color plate 6). Firmly embedded in the scholarly exchange about the plant, Keisuke's use of *shashin* correlates directly to Yōan's use of the same term in translating herbarium (*Kruidboek*) as a necessary component in the study of *materia medica*. Here, we witness the deployment of *shashin* as a

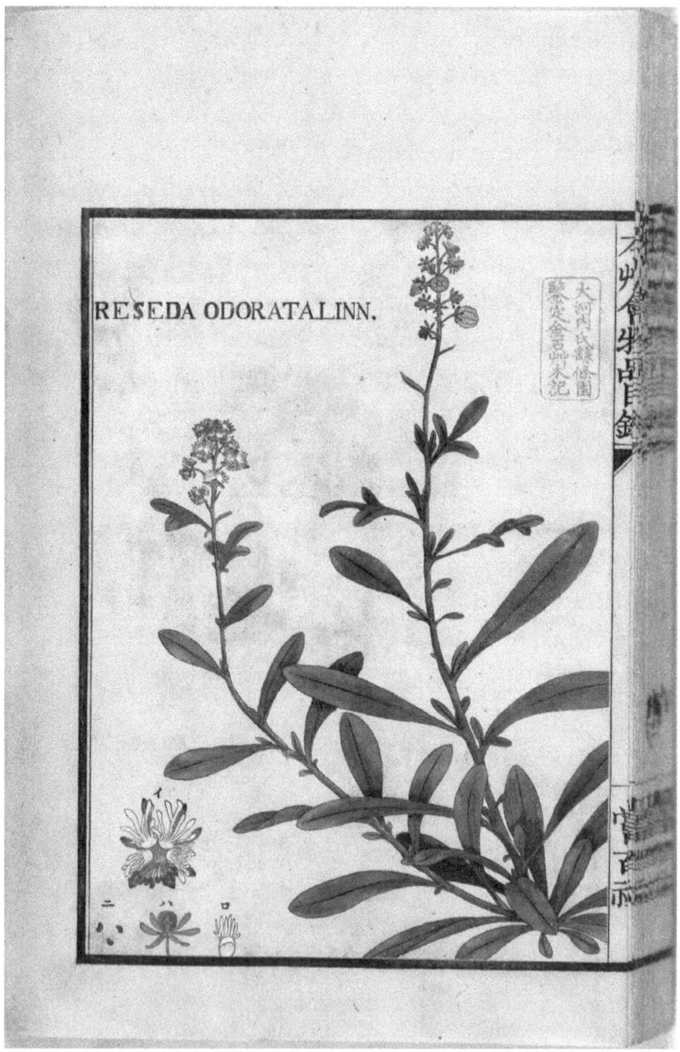

Figure 4.10 Kondō Shūen, *Reseda Odorata (mignonette)*, from *Honzō kai buppin mokuroku*, 1835, copperplate on paper, Hōsa Library

concept that does not conform to the use of camera obscura. Nor is it consistent with apprehending ink rubbing through the concept of *shin*. Rather, Keisuke's deployment of the term here responds directly to the Shōhyakusha's epistemological concern to verify the name and existence of a plant through means of visual representation.

The specimen of mignonette that Keisuke described as "rare" in his 1831 letter to Kaku Sukeyuki was exhibited at a *honzō kai* in 1835. The sample belonged to Keisuke's brother Ōkōchi Zonshin, who noted that the plant was given to him in 1828. Keisuke would have had ample opportunity to verify the existence of the plant and make a pictorial representation after his brother's plant. At the *honzō kai* in 1835, when the plant became more widely known through public exhibition, neither the medicinal attributes nor the proper classification in the Linnaean system were known.[19] That is to say, though the plant's medicinal and scientific particularities were still in flux, Keisuke was able to verify with his own eyes that the mignonette existed. The term *shashin* is mobilized in this case precisely to connect the irrefutable existence of a plant and its pictorial representation.

The technique of copperplate engraving was first introduced to Japan through the works brought by the Jesuit Christian missionaries in the mid-sixteenth century, but it was not until the late eighteenth century that copper-etching printing was practiced in Japan.[20] Shiba Kōkan, the first artist to produce successfully an image using this technique, acquired knowledge of the medium from imported Western books, including Chomel's *Dictionnaire oeconomique*.[21] By the early nineteenth century, the number of artists who produced copper etching was rising, among them Aōdō Denzen (1748–1822) and Maki Bokusen (1775–1824). Although Kōkan preferred to create perspective prints with this technology, transforming the noted places of Edo into a spatial vista with the help of one-point linear perspective, other subjects of copper-etching prints catered mostly to the emerging interest in Western learning (*rangaku*).[22] The scholars of Western learning became interested in learning and using copperplate technology in the late eighteenth century, especially after the shogun's senior counselor, Matsudaira Sadanobu, sought to find an artist who could apply the technique for utilitarian purposes. Denzen was appointed to this role and became the trusted artist who produced world maps as well as illustrations of various types of animals for the shogunate.[23]

Shūen, who produced the copper etching of mignonette for Keisuke, learned the technique from Bokusen, a versatile artist in the Owari domain who studied painting and woodblock printing first with Utamaro and later with Katsushika Hokusai, one of the most productive and popular woodblock artists in late Tokugawa Japan.[24] Shūen's other known work in copperplate etching, a portrait of Carl Peter Thunberg, is included in Keisuke's *Nominal Differentiations* (1829). Shūen's collaboration signals his interest in and familiarity with Keisuke's work and, more broadly, with studies of *materia medica*.[25] Shūen's portrait was based on a portrait of Thunberg that appeared in his travelogue *Voyage de C. P. Thunberg au Japon*, published in 1796 (Figs. 4.11 and 4.12; see also color plate 7). Siebold is said to have torn

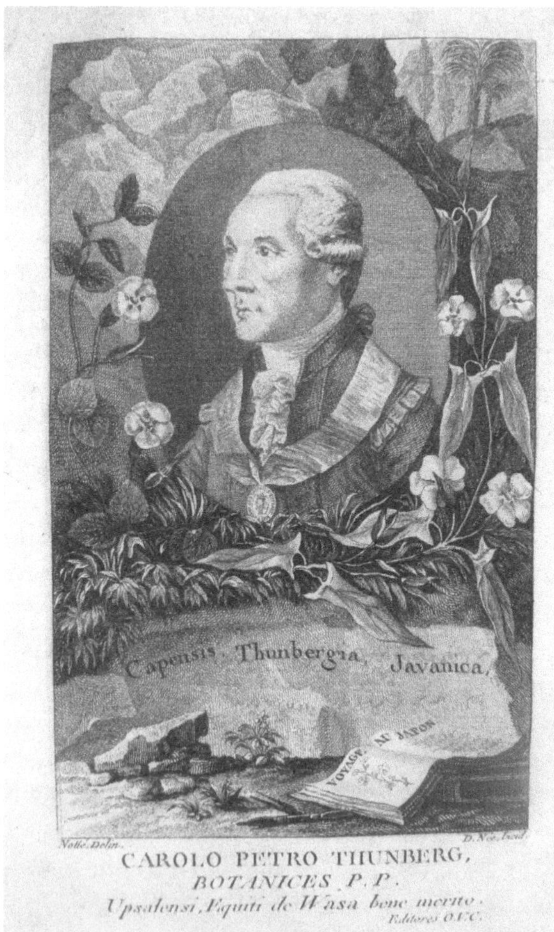

Figure 4.11 Portrait of C. P. Thunberg, *Voyage de C. P. Thunberg au Japon*, 1796, Special Collections Library, University of Michigan

the portrait from his volume of *Voyage de C. P. Thunberg* and given it to Keisuke.[26] Shūen embellished the original by adding a border and decorative framing pattern as well as a caption above the likeness in Chinese characters that read, "Portrait of the notable medical doctor Thunberg."

Shūen added borders around the portrait to mimic the mat and wooden frames that would protect and present the print itself, and he transformed the caption in the original to a label in his print. These additions to the original suggest, in turn, Shūen's keen awareness of material reality of representation. Calling attention to the represented nature of the portrait itself—simulating the framed Western portrait in a two-dimensional representation—Shūen's modifications to the "original" turn the portrait of Thunberg into a self-referential representation about representation. Indeed, prior to Shūen artists had used similar self-referential gestures in copperplate printing. One such example is Aōdō Denzen's series on *Notable Places of Edo* (*Edo meisho zu*), produced in three sizes (large, medium, and small) between

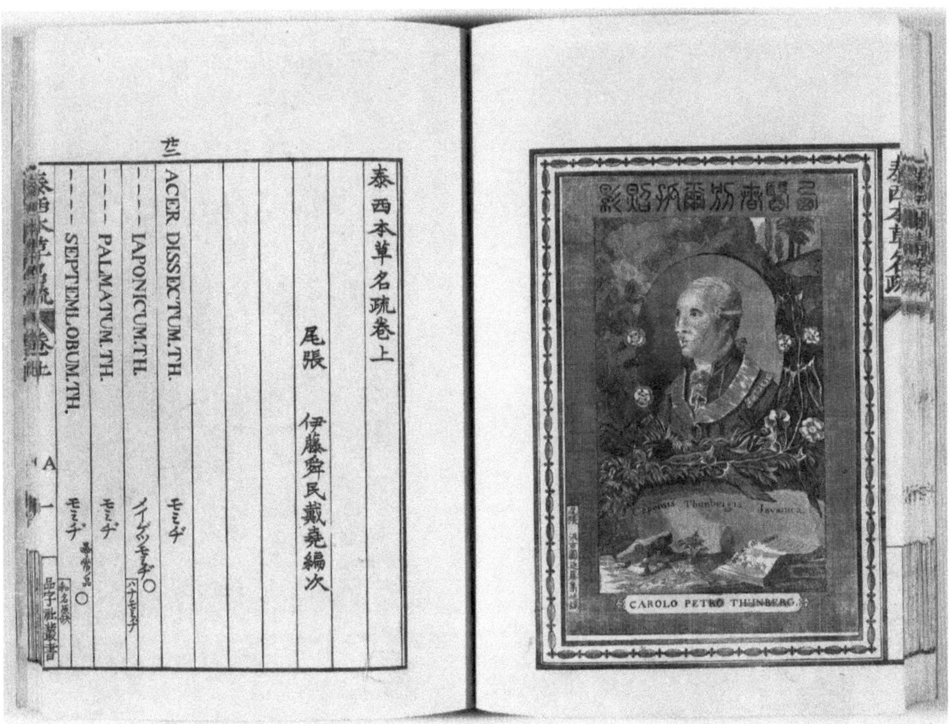

Figure 4.12 Kondō Shūen, Portrait of C. P. Thunberg, from Itō Keisuke, *Taisei honzō meiso*, 1829, copperplate on paper, Hōsa Library

1804 and 1818 (see Fig. 4.13). Denzen played with the pictorial space around the scenes by referencing the picture frame, or the materiality of representation, which was unfamiliar to his Japanese audience. The gap in the actual material on which the images are printed (paper) and the simulacra of fabric around the scene of "Shinagawa," a major inn on the highway in Edo, results in creating a playful and mindful tension within the representation. Denzen's approach moves beyond inserting depictions of threads around the scene. By depicting some threads of the fabric running in various directions, rather than showing all threads uniformly moving toward the edge of the paper, Denzen's simulacra probes the materiality of the print by pointing directly to the gap between the real physical entity (paper) and the representational reference (fabric). Similarly, Shūen represents the physical qualities of the framing mat, conveyed by crossing vertical and horizontal lines. He extends the similar method of self-referencing that points to the materiality of representation in his portrait of Thunberg.

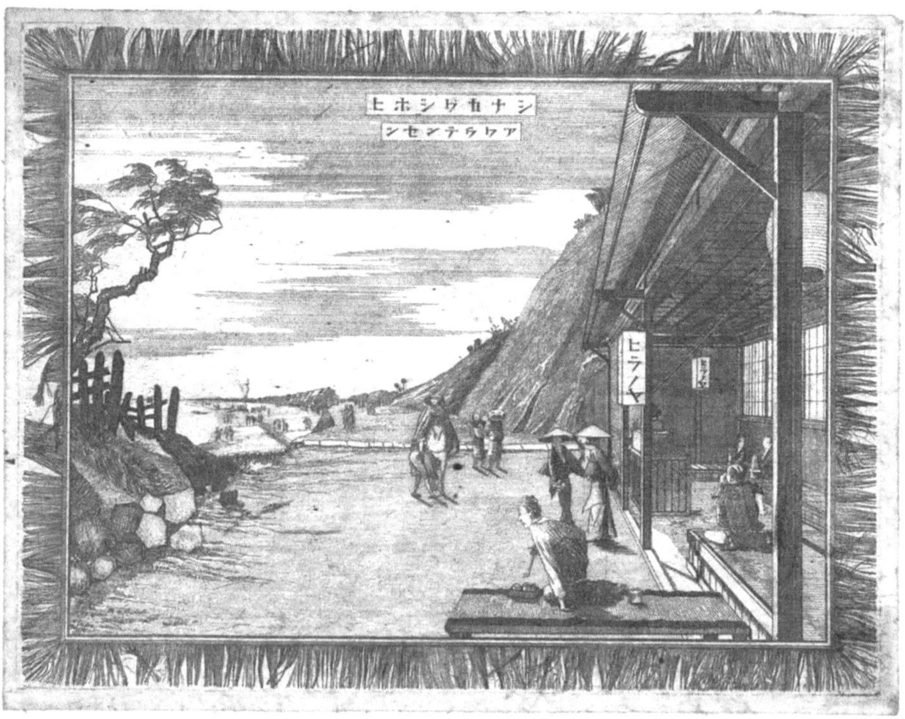

Figure 4.13 Aōdō Denzen, *Shinakawashihoi*, from *Edo meisho zu*, 1804–1818, copperplate on paper, Tokyo National Museum

What differs in Shūen's self-reference to the physicality of the print, however, is his attention to the labels and scripts representing Thunberg's name. Here, the name is expressed by combining four Chinese characters whose sounds mimic the sound of "toonbeig" in Dutch. In other words, they transliterate his name into Chinese characters.[27] Comparing the resulting work to the "original," Shūen's portrait appears crowded and less articulated. Overall, comparison of Shūen's print to its model demonstrates his technical immaturity: the shading on the rocks and leaves behind the profile of Thunberg makes it more difficult to discern the contours of these pictorial elements, tinting the overall portrait with darker hues; and Thunberg's face fails to replicate the sculptural contours of the original, resulting in a flattened and more expressionless impression.

But Shūen's portrait of Thunberg also testifies to the thriving status of *materia medica* in the Owari domain, and acute awareness of representational issues embedded within the study of *materia medica* through representations of two types of script. Interestingly, the added texts in Shūen's portrait combine elements drawn from pictorial traditions of Japan, China, and the West. The inscription on top, reading horizontally from right to left, uses an ancient seal script (Jp: *ten moji*; Ch: *zhuànshū*). The Roman scripts on the bottom appear in all capital letters and are read from left to right, and Shūen reframed the name within a label that evokes those often added in Europe to the frames of oil paintings. Shūen also inscribed his name vertically in the lower left corner in Chinese, noting the name of his city (Owari), his pen name (*go*) "Kyūkoen," and his full name "Kondō Shūen." The combination and coexistence of these three graphic styles in Shūen's portrait of Thunberg visually attests to the intellectual and epistemological stakes of *Nominal Differentiations* as a publication. The content of the book derives from Chinese study of *materia medica*, attempting to refit this body of knowledge into a Japanese context with the aid of a Western nominal scheme.

The zigzagging movement involved in reading the texts on this portrait—reading from right to left, left to right, and top to bottom—mimics the epistemological crisscrossing entailed in understanding and applying the information contained in the book. As we saw in Chapter Two, the Shōhyaku-sha were aware of the visual aspect of Linnaean names. The nominal identification of Thunberg in two types of scripts—in the Roman alphabet on the bottom, and in Chinese characters on top of the portrait—reinforces awareness not only of the scripts as visual media but also of the epistemological impact of placing a name in a different script. The perceptible differences in Thunberg's name in the two scripts point to the pictorial likeness of Thunberg in the center of the page, visually connecting the representation of the person and his names.

Use of several script styles in this portrait illuminates how copper etching became embedded within and articulated through a vocabulary of pictorial style derived from calligraphy in the Owari domain. Calligraphers recognized three degrees of scriptural simplification, known as standard script (*shin*), semicursive (*gyō*), and cursive (*sō*) styles.[28] The term "*shin* style" is written with the same character as the *shin* of *shashin*, while *gyō* and *sō* are rendered with the characters for "path" and "grass," respectively. In 1820, Shūen's teacher of copper plate, Bokusen, took part in the publication of *Illustrated Book of Detailed Surgical Medicine* (*Yōka seisen zukai*), for which he produced two copper-etching prints. There is an intriguing advertisement at the end of *Surgical Medicine* for a forthcoming book by Bokusen with the title "Western *Materia Medica* in *Shin'ei* Styles" ("*Taisei shin'ei honzō*"), which "includes two thousand and forty-eight images in copper-etching reproduced from a book by the Western scholar Carl von Linné."[29] Five years earlier, in 1815, Bokusen advertised another book, "Picture Book in *Shin* and *Sō* Styles" ("*Shinsō gaen*"), which was to contain two illustrations of each plant specimen rendered in two styles. The illustrations in *shin* style were to be executed in minute detail (*seimitsu*) in copper etching, while *sō* illustrations were to be simplified images containing only vital information (*kanyo naru shitsui nomi*).[30] Unfortunately, neither book saw publication as Bokusen passed away in 1824 from accidental ingestion of nitric acid.[31] Nevertheless, how Bokusen mobilized the existing categories of calligraphic styles for differentiating pictorial representations attests to a rich and intertwined relationship that he saw between text and image, and in turn it provides a window onto understanding how the copperplate technique was apprehended in tandem with the styles of scripts.[32] The differences among the three styles in calligraphy are visually manifested in two distinct ways: the fluidity of brushwork and the formal qualities associated with each style. Historically, the *shin* style serves as the foundational style in the practice of calligraphy. In writing in *shin* style, one moves the brush to write one stroke without merging one stroke to the next, "convey[ing] a sense of simplicity, formality, and clarity." The *sō* style is a more interpretive style used to express interiority and movement of energy.[33] Writing in *sō* style often renders more than one stroke into one brush stroke. The legibility of the *sō* style would thus depend largely on the reader's ability to recognize the original character from which the cursive is derived, while legibility is what characterizes the *shin* style.

In 1816, Bokusen and Hokusai published an instructive illustration book, *Manual for Illustration in Three Styles* (*Santei gafu*), the very title of which indicates the pivotal role of the three styles of *shingyōsō* in the organization of the manual. The illustrations in this publication apply *shin*, *gyō*, and *sō* styles to depict a range of subjects that included people of many occupations, mythological gods, plants, insects, birds, fish, animals, and landscapes.[34]

The temporal proximity between publication of *Manual* and the advertisement for "Picture Book"—separated by only one year—suggests that Bokusen engaged in these two projects concurrently.[35] Take, for instance, a series of four flowers—orchid and plum—in *Manual* (Figs. 4.14 and 4.15).

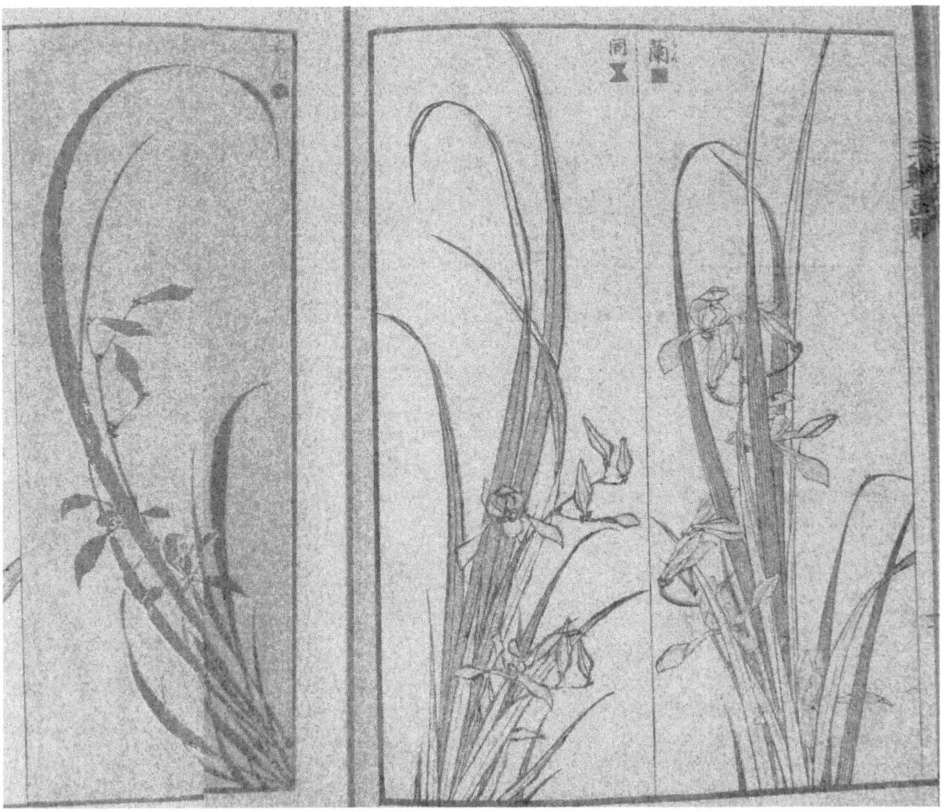

Figure 4.14 Katsushika Hokusai, *Orchid*, from *Santei gafu*, 1816, woodblock print, Rijksmuseum voor Volkenkunde

The arrangement of the styles follows the conventions of reading from right to left: in Figure 4.14, the rectangular frame on the right contains a depiction of an orchid in *shin* style, followed by *gyō* style, then *sō* style in the adjacent rectangular frames. The *shin* style thus serves as the basis of visual comparison and contrast to the styles that follow, directing the viewer's gaze from right to left. The viewers, then, are invited to notice the differences and commonalities among the three styles.

This order of presentation visually demarcates the differences of the three styles through a decreasing degree of detail and intensifying fluidity of brush strokes, indicated by the number of strokes used to depict, for instance, a leaf. In the *shin* style depiction of an orchid, a leaf contains the contour as well as lines indicating the veins of the leaf. Contrastingly,

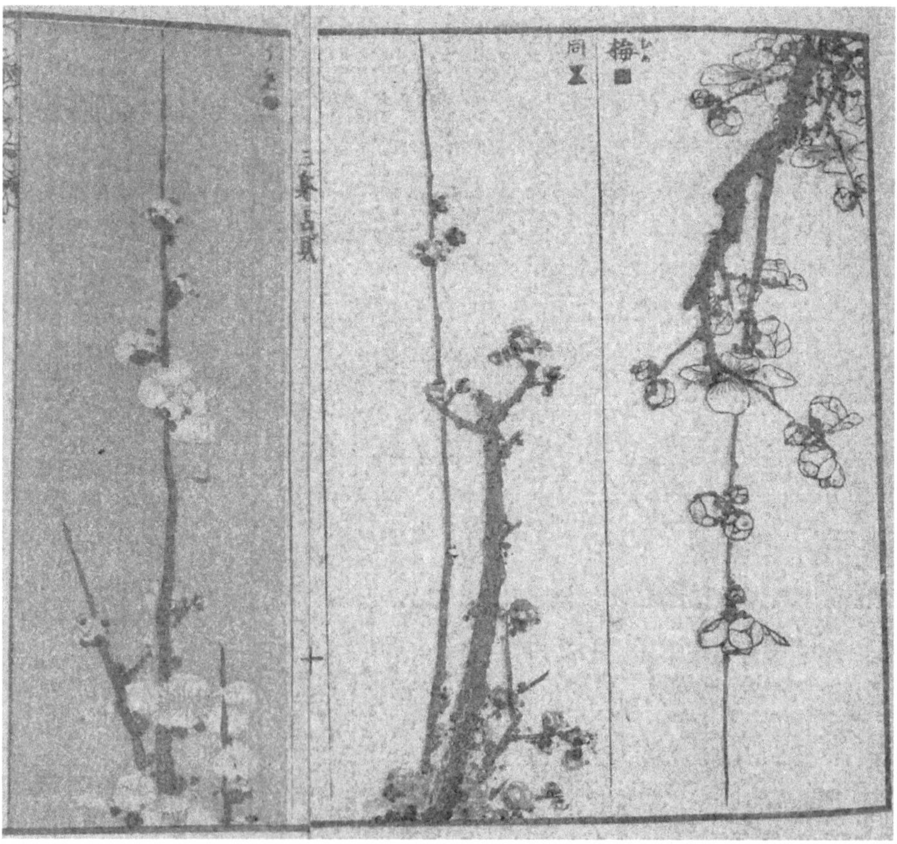

Figure 4.15 Katsushika Hokusai, *Plum*, from *Santei gafu*, 1816, woodblock print, Rijksmuseum voor Volkenkunde

in *sō* style depiction a leaf is rendered in one stroke. Similarly, in *shin* style depiction, each petal of a flower is rendered individually with a crease by noticeable outlining, whereas in *gyō* style the petals are rendered by simple contours of the petals with no visible crease. In *sō* style petals are rendered completely by a black shape that conveys a petal with one brush stroke. The two sides of the leaf are also visibly rendered in *shin* and *gyō* styles, while in *sō* style this concern is visually absent. The number of flowers and leaves accentuates the physical volumes and details of plants in the *shin* style. In representations in the *sō* style, by contrast, the volumetric qualities appear less embellished and the flowers more contracted. The scripts labeling each illustration of the specimens also follow the characteristic differences attributed to each style: *shin* style, being the most formal, is expressed in classic Chinese characters, while *sō* style is rendered in cursive *hiragana*. The organizing framework of this instruction manual presents the three distinct styles of representation and instructively foregrounds and augments their differences for the viewers.[36]

We can see a pictorial and conceptual similarity between this publication and Bokusen's intention in "Picture Book." Bokusen equates the copperplate prints with *shin* style and then contrasts them with existing Chinese and Japanese representations in woodblock prints, which he labels *sō* style. These two categories not only provide him with a shared and accepted vocabulary of styles but also allow him to extend their use from calligraphic discourse to pictorial representation. For Bokusen, the styles of *shin* and *sō* served as efficient didactic tools to showcase different styles of pictorialization. Here too, emphasis is placed on the ways in which the representations convey particular approaches to the pictorial subject. For Bokusen, the primary significance of *shin* style is in the practices of faithful duplication and the formality and legibility already associated with the *shin* style in calligraphy. Conversely, the *shin* style necessitates the existence and availability of an *already* extant image.

Bokusen's other project, the ambitiously planned "In *Shin'ei* Styles," would have contained more than two thousand copperplate prints adopted from Linnaeus's publication. The projected publication would have used the contrastive format of "Picture Book," with the aim of showcasing the differences between traditional illustrations of plants in woodblock prints and illustrations in Western botanical publications. Thus Bokusen's plan for this book would have marked the differences he saw in these pictorial representations, especially in the amount of detail captured by the different media.

More important, the proposed work signaled that Bokusen perceived the *shin* style of the copperplate prints as most suitable for approximating the "original" illustrations. This perception is affirmed by his characterization of his copper-etching prints in *Surgical Medicine* as *shin* style, a designation that emphasizes the fidelity of the illustrations to those in Lorenz Heister's *Heelkundige Onderwyzingen*, on which they were based. It was in this perceived faithfulness to the "original" representation that Bokusen discerned the eye-catching advantage of copper etching and found the term *shashin* fitting for his prints.

Bokusen's own comments on the copperplate prints he himself executed for *Surgical Medicine* further articulate how he evaluated the strength of copper etching and his own ability:

> Koshimura Tokki has carefully translated this *Heelkundige Onderwyzingen* and asked me to copy and pass down all the illustrations included in the book. The original illustrations are incredibly detailed and delicate [*hanahada kuwashiku yūbi nari*] to such an extent that people with elongated eyes would not have been able to depict them. How can I adequately copy the [original] illustrations? I cannot. In addition, there is one copper-etching print included in this book that I carved by hand. When I look at it, I cannot help detesting the print because it is so poorly executed. How, then, can I be exempt from the sneers of others? Moreover, even though the copper-etching prints made by Japanese persons have astonished people, they are far inferior to prints made in the West. I don't know a single person who can make copper-etching prints properly or who fully understands the copper-etching technology itself, and I myself lag far behind in terms of the mastery of this technology. Nevertheless, I have given this technology much thought and made various speculations. I fervently hope that [my] skill advances to the point that my prints have Westerners questioning whether they were made by a Westerner.[37]

Bokusen notes the discrepancies between his work and Western prints in two interrelated and distinguishable ways: the appearance of his reproductions in comparison to the "original" illustrations, and his incomplete mastery of copper-etching technology compared to the European printmakers who labored on *Heelkundige Onderwyzingen*. The possibility that his copied prints might one day be "mistaken" for those of a Western artist becomes Bokusen's ultimate goal, a hopeful projection that comfortably overlaps with the concept of *shin* of *shingyōsō*. Taking a "Western" copper-etching print as the "original" to trace faithfully, Bokusen would have achieved the highest degree of *shin* style if the difference between the two had been negligible. In other words, Bokusen imagined that, to the degree he could refine his copper-etching skill, the resulting images would epitomize the *shin* style. The evaluative qualifications for the *shin* style lay in the exactitude of mimicry that a student demonstrates in copying the master's work.

Closer examination of Shūen's prints suggests an even stronger association of copper-etching technology with the act of replicating an already represented picture. In fact, none of his pictorial compositions were his own invention. The portrait of Thunberg in *Nominal Differentiations* is copied directly from *Voyage de C. P. Thunberg*. The other known copperplate prints by Shūen are based on paintings by Owari painter Yamamoto Bai'itsu (print of a leopard), and Shōhyaku-sha member Okubo Masa'aki (print of a seal).[38]

In the broader context of production and use of copper-etching prints in nineteenth-century Japan, the works by Bokusen and Shūen differ from those of their contemporaries who also employed copper-etching technology, most notably, the pioneers of Western pictorial styles, Kōkan and Denzen. Unlike that of Bokusen and Shūen, Kōkan and Denzen's work in etching is not limited to copying illustrations from imported Western books (although they did copy a fair amount); their original compositions representing Mount Fuji and famous views of Edo attest to their strong interest in applying the technology to familiar subjects to create a new type of picture.[39] The popularity of the novel experience of seeing through a device such as a zograscope no doubt encouraged them to make available single-sheet prints of a well-known scene. In contrast, Bokusen and Shūen did not internalize the technology as a means for making their own compositions. They funneled their interest toward using copperplate technique to reproduce what had already been represented.

Kōkan and Denzen incorporated copper-etching technology as an integral element of an ongoing program to encompass various pictorial traditions of "the West." Ample evidence of this program notably appears in both their images and statements: Kōkan's famous denunciation of the Chinese and Japanese aesthetic tradition as lacking any useful application, and Denzen's own pursuit of one-point linear perspective in his compositions. Bokusen and Shūen, by contrast, evince no such interest in their work. Neither explored or sought to incorporate elements from broader discourses surrounding the imported Western media of oil painting, *vedute* prints, or camera obscura—in all of which Kōkan and Denzen actively engaged.

The underlying motivations for these different approaches become more apparent in light of the broader popular reception of Western imagery. In his eloquent study of the role and meaning of *ran*, a historically and culturally constructed idea of the "West" that the Japanese shaped around their encounter with materials from Holland, Timon Screech coins the term "precisionism" to designate a pictorial quality that enticed Japanese artists and scholars alike around this time to explore and absorb European pictorial representations as a "style."[40] Screech's argument that the notion of "precisionism" emerged from Japanese construction of "the West" enables us to situate the Shōhyaku-sha's engagements with copperplate etching and the association of *shashin* with the technology apart from the context of popular culture and Western learning, thus allowing us to see the group's application of copper-etching more in light of their pursuit of *materia medica*. Copper-etching technology, Screech argues, was understood to possess the valued features of "precisionism," which demanded that picture makers and viewers appreciate the representations *because* the pictures appeared "precise." Screech argues that both Japanese scholars of Western learning and the general population regarded the imported "Dutch" pictorial devices and representations as embodying an acute sense of visuality, with particular emphasis on visualizing the invisible (temperature, distance, time, weight).[41] Screech's concept of "precisionism" proves to be exceptionally useful in exploring the activities of Bokusen, Shūen, and the Shōhyaku-sha within the context of nineteenth-century Owari.

The notion of precision, by definition, requires an established standard by which to measure itself. As explored in previous chapters, the challenges that faced the members of the Shōhyaku-sha were intimately connected to their pursuit of fidelity between names of plants and physical plants. The numerous meetings of *honzō kai*, publication of *Nominal Differentiations*, and continued exchange of information with other scholars of *materia medica* were concrete and active ways by which the group solidified the relationship between what they knew and what they saw. The Shōhyaku-sha sought to secure a verifiable and stable relationship between forms of knowledge transmitted from texts and the various plant specimens that were locally available, a desired relationship whose fidelity they expressed through the notion of the real (*shin*). In other words, the yardsticks for precision—the botanical specimens they strove to understand—were both sources of confusion and objects of their pursuit. By triangulating text, illustration, and the actual entity in the search for knowledge, the Shōhyaku-sha sought to identify the real. But the uncertainty surrounding the real undermined its ability to serve as a stable basis for determining precision of knowledge and representations. Conversely, in examining pictorial representations, whether made by themselves or collected from other sources, the group had to be able to understand *in what way* the pictures were being accurate and *to what* they were being faithful.

That Bokusen and Shūen neither ventured to produce *vedute* prints nor applied their skills beyond copying the "originals" they found in imported books in turn points to the difference in the social context and needs between the Owari domain and the Edo, where Kōkan and Denzen learned their *ran* techniques. Bokusen's ongoing collaborative project with Hokusai in *Manual*, the instructive textbook that co-opted the calligraphic styles of *shingyōsō* as a framing device, which he continued even as he began to experiment with copperplate etching, reveals how the concepts of *shingyōsō* styles derived from calligraphy converged with his understanding of the Western medium. His perception of *shin* style in copper etching hinged, in particular, on the "incredibly detailed and delicate" qualities of the images, which for him ensured faithful imitation of pictorial models through delicate lines and attention to detail. Bokusen's experiments in copper etching, consequently, were guided by his pictorial concern for reproducing an already represented image with a level of detail that no other form of reproduction could achieve. The notion of precision operates here *only insofar as* the resulting representation carries the same attention to detail as the "original." For Bokusen and Shūen the notion of precision was confined to comparing the reproduction to the "original." The yardstick for them was not a general adherence to things from the "West" as an embodiment of imaginary precision, but the actual visual correspondence between the two. The permeating association of *ran* with "precisionism" that Screech finds central to representations of *ran* in popular culture eluded Bokusen and Shūen.

In addition, the meaning of *shashin* that circulated in Owari domain was not subsumed within the aesthetic discourse promoted by Kōkan. Rather, the concept proved important as a response within a specific regional and intellectual environment that fostered the collaboration of Bokusen, Hokusai, and Shūen with the Shōhyaku-sha. Whereas Kōkan aimed to absorb and champion Western pictorial techniques, which included perspective pictures and etchings, by heralding their ability to foreground the visible aspects of the pictorial subject, the Shōhyaku-sha's concern with pictorial representation pivoted around a different primary concern: How do we secure and stabilize the relationships linking our knowledge of a plant, its representation, and the plant in question? Their concerns could not be satisfied simply by assuming that all "Western" representations were faithful depictions of the visible world. What was crucial for the Shōhyaku-sha was to ascertain that the represented subject in a given illustration existed.

Here we see the overlapping, on a conceptual level, of Bokusen's approach to copper etching and Yōan's characterization of ink rubbing. The capacity of copper-etching technology to reproduce the original down to the most delicate detail and the necessary contact with the physical "original" in ink-rubbing prints coalesce with one another as both point to their maintaining "*shin* form," the real form of the represented plant. Neither embellished nor modified by artists, the image produced through these two processes of picture making guaranteed fidelity in representation. Comparisons with other contemporary studies and publications on plants in the final section of this chapter highlight these regionally and intellectually specific characteristics of *shin* and *shashin* in representation.

The Concept of Shin *in Contemporary Illustrations*

Kawahara Keiga, an artist in Nagasaki, produced a series of plant illustrations under the title *Plant Transpositions of the Real by Keiga* (*Keiga shashin sō*) in 1836 (Fig. 4.16).[42] Working in Dejima under Jan Cock Blomhoff, the chief trader of the Dutch factory from 1809 to 1813 and again from 1817 to 1823; J. F. van Overmeer Fischer, a secretary and an accountant between 1819 and 1829; and Phillip Franz von Siebold, Keiga left an enormous body of pictorial representations that range from domestic scenes to depictions of surgical tools.

The two surviving volumes of *Plant Transpositions* include fifty-seven illustrations of various plants with a short description of their blossoming seasons and medicinal benefits. It also contains close-up representations of the pistil and stamen, and, for some plants, the ascribed Linnaean names. In the introduction, Keiga's contemporary Ishizaki Yūshi stated:

> All pictures are not the same. When a picture captures the spirit and the elegance [*ki'in*] of the subject, the movement of the brush [*hippō*] is absent, or when a picture is executed with wonderful brushwork, it lacks a balanced

> composition [*ichi*]. It is rare to find a good picture that contains the delicate balance of these aspects. The person who can create such a picture is Kawahara Keiga.... The foreigners at the port of Nagasaki were astonished by Keiga's ability to depict scenes of water and land, as well as the changing state of heaven-made creatures. Keiga created images by observing rare animals, differentiating plants, and clarifying the varied kinds of fish. Many foreigners asked Keiga to make copies of the illustrations for them.... One year, someone gave me pictures by Keiga. In my eyes, his images are outstanding in their quality, in the detailing of the shades of objects, and in providing ample consideration to both the exterior and interior of the depicted objects, transferring the spirit [*shin*] of the object to the paper. His brushwork is fast and clear and he is not necessarily confined to the established methods of picture making.[43]

What stands out particularly in this evaluation is how Yūshi incorporates the aesthetic concepts prevalent among literati painters, especially among those of the Southern School (Jp: *nanga* or *bunjinga*), in praising Keiga's work. The concepts of *hippō* and *ki'in* that Yūshi incorporates into his remarks were derived from the theories of painting put forth originally by Xie He in *Guhua pinlu* in sixth-century China.[44] The stylistic characteristics of this school were fine-tuned and interpreted by various artists throughout Japan. In particular, artists and critics struggled to balance what appeared to be opposing forces that were critical in the paintings and criticisms of the Southern School: portraying the interiority of the subject or "spirit resonance" was often in tension with depicting visible forms, the "correspondence to objects."[45] Yūshi's appraisal of Keiga's work as conveying the interiority *as well as* the exteriority of the subject also points to a stunning merger of artistic genres of painting and scientific illustration that Keiga achieved in this book.

Yūshi's statement witnesses the integration of aesthetic concepts into images produced to serve a function beyond the norms of artistic appreciation. Certainly, a tradition depicting birds, flowers, and plants existed long before Keiga, and it worked within the vocabularies and aesthetic standards of an established hierarchy.[46] The aim of Keiga's publication, however, differs from this tradition in that it was to accompany text with the goal of imparting knowledge. In this regard, the purpose is similar to *Compendium of Materia Medica* (*Bencao gangmu*) and other *materia medica* publications that aimed to instill and disseminate botanical knowledge essential to the study of medicine. In traditions of Southern School painting, the finished pictures functioned and appealed as pictorialized expression through which a viewer could access the interiority of the painter himself. In Keiga's work the pictures were presumed to function as a knowledge-informed representation.[47] The woodblock images were a visualized source of knowledge accompanied by complementary text that functioned as an integral source of verified and reliable information. Yūshi nevertheless employed the vocabulary of aesthetics to describe and characterize the images regardless of the difference in the intended purpose of the pictures.

Keiga's illustration of a magnolia (Fig. 4.16; Jp: *kobushi*; Ch: *xin yi*; see also color plate 8) demonstrates how the meanings attached to the concept of *shin*, the "real," in his prints remained rooted in aesthetic traditions. Keiga consistently included details of pistils and stamens, key morphological features in the Linnaean taxonomy, in the lower left corner of the

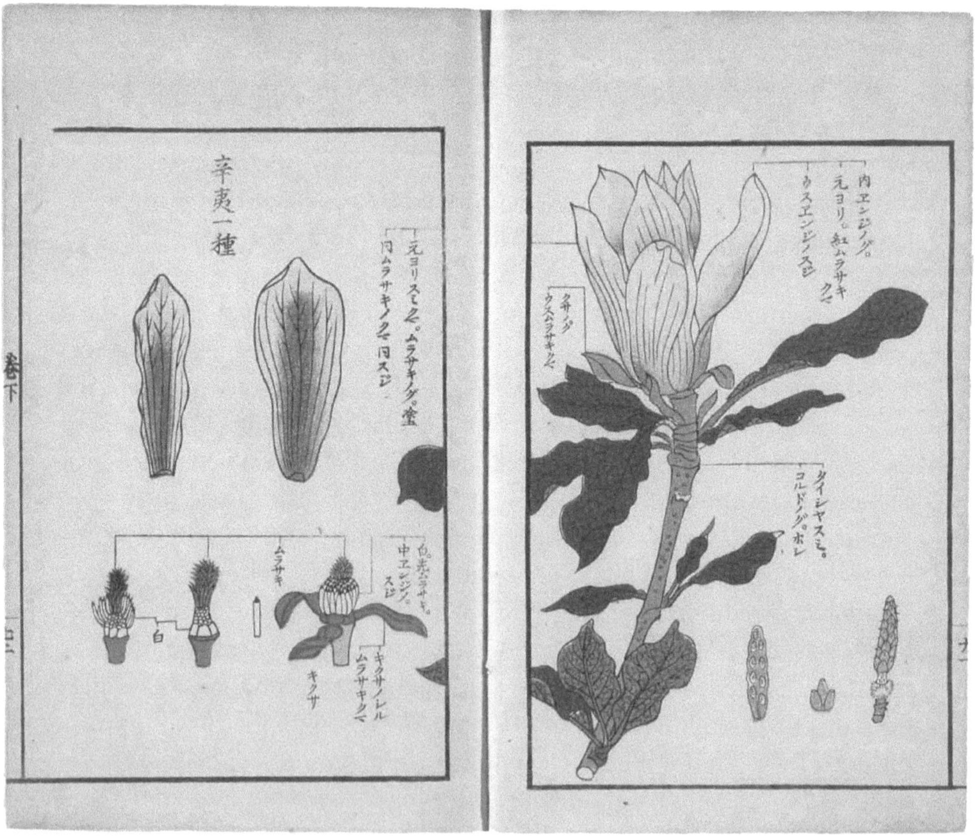

Figure 4.16 Kawahara Keiga, *Magnolia*, based on *Keiga shashin sō*, republished with colors in *Sōmoku kajitsu shashin zufu*, 1868, woodblock print, Waseda University Library

illustrations in his publication. He omits, however, the number of pistils and stamens for each plant species, another piece of vital information in applying the Linnaean method. Instead, the short descriptions of the medicinal properties and uses of each species, offset by the rectangular framing that surrounds the pages, are given as accompanying textual information.[48] The collaborative relationship between text and image on these pages thus signals the intention to compile an illustrated *materia medica* resource that does not rely on a particular classifying principle.[49] The use of the concept of *shashin* in the title of this book is particularly telling in this regard; it demonstrates use of a conceptual term from the pictorial tradition of the Southern School to describe images with a new, decidedly different scientific function. In the copperplate etchings of Bokusen and Shūen, by comparison, the notion of the artist's interiority did not enter the discussion. Their concern was directed toward accurately representing the external and visible elements of objects: the smaller the gap between the existing "original" and their reproduction, the more "successful" their prints were deemed. The concept of *shashin* in Keiga's case is owed to the merger of the interiority of the artist and the exterior visible shape of the subject of the depiction. In the case of Bokusen and Shūen, the term served to designate the degree of correspondence between the "original" representations and their own.

Another popular illustrated book of *materia medica* from this period is *Illustrated Book on Materia Medica* (*Honzō zufu*), published between 1830 and 1844 by Iwasaki Kan'en, a *materia medica* scholar in Edo. An extensive resource consisting of ninety-four volumes, *Illustrated Book* was praised for the details and elegance of illustrations by contemporary *materia medica* scholars; it was characterized by one as the encyclopedic resource that allowed diffusion of knowledge without one's having to travel to distant places and lands.[50] Consideration of Kan'en's work illuminates the representational issues shared by the scholars of *materia medica*, and the challenges they faced in producing more reliable illustrations that embodied *shin*. The draft of the manuscript was completed in 1828, and the first five volumes were published in 1830. The illustrations in these volumes were in woodblocks, but those in subsequent volumes were copied by artisans by hand.[51]

The illustration of a *Chōsen asagao* (*Datula alba*) is included in volume 12 of a copy of *Illustrated Book* and offers a helpful point of contrast with how the concept of *shin* and botanical illustrations coalesced in works by Bokusen, Shūen, and Kan'en. Kan'en's illustration, like Keiga's, is indifferent to the number of pistils and stamens (Fig. 4.17 and color plate 9). Instead, the illustration looks at the plant from a broader perspective that includes the larger structure of the plant. The leaves, branches, buds, flowers, and its fruit are all present in this illustration, which focuses on

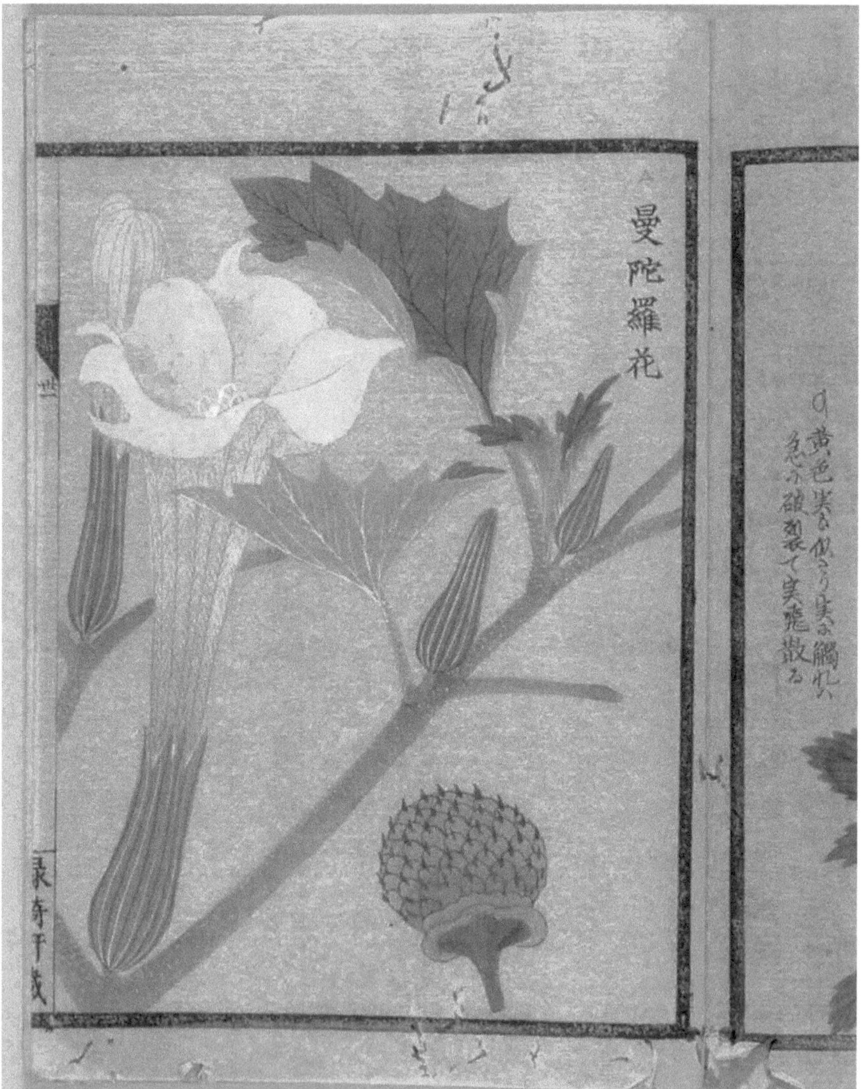

Figure 4.17 Iwasaki Kan'en, *Chōsen asagao*, from *Honzō zufu*, before 1830, woodblock print, Tokyo National Museum

the temporal stages of the blossoming, an aspect completely absent from Keisuke's pamphlet on oleander.

In fact, Kan'en employs the concept of *shin* several times in an explanatory note, given after the introduction, that situates his own illustrations in relation to others available to him:

> There are illustrations on *materia medica* such as *Beiji bencao*, *Erya*, *Sancai tuhui* and *Hiden kakyō*.[52] However, they are rough and abbreviated, and thus do not contain details. That is to say, these illustrations fail to convey *shin*. Including books such as *Sansō noyama sōkai*, only a handful of illustrated books contain color and even when they are colored, they do not capture *shin*. They all fall short of being reliable. Illustrations in this book, which includes varied shapes and variations of the same plant, are all detailed and sketched. I have used the method of *sōkō* and asked the wood carvers to carve them with care and skill. After this initial printing was finished, I have asked another artisan to carve out the blocks for coloring. I have examined the resulting images very carefully, and they have not lost their *shin*.[53]

Unlike the case of Keiga's images, Kan'en's use of the concept of *shin* here is not derived from the aesthetic terms of Chinese painting traditions. Rather, *shin* originates from within the discourse of *materia medica*, which located *shin* in the reliability of the illustrations. Kan'en summarizes the inadequacies of previous illustrations as a "lack of *shin*." Kan'en attributes the *shin* of the illustrations to their details in particular, much as details were seen as the source of *shin* by Bokusen and Shūen in their "original" illustrations from the West and by Udagawa in ink-rubbing prints.

The relationship between these different techniques indicated by their common association with *shin* is underscored by consideration of the technique of *sōkōtenboku*, a method of copying from the original extensively used in calligraphic practice, which Kan'en used to make the illustrations. The technique entails laying a thin piece of paper over the original calligraphy, and then outlining the contours of the letters with a brush. The paper would then be removed and the outlines of the original calligraphy filled in to duplicate the original script. The technique was used to preserve original calligraphy, and it served as a pedagogical tool to instill in students of calligraphy the model calligraphic styles of masters. Kan'en claims in his book to have adapted this method to the copying of woodblock prints. He first placed a thin piece of paper over the original illustration and traced the contour of the depicted plants. Then he glued the copied contour onto a woodblock and turned to wood carvers, who excised the areas outside of the copied contour. Close examination of the illustrations confirms Kan'en's innovative adaptation of this technique. The two leaves at the center of the illustration, of *Chōsen asagao* for example, lack the heavy black outlining of traditional woodblock prints (Fig. 4.17 and color plate 8). By using the *sōtō* technique, Kan'en eliminated this pictorial element that predominated and characterized woodblock illustrations of plants in *materia medica* (see for instance Figs. 1.6 and 4.5). The leaves in Figure 4.17 were produced by two blocks—one in which the areas in lighter green remain uncarved, and the other in which the leaf veins were left untouched. The woodblock prints remained indexical to the hand-sketched illustrations made by Kan'en as well as the original prints. It was this fidelity that constituted the *shin* of Kan'en's woodblock prints.

As someone with direct access to the depicted objects, Kan'en could verify how the resulting images actually conveyed the realness of the plants. In this sense, what is believed to be present in Kan'en's illustrations is the extent of coverage in illustrations derived from repeated observation and sketching by the illustrator himself. Kan'en explains in a note that the degree of reliability of his books is derived from his relationship to the illustrations, thus foregrounding the illustrator himself as the reliable source who could vouch for the integrity of the knowledge provided in the book. By identifying himself as both the trustworthy illustrator and the scholar of *materia medica*, Kan'en collapses the gap between the two that Keisuke and Yōan had lamented previously. Here, we witness the triangulated relationship of text, actual object, and illustrations applied with slightly different emphasis on illustrations. Rather than illustrations being a shared source of agreement, Kan'en produced his illustrations as a testament to his own observation. The reliability of the images stems not from Kan'en's resolution of problems with naming but from his own experience of attentively observing plants in all their detail over years. The Shōhyaku-sha, too, turned to observation and produced their pictorial representations to locate and grasp the *shin* of plants. Their botanical explorations, however, focused upon finding stable names for plants; this was then believed to lead to better medical practice.

In the introduction to Kan'en's book, Kitamura Naohiro, a medical doctor and teacher employed by the shogunal medical academy, acknowledges the pivotal role of *Compendium* in his profession. He notes that regrettably, through multiple copying and reprinting, the illustrations of *Compendium* lost their original appearance altogether, resulting in misunderstandings and misinformation. Kitamura continues:

> Iwasaki [Kan'en] planted over two thousand herbs and flowers in his garden at home and observed them daily. Through this experience, which extended over thirty years, he was able to gain a great depth of knowledge on each plant. He took a brush to his hand and sketched these plants. Therefore, his illustrations contain sophistication and detailed observation. This publication includes illustrations that show the similarities and differences of medicinal herbs from varied regions, variations of a plant in distant mountains as well as those imported from abroad. That is to say, one can gain knowledge of illustrated plants without having to travel to these places where they grow. The effort that [Kan'en] took to complete this project is extraordinary. The illustrations and textual explanations should form an organic unity that binds the two, but the illustration by Li Kengen [the illustrator for *Compendium*] did not adequately fulfill the role. The readers of *Compendium* are thus left disappointed. In compiling this publication, [Kan'en] shows us his detailed categories and unprecedented knowledge. Not only did he show his supreme knowledge but he also provided merits to later generations. It is not exaggeration to characterize [Kan'en] as the meritorious retainer of Li Shizen.[54]

The emphasis on the fact that Kan'en had direct access to the depicted objects—because he himself had grown them—and on the representations that result from direct observation is similar to the concerns of the Shōhyaku-sha. What differs here, critically, is the importance of the concept of detail (expressed in the character of *sei*, read as *kuwashiki*) in the case of Kan'en and the concept of the real (*shin*) articulated in relation to pictorial representations by the Shōhyaku-sha. What was at stake for Kan'en was the standard against which one discriminated the level of "detail," while for the Shōhyaku-sha the concern with detail was not the priority. Characterized as the shadow of the real, the ink-rubbing prints attested to and accounted for the existence of particular plants, while the copperplate prints were understood as a way to faithfully replicate the already represented illustration. For Kan'en, such evaluation of the real relied on his own vetting process, which was informed by his extensive observations. The central criterion was whether the resulting representations in his book embodied the plants in the way that was faithful to his own repeated observations. The use of *sōtō* guaranteed technical fidelity between his sketches and the woodblock illustrations. Kan'en's confidence in his experience and abilities of perception thus formed, in his view, the foundation of an authoritative botanical knowledge that, in turn, guaranteed the fidelity and utility of his book.

In Kan'en's case, the idea of detail in his own illustrations is based on his unmediated observation of the plants and the technique used in creating the woodblocks derived from his illustrations: the "details" emerged from an entirely personal mode of observation and filtering, and thus the accountability of details rests ultimately on his judgment. In addition, by using the *sōtō* technique, which allowed woodblock prints to maintain an indexical relationship to his sketches, Kan'en claims that the woodblock illustrations preserved the *shin* of the illustrations he created for each plant. In other words, though Kan'en's detailed observations and representations are heralded as worthy and even superior to what was circulating previously, the details of his illustrations evince his personal authoritative knowledge rather than belonging to a shared system for evaluating the fidelity of images. Kan'en's details stem from his ability to bring a thoroughness and breadth in his presentation of plant species both in pictures and words, but this ability remains at an ultimately individual level.

On the contrary, how ink-rubbing prints and copperplate prints were characterized and used by the Shōhyaku-sha suggests a different approach to illustration. The verified existence of the depicted plants, the ascribed names, and the encounter with actual plants all play pivotal roles in the Shōhyaku-sha's incorporation of these pictorial media that differs from representational concerns expressed by Kan'en. Rather, the *shashin* picture of the mignonette and the "shadows of the real" prints, by means of pictorialization, pointed to the very encounter with plant specimens and *already visible* entities. The concept of the real was applied to these illustrations not to designate pictorial resemblance or proximity between image and object but to indicate the direct relationship between the illustrations and the "original." In other words, while Kan'en argued that his extensive knowledge and experience of individual plants gave him the ability to differentiate the *shin* of plants, and he found technical ways to maintain the *shin* in his illustrations, the Shōhyaku-sha used these *shin* pictures as the observable real thing *because* their existence was verified.

The entanglement of the concept of the real in these publications requires us to attend not only to these differences but also to the common meanings and usages of the term *shin* as well. For both the Shōhyaku-sha and Kan'en, the concept of *shin* was critical to identifying the real plant and thereby eliminating the possibility of an error in prescribing medicine and thus wrong treatment of an illness. In Keiga's publication, his illustrations were heralded through application of aesthetic categories. Kan'en further focuses on the significance of illustration and its role in aiding physicians to recognize plants with medicinal value. He notes that physicians and pharmacists are not only incapable of visually discriminating between the two, but they are indifferent to their own inability. With acrimonious feeling, Kan'en laments the fact that people are dying.[55] This concern, indeed, resonates deeply with the shared urgency among *materia medica* scholars described in previous chapters. The ardent quest for the real thing (*shin butsu*) to eliminate misidentification and thus faulty prescription was indeed a concern that inspired a variety of publications and explorations of *materia medica* in the late Tokugawa period.

In Kan'en's case, he was compelled to grow the plants at home to make illustrations after reaching the boiling point with his frustration with the illustrations of *materia medica* publications, and concern for commonplace misidentification of plants on the part of physicians. Similar factors motivated the Shōhyaku-sha. The critical difference between the two is in the fact that while Kan'en sought to qualify his illustrations as containing *shin* by authenticating himself as a reliable and experienced scholar of *materia medica*, the Shōhyaku-sha tried to eliminate the distancing effects of mediation.

At the same time, the introductions to their publications written by contemporaries offer insight into how these illustrations were viewed and what about them was particularly notable. Appraisal of Keiga's work through the ideals of the aesthetics of Chinese painting, for instance, signals what contemporary readers recognized. From these texts, we can deduce the outlines of the operative standard of evaluating *materia medica* illustrations in this era. But nowhere in these descriptions are the images characterized as realistic, or illusionistic for that matter. These illustrations did not stop at forming symbiotic relationships with the eyes of viewers alone; rather, they were created to address and ease particular epistemological tensions. Consideration of the intended functions of these illustrations points precisely to how the modern impulse to characterize them through the language of realism has failed to understand these images historically. The standard of evaluation shared by the image makers and the viewers did not concern itself with the realism of the illustrations per se. Rather, the yardstick by which they were measured was formed by the kinds of knowledge that the illustrations, as epistemological resources, were able to convey.

FIVE

Shashin in the Capital

The Last Stage of Metamorphosis

> The pivotal role of pictures is to bridge over the time between thousands of years ago and today, and to shrink the hundred thousands of miles that lay between places. That is to say, if the shape of objects is transferred precisely and its spirit maintained, there is not even the slightest dissatisfaction in pictures.
>
> Yanagawa Shunsan, *Illustrated Book on Photography* (*Shashinkyō zusetsu*, 1868)

In 1868 and 1869, Yanagawa Shunsan (1832–1870) published the first book to systematically introduce the photographic process in Japanese, *Illustrated Book on Photography* (*Shashinkyō zusetsu*).[1] Trained as a physician, Shunsan became an assistant to Itō Keisuke in Nagoya and was an active member of the Shōhyaku-sha before moving to Edo in 1856, to continue his studies of European languages and Western learning with the assistance of the Katsuragawa family.[2] *Photography* is itself largely based on the writings of the French photographer and inventor René Dagron, but it also incorporates information Shunsan learned from his reading of other European publications on photography.[3]

It was there in the city of Edo, as it was being renamed Tokyo amidst the political and social upheavals of the Meiji restoration (1868), that the genealogy of the Shōhyaku-sha's use of the term *shashin* and their fascination with transposition of the real were first connected with the practices of photography.[4] There is thus an intersection of two discourses on *shashin* that takes place in the years surrounding the Restoration, as the lineage of the Shōhyaku-sha's concern with the connection between images and reality merged with a separate preoccupation with pictorial techniques capable of producing illusory experiences of reality.

In this light, it is no coincidence that many of photographic images produced around this time were portrait photographs accompanied by a textual rendition of the sitter's name. The soldiers of *bakufu* and anti-*bakufu* forces began to appear at photographic studios to have their portraits taken (Figs. 5.1 and 5.2). These ambrotype portraits were produced by covering one side of glass with collodion emulsion, and then setting the still-wet glass in the camera for exposure. The exposure time was drastically shorter than that of daguerreotype, less than twenty seconds, but like a daguerreotype it

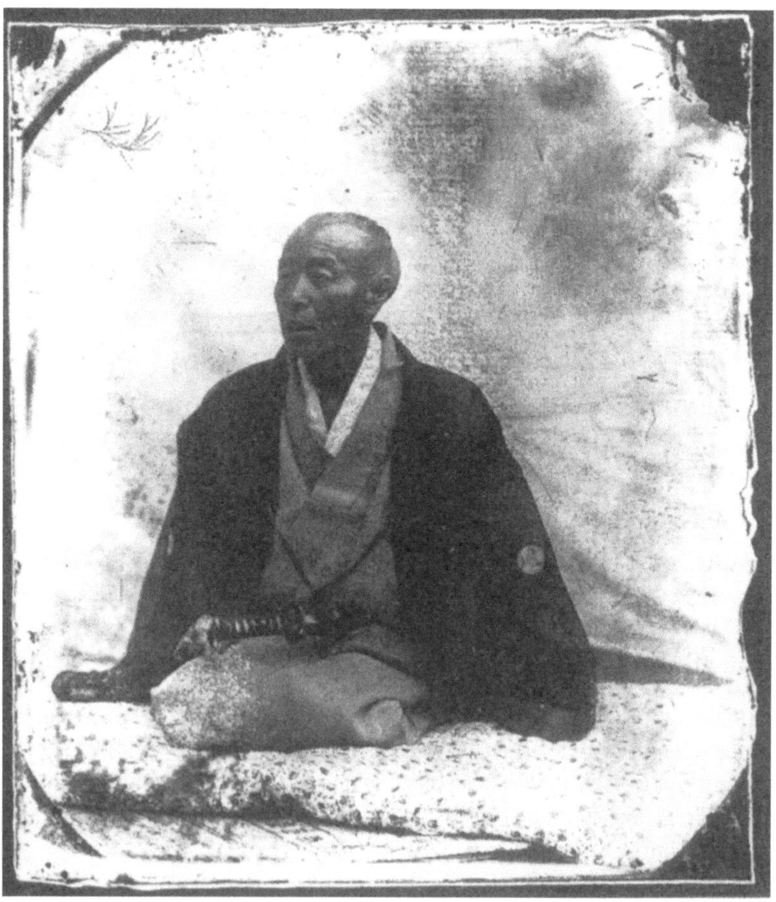

Figure 5.1 Shimo'oka Renjō, Portrait of Kimura Masanobu, 1862, ambrotype, Tokyo Metropolitan Museum of Photography

produced a positive image on the plate, rather than a negative from which copies could be made. The glass positive was encased, emulsion-side down, in a box made of paulownia (*kiri*), a kind of wood typically used in Japan to protect valuable objects. The portable size of the box (about 8 x 7 cm) made it small enough to hold the encased portrait in one hand. The mechanism of viewing the portrait—one has to remove the top cover in order to see the image kept inside—connects the viewer intimately and personally to the photographed subject through revelation. The sitter's name in handwritten script on the boxes intensifies this physical and personal connection between the box and the viewer by adding another layer of per-

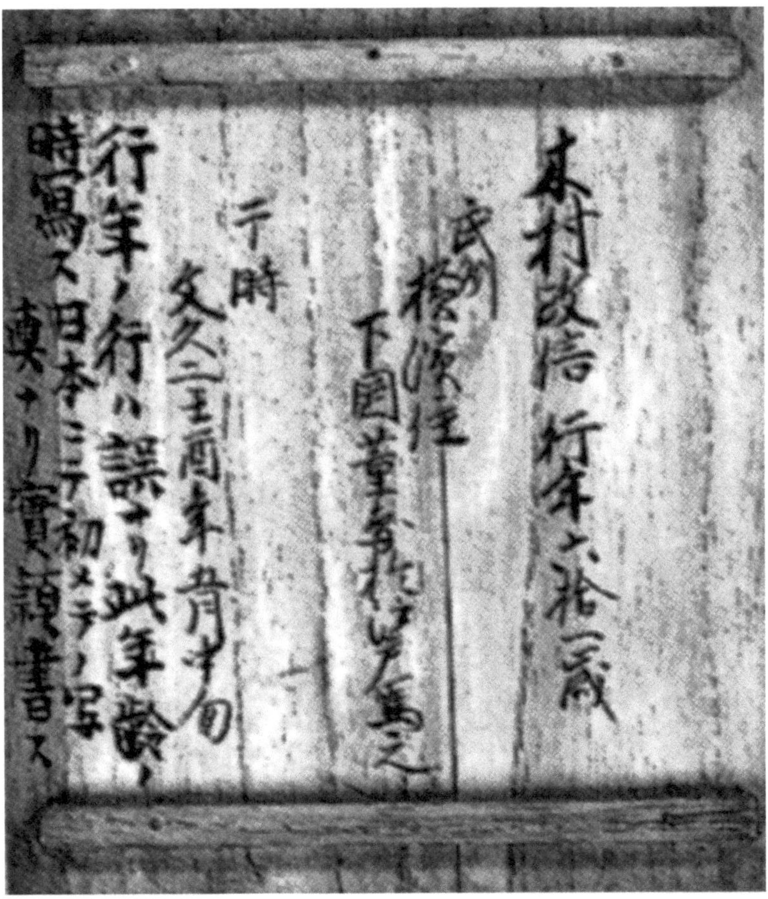

Figure 5.2 Inside of the box covering the portrait of Kimura Masanobu, 1862, calligraphy on wood, Tokyo Metropolitan Museum of Photography

sonal identification. Indeed, these "one-off" *kiri* box portraits often served as *memento mori* for the *samurai*'s family.[5] These customized portraits link the photographed subject and their names directly and intimately through a proper noun. Unlike the debate over various names of plants in *materia medica*, the name given to an individual remains constant and shared, and the hermeneutic link established between the elements that constitute the

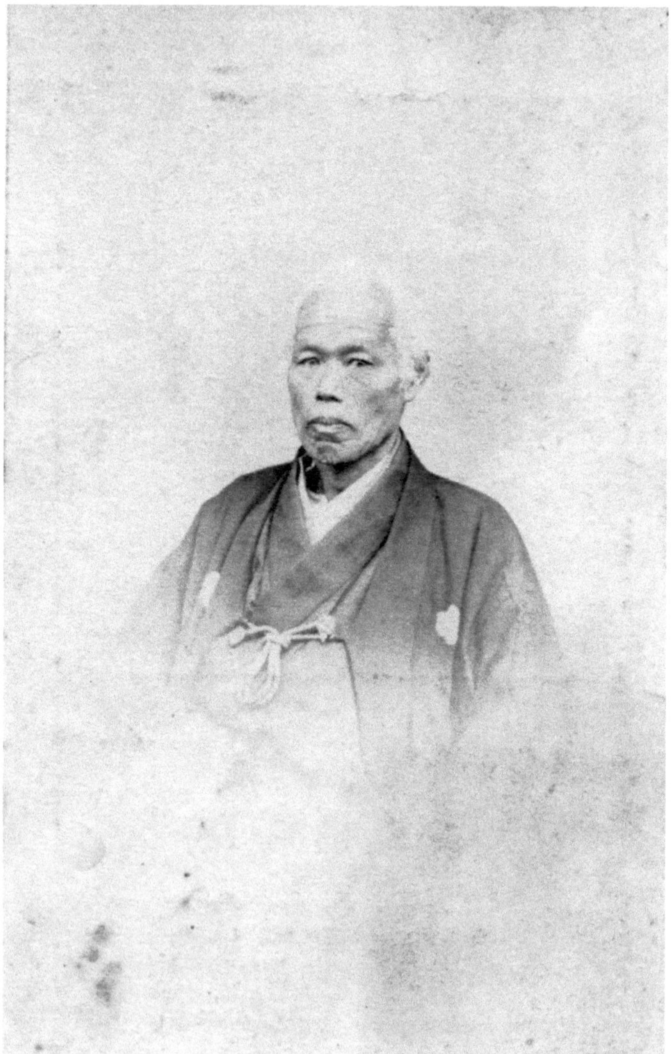

Figure 5.3 Uchida Kuichi, Portrait of Itō Keisuke, 1871, collodion process on paper, Itō Keisuke Collection, Higashiyama Botanical Garden

boxed portraits—the photographed subject and the written script—leaves no room for doubt of fidelity. The photographic portraits of Keisuke from this period are also accompanied by his name, written in his own hand, and the date when the image was taken (Figs. 5.3 and 5.4; see also color plate 10). Although not in *kiri* box, these portraits similarly strengthen fidelity of the relationship between the photographed subject and his writing.

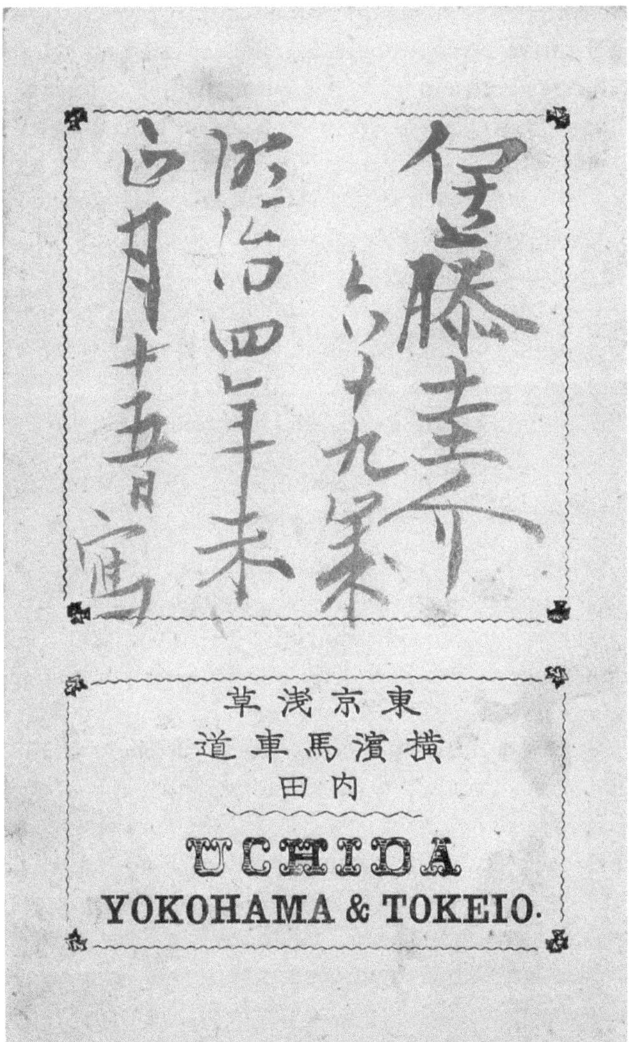

Figure 5.4 Recto, Uchida Kuichi, Portrait of Itō Keisuke, 1871, paper, Itō Keisuke Collection, Higashiyama Botanical Garden

In this concluding chapter, I first explore the relationship between studies of plants and photography through the figure of Iinuma Yokusai and his approach to photographic technology as a way to concretize how these two seemingly distant interests intersected historically. I then shift the focus to trace the process through which the Shōhyaku-sha's interest in *shashin* became a central concern for the official scientific institutions of the political capital of Edo/Tokyo in the decades surrounding the Meiji Restoration. This process emerges most clearly in the work of three individuals in the city of Edo at this important political juncture: Keisuke's work on natural history for the shogunal and Meiji governments; the work of Keisuke's colleague Takahashi Yuichi, a painter trained in the official Kanō school and a pioneer of oil painting in Japan, in his new governmental appointment; and the photographic experiments of Shunsan, who knew both Keisuke and Yuichi. Their work in the new institutional spaces of the scientific establishment reveals an emerging synthesis of the conceptual meanings of *shashin* and the technology of photography, driven by a persistent concern for correlating names and representations. Thus *shashin*, even as it became "photography" in these years, carried with it a reminder of its earlier uses, notably its fixation on questions of verifiability and of fidelity.

Iinuma Yokusai and Photography

In fact, the interest in and use of photography by Shōhyaku-sha members such as Keisuke and Shunsan in Edo was prefigured by a curiosity about the new technology of representation that extended back to their time in Owari. The practice of photography around Owari can be traced back to Iinuma Yokusai, a close collaborator of the Shōhyaku-sha though not himself a member.[6] Yokusai became interested in photography in 1852 when drafting his magnum opus, *Illustrated Explanations of Plants and Trees of Japan: Grasses* (*Sōmoku zusetsu*, 1856–1862).[7] The surviving evidence of his photographic experimentation, though limited, is nevertheless significant, as his use of the term *shashin* lens (*shashin kyō*) to refer to a photographic camera suggests. Although this term had been employed to refer to camera obscura since the late eighteenth century, a variety of other terms were used in the 1850s to refer to daguerreotype photography. Kawamoto Kōmin, a scholar from the Satsuma domain on the southern island of Kyūshū who had a working relationship with the Shōhyaku-sha, called daguerreotype photography *choku sha'ei kyō* (literally, "directly transcribing shadow lens").[8]

The famous innovator and political activist Sakuma Shōzan, by comparison, used the term *rui'ei kyō* (literally, "halting shadow lens"). Yokusai's choice of the term *shashin* to designate a photographic camera was the first clear indication of the converging intellectual currents examined in this chapter.

The earliest instances that refer to photography in Japan date to 1848, when Ueno Toshinojō, a watch merchant in Nagasaki, purchased a set of daguerreotype equipment and presented it to the lord Shimazu Nariakira of the Satsuma domain. Nariakira referred to daguerreotype photography as *in'ei kyō* (literally, "impressing shadow lens").[9] Other wealthy domainal lords, such as Kuroda of the Fukuoka domain, and the Tokugawa of the Owari domain—including the last shogun himself, Tokugawa Yoshinobu—experimented with the medium with the aid of their retainers.[10] It was one of these servants or courtiers who took the first known photograph made by a Japanese, a daguerreotype portrait of Lord Shimazu Nariakira that dates to 1857.[11]

The political shifts that occurred following the arrival of the American Commodore Matthew Perry's Expedition in 1853 also afforded two kinds of experience of being photographed to the Japanese. The daguerreotypists who came with foreign delegations to Japan, such as Eliphalet Brown, Jr., of America and Alexander Mozhayskiy of Russia, made portraits of *bakufu* officials with whom they came into contact during their duties. Members of Japanese delegations sent abroad by the *bakufu* in turn were photographed on their tours by Félix Nadar in Paris and by Antonio Beato in Egypt.[12] By 1862, three professional Japanese photographers had set up their own studio—one in Nagasaki run by Ueno Hikoma (the fourth son of Ueno Toshinojō), and two in Yokohama operated by Shimo'oka Renjō and by Ugai Gyokusen. These three photographers dominated production of commercial photographic prints into the early years of the Meiji period.[13]

Yokusai acquired his knowledge of photography from his third son, Kōsai, who was adopted in 1843 by the scholar of Western learning Udagawa Yōan, whose translations of works on botany have already been discussed.[14] Living in the capitol Edo as a scholar of Western learning (*rangaku*), Kōsai had access to the latest news and imported objects from the West.[15] Around 1860, Kōsai translated and left a manuscript that explains the process for making albumen prints: he titled the manuscript simply "Potokarahī," a transliteration of the Dutch term *fotografie*. This manuscript attests to his interest in and familiarity with photographic technology, while it also highlights the possibility that he was trying to relay knowledge of the albumen technique to his biological father Yokusai in Ogaki.[16]

Although Yokusai mentions photography in letters to his son and other correspondents several times, only three glass negatives that Yokusai made using the wet collodion process survive today.[17] In a portrait dating from 1863 (Fig. 5.5), the eighty-one-year-old Yokusai sits in a formal *seiza* position with his eyes focused to the right of the camera and his legs slightly slid to the right to protect his failing knees. The sword on Yokusai's left, which rests with its hilt across his lap, symbolizes his status. His gray hair and the wrinkles on his face indicate his age, a topic on which he com-

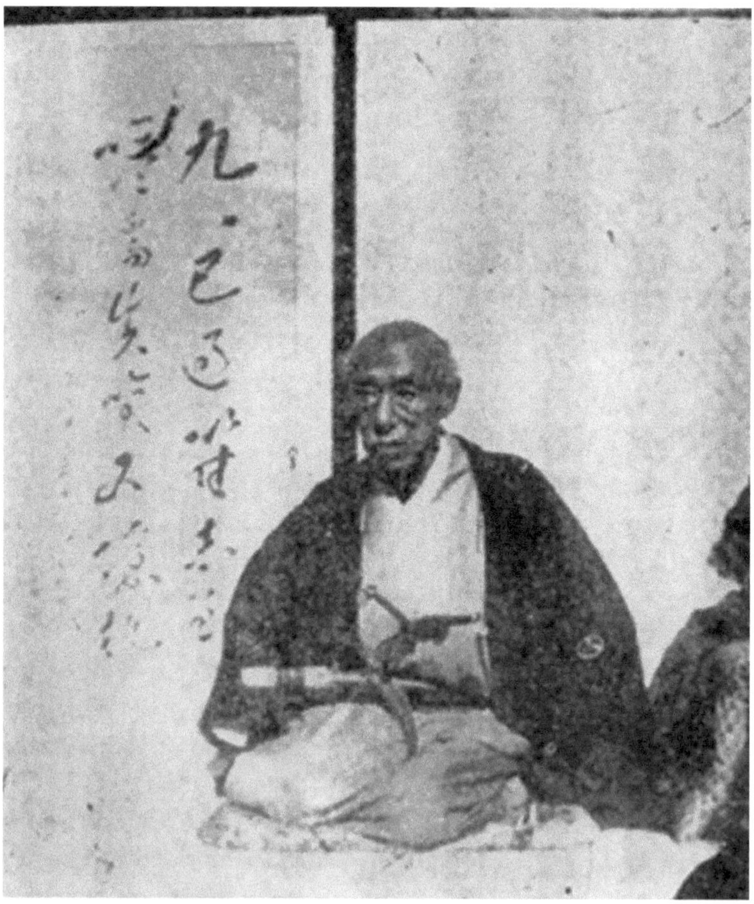

Figure 5.5 Portrait of Iinuma Yokusai, 1863, ambrotype, Waseda University Library

ments in the beginning of the poem that appears on the hanging scroll to his right, which is legible in the photograph.[18] Yokusai composed the poem in classical Chinese and wrote the scroll during the winter of 1863, shortly after he published the first set of *Plants and Trees*, the twenty volumes on grass, on which he had labored for three decades. He had yet to find a way to publish the second set on trees. In the second half of the poem, Yokusai steps back from his own frustration and comments on the incidental nature of everything, laughing at his own feeling of urgency. This portrait invites the viewer to make a visual connection between the signed poem, written in Yokusai's calligraphy, and his photographed body. The image thus stages a particular relationship between the subject of representation and the existence of the subject, which resonates with the ideals of representation that shaped Yokusai's *Plants and Trees*. Indeed, Yokusai had practiced a pictorial logic similar to the photograph repeatedly and assiduously in his study of *materia medica*.

Plants and Trees marked the first instance in Japan in which a publication comprehensively systematized indigenous plant specimens in accord with the Linnaean binominal system. It extended the scope of Keisuke's *Nominal Differentiations in Western Materia Medica* (*Taisei honzō meiso*, 1829), which introduced Linnaean nomenclature and listed the names of known plants, by identifying available plants in Japan through Linnaean taxonomy, by providing illustrations and known Chinese and Japanese names for each plant, and by describing the physical attributes of the plants. In the introduction of the book, Yokusai remarked that he himself had drawn every one of 1,259 specimens, most of which were grown by him in his garden. The Shōhyaku-sha's repeated attribution of primacy and authority to direct observation evident in their exhibition practices is echoed in Yokusai's insistence on the experience and encounter with the plants as critical to botanical knowledge. At the same time, his articulation of the need for observation-based illustration differs significantly from that of Iwasaki Kan'en, who understood his own repeated observations as the source of his confidence in the extent and amount of illustrated "details." Yokusai, by contrast, relies on the Linnaean classificatory system as the guiding standard to shape his observation. In other words, the primary "details" included in his illustrations are in accord with the "details"—notably the number of stamens and pistils—required to apply the Linnaean nomenclature. For Yokusai, then, the confidence of his illustrations draws from two sources: his own experience of growing the plants and trust in the Linnaean taxonomy as a stable schema of organization. Yokusai then comments on the complementary roles of texts and images in recording and disseminating knowledge:

> In the conveyance of an object, it is not possible to reveal detailed knowledge of it unless you provide both illustrations and texts. In the West, this [format of revealing knowledge] is commonly understood, and thus most of the books on *materia medica* [from the West] are accompanied by illustrations.... If no illustration was available for the plant that he wished to write on, he would make his own and attach it to the text.[19]

Yokusai's central concern was to provide illustrations and text together, under the presumption that this approach made knowledge related to the object—its name, efficacy and existence in his region, those constitutive aspects of the real enumerated by Keisuke—available in a more shareable and informing format. Indeed, in the introduction to *Plants and Trees*, Keisuke himself complemented Yokusai on his achievement:

> When I was twenty-five or six years old, I wrote and published the three-volume book called *Nominal Differences*. At that time, I thought that I should spend more time on this book, that is to say, I should write a fuller book called "Honzō taisei," and make the knowledge available to the public. After its publication, however, I was often sick and unable to further my study. I have postponed this project for almost thirty years, and I am still unable to actualize my intention. I have constantly regretted this fact. Iinuma Yokusai, however, has always enjoyed pursuing *materia medica*, and as he aged his involvement with it deepened. He researched and overcame many obstacles to explain plants in many areas. In other words, he had traced the study of western botany, and wrote this book entitled *Plants and Trees* in accord with the Linnaean system of class and order. His explanations are consistently detailed, his illustrations are finely executed, and this work encompasses all plants and trees in this country, leaving none unaccounted for.[20]

Keisuke situates *Plants and Trees* within the extended line of the genealogy of botany that includes *Nominal Differences* as its predecessor. He implies that Yokusai's *Plants and Trees* is tantamount to his envisioned and regrettably unrealized publication project, "Honzō taisei." In Keisuke's mind, Yokusai's *Plants and Trees* completed the correlation of local flora to their Linnaean names that Keisuke had sought to undertake. Keisuke's effort to provide Linnaean names along with Chinese and Japanese names in *Nominal Differences* did not include illustrations or descriptions of the physical attributes of the plants. Yokusai produced his own illustrations from the very plants that he himself had grown in Ogaki and provided the name (Linnaean, Chinese, and various Japanese) and description for each plant included in *Plants and Trees*. Yokusai's *Plants and Trees* thus concretized the triangulated relationship among texts, illustrations, and the actual existing plants in the region more fully and comprehensively. That Yokusai experimented with photography suggests a similar incentive might have motivated his study of the medium. His photographic portrait from 1863 combines text, name, and visual representation of the subject in a mutually informative relationship guided by the same principles for portraying and conveying knowledge found in his botanical magnum opus. Yokusai began his experimentation with photography during the period in which he was drafting *Plants and Trees*. It would have been technically impossible to photograph plant specimens, since the available lens would not have permitted short focal depth, but Yokusai's deep concern for producing reliable illustrations for *Plants and Trees*, based on his observation and how his portrait combines the very constitutive elements he prioritized in his drawings of plants, signals a suggestive overlap that illuminates Yokusai's interest in photography.

Shashin as Envisioned by the Bakufu

In the very years in which Yokusai was engaged with his project, Keisuke began work on a similar project as a newly appointed official in the academy operated by the *bakufu* in the city of Edo. Keisuke's project can best be understood by way of a manuscript, "Miscellaneous Collections of Natural History by Kaiunrō" ("*Kaiunrō hakubutsu zassan*"), likely compiled between 1862 and 1865.[21] To historians interested in the nature of image making in the final years of the *bakufu* and the opening years of the Meiji period, the "Miscellaneous Collections" attracts particular interest because it was a collaborative endeavor that Keisuke undertook with Takahashi Yuichi, who was then a student at the newly established Bureau of Pictorial Studies (*Gagaku kyoku*). Yuichi would later become one of the strongest proponents of Western style painting while advocating didactic use of oil paintings in public life.[22] Indeed, the collaboration between Keisuke and Yuichi on this manuscript, which contains illustrations of domestic flora, represented not only an intersection between two individuals but also a point at which genealogies of *shashin* began to merge in a way that would guide the approaches to photographic practice in following decades.

Keisuke's preoccupations in the opening years of the 1860s strongly shaped the manuscript. The year 1861 was critical for the fifty-nine-year-old Keisuke. In that year, he was summoned to Edo as an employee of the Academy for Western Studies (*Bansho shirabejo*) operated by the *bakufu*, and he would meet Siebold for the second time, three decades after their first encounter in Owari. The circumstances that enabled this second meeting depended on the institutional transformation of the shogunate's system of academies that followed in the wake of Perry's arrival in 1853.

In response to this new foreign contact, the *bakufu* was pressured internally by their bureaucrats to expand the capacity and power of the Branch for Translating Barbaric Books (*Bansho wage goyō gata*) to better address the current political situation.[23] By 1856 this office, renamed to the Academy for Western Studies, was primarily devoted to gathering and analyzing information about foreign countries, and it was particularly concerned with foreign access to and control of information about Japan. To meet the pressing need for more knowledge about the West and to prepare for imminent expanded economic and political intercourse with Western nations, the *bakufu* recruited Western learning (*rangaku*) scholars such as Keisuke from across the country to operate the institution. In January 1857 the institute began to accept students by enrolling 119 children of direct retainers of *bakufu*, and it continued to train students after the Meiji government absorbed the institution in 1868.[24]

In 1861, the year in which Keisuke would meet Siebold for the second time, two new offices within the Academy were established: the Office for the Study of Material Goods (*Bussan gata*) and the Bureau of Pictorial Studies (*Gagaku kyoku*). Keisuke headed the former and worked closely with the latter, a connection that throws into high relief the discourse surrounding *shashin* and image making in the decade of the 1860s and its broad political implications. In the draft of an opinion note he wrote to submit to the *bakufu*, Keisuke revealingly described his own understanding of the meaning of "material goods" (*bussan*):

> The study of material goods [*bussan gaku*] is one that involves overarching and expansive objects and considers everything such as plants, trees, birds, animals, insects, fish, minerals and stones. Each item is evaluated via comparisons to already published materials and to the actual sample objects. Unless we gather adequate number of foreign books as well as sample objects, our inquiries in the study of material goods cannot be [perceived to be] answered in a satisfying manner.... For the time being, the Office for

> the Study of Material Goods relies on the books on material goods from the West owned by the highest-ranked scholars of English and Dutch studies to sort out the names of objects that are difficult to identify. People bring objects to the Office for the Study of Material Goods to be identified and evaluated. And although not on the same scale as the contents of Western books, for objects that originated in Japan we not only research their various Japanese names but also study their shapes and conditions. If the object's Chinese name is known, we make a note of it as well and keep the actual object for future assessment and research... As more people become equipped to conduct comparative research on these objects, we will be able to add to the national profit [*gokoku'eki*] gradually while also revealing the attributes of objects, by so doing, we will fulfill the rationale for establishing the Office for the Study of Material Goods.[25]

Keisuke's assessment of the study of material goods highlights a continuation of the familiar practices from the activities of the Shōhyaku-sha such as hosting regular *honzō kai* and borrowing and copying imported books. In this note Keisuke proposed that the Office for the Study of Material Goods host exhibitions to display recent acquisitions and objects submitted by instructors at the office for viewing by any interested person, in addition to hosting smaller monthly exhibits. The goals of these displays outlined in the note tellingly echo the intended goals of the *honzō kai* described in the 1830 flier discussed earlier. The same underlying principles that shaped the Shōhyaku-sha's *honzō kai* also direct Keisuke's proposed gatherings at the Office for the Study of Material Goods, which similarly rely on strategized observation and public participation to clarify and certify the relationship between names and the objects.

The evocation of "national profit" (*gokoku'eki*) is the most notable difference in Keisuke's language separating his conception of the role of the Office for the Study of Material Goods from the *honzō kai* gatherings in Owari. Art historian Suzuki Hiroyuki delineates the broader change in governmental and official attitudes toward "objects" in the *bakufu* and Meiji governments that underpins this shift in Keisuke's thought. Exploring how "old objects" had been evaluated, protected, and appraised in the late Tokugawa and early Meiji periods, Suzuki asserts:

> What changed first in the Meiji period is the fact that various environments that surrounded old objects came to have intimate relations with the infrastructure of the state. During the Tokugawa period, [the government] encouraged the study of material goods from the perspective of social utility, but the scale and content [of the study of material goods] changed vastly

> during the Meiji period. In the Meiji, old objects came to be understood as that which ought to be protected legally, to be exhibited in public, and to be researched in order to be exhibited.[26]

Similar to the way in which the category of "nation" assumed significance in the rhetoric and tactics of the Meiji government in relation to "old objects," material goods came to be assessed in terms of their value to the state.

Just as images and image making played an important role in Keisuke's activities in the Shōhyaku-sha, so too would Keisuke continue to have a deep interest in pictures and their role in the newly formed Office for the Study of Material Goods. As he wrote from Edo to his family in Owari on February 28, 1863:

> Today, I went to the Academy for Western Studies to study England and to learn their language gradually. There is also an office to practice methods of picture-making within the Academy, and a highest-ranking official [*kyōju*] assigned to this office. In pursuing material goods, the lack of familiarity with picture-making impedes [the study] rather severely. Therefore, I send my apprentice, including Yoshio, to this office to be admitted. The office not only accepted them but also gave some sample examples of the Dutch method of picture-making [*oranda ryū ga no kakikata*] to them. I am thankful that we can study whatever field of study [within the Academy] and that we do not need to pay for this service.[27]

Here Keisuke explains the goal of developing a relationship between the Office for the Study of Material Goods and the Bureau of Pictorial Studies; he deems understanding of what he calls the "Dutch method of picture-making" to be indispensable in pursuing the study of material goods. What needs to be emphasized, moreover, is the fact that this was not a novel methodological innovation that Keisuke arrived at in Edo, but rather a practice that emerged as an extension of the familiar approach to *materia medica* that Keisuke and other members of the Shōhyaku-sha had taken in Owari.

In January 1861 Koga Kin'ichirō, a *bakufu* bureaucrat working for the chief senior counselor Abe Masahiro, petitioned to reform the already existing Office of Pictorial Illustrations (*Ezu shirabe gata*) and appoint two full-time teachers.[28] This petition stirred a contentious discussion among *bakufu* bureaucrats: the central issue was the expediency of pictures in various branches of knowledge. The debate surrounding establishment of the Bureau of Pictorial Studies demonstrates a conflict that permeated understanding of the roles of pictorial representations. Significantly, the concept of *shashin* lay at the crux of the dispute. In his original petition, Kin'ichirō argued:

> As I mentioned in an earlier notice, because the offices dealing with technologies and techniques of the West are overwhelmingly busy and the study of pictorial illustration must be pursued in the coming future, Office of Pictorial Illustrations is lately exhausted from their practice and study. Steadily, the number of students is increasing and for now we do not have expert teachers on this subject and thus cannot provide adequate teaching instruction. I petition to assign two highest-ranking officials to this office.[29]

In order to meet the demands of research and teaching at the Office of Pictorial Illustrations Kin'ichirō recommends expanding its capacity by appointing two full-time instructors. As part of standard bureaucratic procedure, the *bakufu* officials then passed his petition on for approval to the Magistrate's Office Supervising Inspections (*Daishō metsuke bugyō*), Magistrate's Office for Foreign Affairs (*Gaikoku bugyō*), and Magistrate's Office of Financial Supervision (*Kanjō bugyō*). The first two supported the petition, whereas the Office of Financial Supervision protested, essentially claiming that the study of pictures amounted to nothing more than a hobby. Moreover, since the Academy was not equipped with an office of Chinese picture making, they argued, there was no reason to establish an office that promoted Western picture making.[30]

At issue in this bureaucratic dispute was a fundamental disagreement on the role of pictorial illustrations within the institutional settings. The *bakufu* officials at the Office of Financial Supervision saw any type of picture making and its application as irrelevant to the institutional goal of the Academy. Instead, they insisted that *all* types of picture making, including Chinese, Japanese, and Western traditions and media, belonged to the category of leisure as individual pursuits that did not fulfill larger sociopolitical needs.

Shashin furnished the critical ideological centerpiece in the counterargument presented by the two offices that supported the petition. The term was employed to underscore what were perceived to be profound conceptual differences in the techniques and uses of pictures from the West (*yōga*) and those from China (*tōga*). In the response to the officials from the Office of Financial Supervision, "pictures from the West" were described as "pictures of actual use" (*jitsuyō no shina*), whereas Chinese pictures were characterized as "playthings, in short" (*hikkyō ganrōbutsu*).

> In the previous note on hiring highest-ranking officials for Western pictorial studies, the Magistrate's Office of Financial Supervision had interpreted Western pictures to be entertainment and argued that because there is no objection in not having Chinese pictorial studies in our institution, equally, not having Western pictorial studies would also pose no hindrances. There is an enormous difference between these two pictorial traditions. Pictorial studies of our country and China aim to produce pictures for amusement, and thus pictures are not of utility. The methods of picture-making in the West center on its utility, and thus constitutes one distinct discipline in education. That is to say, there exists no practice to regard the mediocre picture as a proper talent in the Western pictorial tradition. In general, the Western method of picture-making has detailed rules as exemplified in survey drawings. Furthermore, other *shashin* methods of picture-making are fitting as part of an introduction to the study of material goods and scientific [*kyūri*] studies. These [*shashin* methods] serve as goals of military science and the foundations of shipbuilding. One who is not familiar with the above mentioned method is incapable of understanding pictures done in *shashin* methods. For this reason, we now establish the study of pictures [*gagaku*] and the study of science [*kyūri*] as introductory foundational classes to various disciplines so that when the young students enter our school, they will enroll in these two classes first.[31]

This supporting petition fully discloses the perceived relationship between Western visual representations and knowledge that initially motivated Kin'ichirō to request the appointment of two teachers of Western pictorial traditions at the Office of Pictorial Illustrations. As an institution that accumulated and analyzed information from the West, the Office of Pictorial Illustrations was compelled to advance the study of Western picture making as an integral skill that ultimately enabled students to understand and internalize the objects and information keenly needed by *bakufu* at that time.

Imbedded in the core of this argument is Shiba Kōkan's ideation of *shashin* from the late eighteenth century that equated it with pictures using Western pictorial conventions. Kōkan used the term *shashin* to designate what he called "methods of Western pictures," specifically referring to techniques of depiction that render the observable sense of light (shading) and space (linear perspective). The "*shashin* methods" here point to *specific* rules and techniques of picture making from the West, a "how-to" guide to making images in a Western idiom. As a concept encompassing the origin ("the West") and utility of imagery, the implications of the concept of *shashin* here resound with the uses of the same concept by Kōkan more than sixty years prior to this bureaucratic dispute. For instance, in his *Theories of Heavens from Holland* (*Oranda tensetsu*, 1796) he employed the concept of *shashin* in this way:

> The books of the Western countries explain things that could not be explained in language by pictorial representations. I follow their step and treat language and pictorial representations as similar systems. I want to use these systems in this country also. The representations in the West use the method of *shashin* [*shashin hō*] unlike pictures made in China or Japan, and do not use brushes. Therefore, there is no discussion of theory of the use of brush or momentum of the brush. Simply they add shadow and express three-dimensionality.[32]

In his capacious *Discussions of Western Pictures* (*Seiyō gadan*, 1799) Kōkan denigrates Chinese and Japanese pictures as entertaining trifles compared with Western paintings (*shashin*) that capture the real:

> [The Japanese and Chinese method of painting] is similar to a doctor trying to cure an illness by using good medicine but not knowing why the illness started. Pictures from the West simply absorb the laws of nature and heaven. Japanese and Chinese pictures do not serve such a function since they are simply entertainment. According to the Western method of painting [*shashin*], darkness, lightness, convex and concave surfaces, farness, nearness, and depth and shallowness are expressed by employing shading, and thus imitating the state of *shin* of the depicted things. The functions of writing and painting are similar, but regarding the form and the state of things, one cannot easily distinguish them unless it is depicted in painting [*benjigatashi*]. Therefore, the method of *shashin* is a technique of real utility and a tool to govern the people.... When looking at deteriorating things, such as those that are losing *shin* as they age, there is no way of knowing about their state if there is not any painting of them. Thus if the painting does not capture *shin*, it does not serve the purpose.[33]

Kōkan then explains further the method of *shashin*:

> What I mean by *shashin* method is a way of picture-making that, in depicting scenes and subjects like landscape, birds, animals, insects and the like, gives an illusion to the viewers that the depicted subjects are about to move out of the picture whenever one looks at it. It is impossible to create this effect pictorially unless one uses Western methods of picture-making.

> Therefore, seen from the point of view of a person who is capable of *shashin* method, pictures from China and Japan are similar to children's play, and thus fall short of being pictures.[34]

Written approximately sixty years prior to the conflict over the merit of the Bureau of Pictorial Studies among the *bakufu* bureaucrats, there is a distinct and intimate echoing of Kōkan's use of the concept of *shashin* in the counterargument presented by Kin'ichirō against the Office of Financial Supervision. For Kin'ichirō and others who favored establishing the Bureau of Pictorial Studies at the *bakufu* institution, the concept of *shashin* denoted a system of communication uniquely different from language, albeit equally efficient and capable of articulating the states of depicted subjects. The drawing techniques, then, were seen to serve a utilitarian purpose in enabling the accumulation of knowledge about the West with which the institution was charged: machine builders, for instance, would be able to build a smelting machine, and those studying geography (*chirigaku*) would be able to decode Western maps. No doubt Kin'ichirō envisioned the Bureau of Pictorial Studies also assisting in other studies offered at the Academy such as the study of machinery (*kikaigaku*) and the study of smelting (*seirengaku*) that were initiated the previous year. They probably also anticipated establishment of the Office for the Study of Material Goods five months later. In their petition, the expanded and fully staffed bureau, an office that promoted the *shashin* method of picture making, was situated as integral and foundational to furthering already established fields of study at the Academy.

These contentious discussions illuminate Keisuke's 1863 letter to his family in which he talks about the uses of picture making within the Academy. The value assigned to Western picture making as a tool of knowledge in particular provides a better gauge for Keisuke's own interest in the work of his colleagues in the Bureau of Pictorial Studies and how he conceived of the role of the image in his own work on material goods. It is revealing that in this letter Keisuke does not use *shashin* to refer to the Western method of picture making, especially when Kōkan's definition of *shashin* as the techniques and media of Western pictorial tradition had been so deeply internalized by the *bakufu* that bureaucrats were evoking it to defend Western pictures. Keisuke, however, does not follow Kin'ichirō and the others in equating the concept of *shashin* with Western pictorial methods. By rendering the Dutch method of picture making as "*oranda ryū*," literally "Dutch-style," Keisuke pointedly reserves the term *shashin* for a different purpose.

Keisuke's understanding of the term, with its implications for knowledge and practices of representation, emerges in his second meeting with Siebold and how *shashin* surfaces in the encounter.

Shashin *and Keisuke's Second Meeting with Siebold*

Keisuke and Siebold met for the second and last time in Yokohama on November 10 and 11, 1861. Thirty years had passed since their first meeting, and by then Siebold was sixty-five years old and Keisuke fifty-nine. Siebold had returned to Japan to work for a trading company a year after the Japan-Holland Treaty of Amity and Commerce was signed in 1858. Keisuke had begun working for the Office for the Study of Material Goods just two months earlier in September 1861. To meet Siebold, Keisuke traveled about twenty miles from Edo to Yokohama with his apprentice Tanaka Yoshio, who recorded and recollected the meeting in the manuscript "Miscellaneous Notes from Yokohama."[35] Significantly, the term *shashin* is used in this document in a completely different way from how *bakufu* officials used it. In the manuscript, *shashin* is employed both to point to illustrations in natural history books and to sketch a portrait of Keisuke. Viewed from the perspective of modern academic disciplines, these two types of work appear to be completely separate categories of illustration. For Keisuke and Yoshio, however, these differing pictorial works shared a deeper commonality as visual records of the relationship between the depicted subject and its proper name. Their use of the term *shashin* for these images rested not on an association with Western picture making but on these pictures' ability to attest to a concrete and verifiable relationship between a name and a representation.

As he had at their first meeting in 1830, Keisuke brought numerous specimens of minerals and plants so that Siebold could verify their Linnaean names. The abridged record of the conversation between the two reveals that even after thirty years of study Keisuke continued facing difficulty in identifying natural objects within the system of Western taxonomy. On seeing the vast number of Keisuke's dried specimens and realizing the limits of time, Siebold suggested that Keisuke send them to Nagasaki, Siebold's next destination. Siebold assured Keisuke that he would verify their names there and send them back to Keisuke along with a few mineral specimens. Keisuke then asked Siebold about the success of his publication projects. Siebold responded by bringing out four manuscripts on natural history.[36]

It is in describing these illustrations from manuscripts that Yoshio employs the term *shashin*.

> After this remark, Siebold brought four large Western manuscripts to Keisuke. One was on the subject of fish, the second on birds, the third on crabs, and the last on trees. All the manuscripts were of specimens originating in Japan, and [the illustrations] were beautifully colored, so beautiful that they surprised the eyes. According to Siebold, these illustrations were *shashined* by the Nagasaki artisan Toyosuke [Kawahara Keiga] and a Dutch artisan named de Villeneuve.[37]

Yoshio uses *shashin* here as a verb. He also employs the term in the description of the meeting on the next day when Siebold showed Keisuke a manuscript containing *shashin* pictures of lilies, fir trees, and other objects. Although in this second instance *shashin* functions as a noun, the use of the term in relation to Western illustrations remains conceptually consistent. In both cases, *shashin* signifies the process or the product of making pictures after existing specimens. The usage of the term thus emphasizes the direct nature of the relationship between the representation and object, which grants the images an authority to verify and authenticate the names ascribed to the specimens.

Even more revealing is Yoshio's use of *shashin* in relation to the sketched portrait of Keisuke made on the second day of their meeting. During their discussion an unnamed Dutch person suggested that Siebold "*shashin*" Keisuke. Liking the idea, Siebold called "a British artisan" (*eikoku no gakō*), who then sketched Keisuke in pencil while he sat on a chair, wearing a long sword on his left. Keisuke was asked to focus on one spot and not move while the artisan made the sketch (Fig. 5.6).[38] The sketch artisan was Charles Wirgman, who came to Japan as a reporter for the *Illustrated London News*, along with the British diplomat Harry Smith Parks in 1861.[39] Siebold later sent the sketch to Keisuke along with a sketch of himself.

Yoshio thus equated Wirgman's portrait of Keisuke with the illustrations of natural specimens in Siebold's manuscripts through use of the term *shashin*. However different the portrait and scientific illustrations were in subject matter and medium, Yoshio's choice of words pointed to an underlying commonality between the images in which the represented subjects and their names were already verified to be existent. What is more, these portraits were made to commemorate and record the meeting of Siebold and Keisuke and are thus results and evidence of the direct encounter among Wirgman, Siebold, and Keisuke. Here, we witness the proximity of meaning working through the concept of *shin* that resonates with the Shōhyakusha's use of ink-rubbing prints (*shin'ei*). Both pictorial representations were produced to serve as visual documents of historical encounters. Wirgman and Siebold were present with Keisuke when the portrait was made. Similarly, Siebold had verified the Linnaean names and handled the plants that Kawahara illustrated in the manuscripts Siebold showed to Keisuke. In other words, to Yoshio, the fidelity between the name of the subject of the

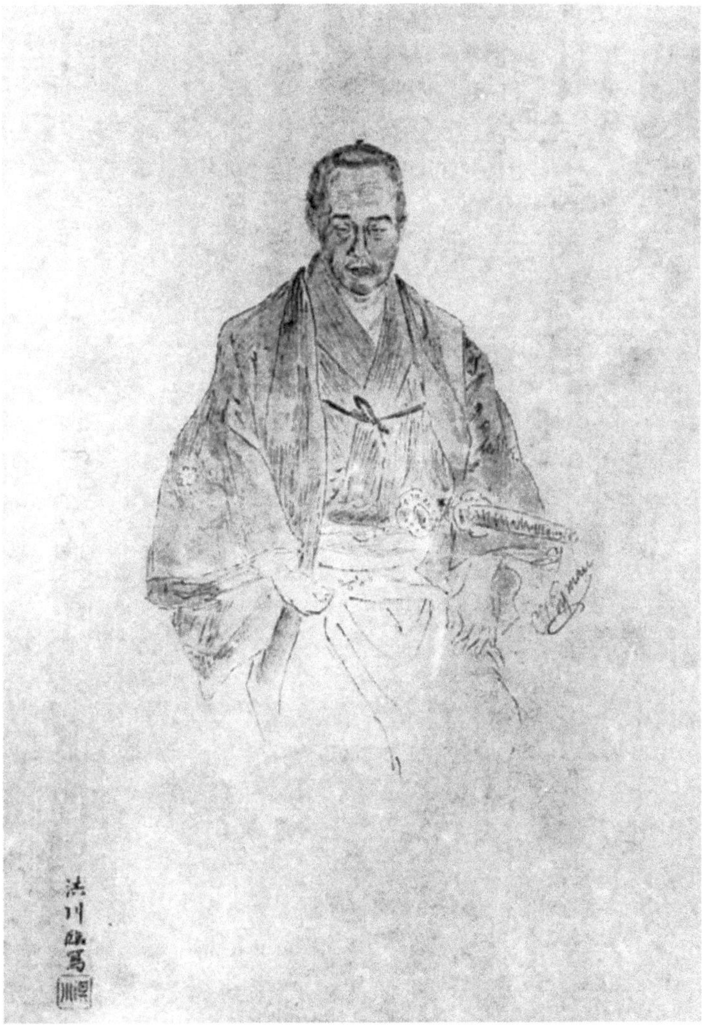

Figure 5.6 Shimizu Kōsui, copied after Charles Wirgman, Portrait of Itō Keisuke, 1861, ink on paper

representation and the represented subject was uncontestable. The images are authoritative sources of knowledge about their subjects. What had been the driving question for the Shōhyaku-sha's study of *materia medica*—establishment of a definitive and accurate relationship between names and representations—finds concise expression in Yoshio's use of the term *shashin*.

The understanding of *shashin* revealed in his account of Keisuke and Siebold's meeting encompasses the conceptual approaches the Shōhyaku-sha members would take in identifying photographic technology as *shashin*. In the decades that followed, this resonance would become gradually flattened out under the association of *shashin* with Western pictorial method. Yoshio's identification of these two representations as *shashin* affirms the transmission of the Shōhyaku-sha's earlier conceptualization of the term to the metropole of Edo, and it demonstrates that their sense of the term distinctly differed from the *bakufu* bureaucrat's use of *shashin* to designate a conglomerate of Western pictorial techniques.

In much the same way that Keisuke described Western pictorial techniques as Dutch-style, or *oranda ryū*, in his 1863 letter to his family, thereby disassociating the term from its prevalent use by the *bakufu* as a term connoting specific Western pictorial techniques, Yoshio's use of the term *shashin* also singles out how it diverged from received use of the term by the *bakufu*. Indeed, Yoshio deploys the term precisely as a picture that concretizes the relationship among text, illustration, and the existence of the subject, the very relationship that the Shōhyaku-sha had sought to stabilize and verify through their activities in Owari.[40] The illustrations Siebold showed to Keisuke were understood to be representations of actual plants that Siebold himself had handled, and he attached names verified by him to the plants. Similarly, the portrait of Siebold and Keisuke represented two existing subjects with proper names attached.

In December 1863 Keisuke left his position in Edo because of his concern for family members living during a cholera epidemic in Nagoya, returning there while his apprentices remained in Edo.[41] Keisuke spent the next seven years working for the Institute of Western Study (*Yōgaku sho*) in Owari, and publishing books such as *Information for Nonspecialists in Regard to Cholera* (*Bakushabyō teate shirōto kokoroe sho*, 1863) during the cholera epidemic. In 1870, the new Meiji government asked Keisuke to join the faculty of its professional school (*Daigaku kō*), reconfigured from the *bakufu*-operated Academy of Developing Intelligence for Successful Enterprises (*Kaiseijo*). Sixty-eight-year-old Keisuke moved to the capital again.

During Keisuke's absence from Edo/Tokyo, his fellow Shōhyaku-sha members Yanagawa Shunsan and Tanaka Yoshio came to play prominent roles in the establishment and reconfiguration of the new Meiji cultural institutions and practices. In response to the changing needs of the *bakufu*, Shunsan and Yoshio engaged in numerous new projects, including compilation of an English-Japanese dictionary, participation in and preparation for the Paris World Exposition in 1868, and translation and revision of a book on photography in Edo. In the process of realizing these projects, the term *shashin* appears again. Sifting through its meanings in this context, so decidedly different from the study of *materia medica* in Owari, reveals the process through which ideas and associations already in circulation were negotiated and oriented around the concept of *shashin*.

Yuichi's Rendition of the Concept of Shashin *in 1865*

Returning once more to the manuscript "Miscellaneous Collections of Natural History by Kaiunrō," a compilation of botanical illustrations dating between 1862 and 1863 made collaboratively by the Bureau of Pictorial Studies and the Office for the Study of Material Goods where Keisuke worked, we witness the two genealogies of *shashin* meeting in practice. Keisuke brought to the manuscript, on the one hand, the understanding of *shashin* that he had cultivated studying *materia medica* with other members of the Shōhyaku-sha in Owari. In the bureaucratic discourses that he encountered in Edo, on the other hand, *shashin* was used chiefly as a term to designate "Western techniques of picture making." These two strands of meaning intersect in "Miscellaneous Collections," an album of illustrations for which Takahashi Yuichi drew sketches for the Bureau of Pictorial Studies. The context in which Yuichi came to pursue picture making sheds light on both the historical climate of the end of the Tokugawa period as well as the intensity of Yuichi's investment in acquiring skills to create pictures in accord with Western techniques and media.[42] Yuichi's inspiration to pursue "Western-style paintings" is recounted in "The Résumé of Takahashi Yuichi" ("*Takahashi Yuichi rirekisho*," 1891), as transcribed by his son when the sixty-four-year-old Yuichi recounted his life during his illness.

> During the years of Ka'e'i (1848–1854), I borrowed a lithographic print that originated in the West from a friend, and discovered that beyond the ability of [Western-style pictures] to thoroughly approach the real (*shin*) of the depicted subject, there is a particular flavor in pictures from the West. On

> the spot, I felt the urge to acquire the knowledge about [Western pictures]. However, it was difficult to find a way to learn Western picture-making, and this obstacle worried me day and night.... Mr. Mashimo continued that there was in fact the Bureau of Pictorial Studies that was established earlier within the Academy for Western Studies and a senior official that he knows teaches there. Mr. Mashimo urged me to submit an application immediately. Jubilant and thrilled with such a reply, I felt as if I were ascending to the heavens. On September 5, 1862, I obtained the permission to study at the Bureau and started my pursuit under the guidance of a faculty member named Kawakami Bannojō [Tōgai].[43]

It remains uncertain what kind of lithographic print Yuichi saw, but his instantaneous, almost compulsive reaction to the lithographic image demonstrates that, beyond technical differences, Yuichi was responding to a particular "flavor" he intuitively sensed, a flavor absent for him in the familiar Chinese and Japanese pictorial representations he had once studied.[44] Yuichi's use of *shashin* in this recollection reveals the continued metamorphosis and rearticulation of the concept.[45] In this document, Yuichi fully adopts Kōkan's ideation of *shashin* as a technique of and approach to representation that produces pictures for utilitarian purposes, and Yuichi decidedly extends it by directly naming governance as one of the uses of *shashin*. More important, Yuichi's concept of *shashin* further develops Kōkan's notion of it by expanding the term to mean a type of expression that joins textual and pictorial representations.[46]

Yuichi's 1865 credo entitled "Statements for the Office of Western Pictures" ("*Yōgakyoku tekigen*") clearly demonstrates the evolution of these ideas.[47] (See Appendix.) Yuichi recollected almost three decades later in "The Résumé" (1891) that he wrote the credo and put it on the wall of the bureau because he wanted to "bring prosperous spirit to the Bureau of Pictorial Studies" (*gakyoku no ryūsei wo hakaranto hosshi*).[48] Throughout this credo, Yuichi asserts that the central purpose of the Bureau of Pictorial Studies (and, by extension, his own role as an official in this organization) is to serve the national interest, and he argues that the media of text and illustration were crucial to better governance. As quoted earlier, Kōkan argued similarly in his *Western Pictures*, stating that "the function of writing and painting are similar, but regarding the form and the state of things, one cannot easily distinguish them unless it is depicted in painting [*benjigatashi*]. Therefore, the method of *shashin* is a technique of real utility and a tool to govern the people."[49] He listed the cases of a "mermaid's bone," "Jan Jonston's table of living organisms," and "portraits of sages and prestige for prosperity" as exemplary demonstrations of such uses.[50]

Yuichi recommends in the text several exemplary uses for *shashin* pictures that a country ought to take advantage of: portraits of virtuous figures, scenes of joy for peaceful periods, and depictions of military power during times of war. Yuichi also endorses the combination of texts and illustrations as the most efficient way of recording one's life for posterity.

These remarks strongly suggest, moreover, that the illustrated Western books available to Yuichi at the Bureau of Pictorial Studies consisted primarily of works with didactic images accompanied by explanatory captions. Among these books at the bureau, for example, was François Pierre Chaumeton's 1817 *Flore Médicale* (Fig. 5.7). The caption of each illustration names the object of depiction in French, demanding that readers identify the depicted plant with its proper name. Each illustration follows the same formula of composition: the bottom of the page includes a row of images detailing the sexual organs of the specimen, while the space above is dedicated to the depiction of its flowers and a branch or stem. The illustrations often include a monochromatic rendering of a leaf of a plant in the background behind the flower or fruit in order to indicate the scale of the plant. This systematic and consistent format of illustrations orchestrates a sense of uniformity and cohesion to the botanical knowledge gathered in the book.

Unfamiliar with knowledge of *materia medica*, and thus the ongoing concern to stabilize the relationship among plant names, illustrations, and the existing plants, Yuichi had no reason or incentive to question the fidelity of the implied relationship between the given names and the representations. This consistent relationship between the name and illustration presented in *Flore Médicale* thus was devoid of any degree of figurative implications for Yuichi. In Yuichi's view, the relationship between the caption and the illustration becomes fixed, permanent, and accepted. Compared to the Shōhyaku-sha, whose skepticism about the incomplete state of knowledge of *materia medica* also led them to mistrust texts and images, Yuichi had no epistemological grounds for doubting the representational logic he encountered in Western illustrated books. His ignorance of the debates that engrossed the Shōhyaku-sha allowed Yuichi to presume, and indeed believe, that the coordination of names and illustrations in the natural history books was correct. The concept of *shashin*, in this sense, emerges from a very different conviction from that of the Shōhyaku-sha's. For the members of the Shōhyaku-sha, the shared knowledge of *materia medica* formed the basis for supporting or challenging

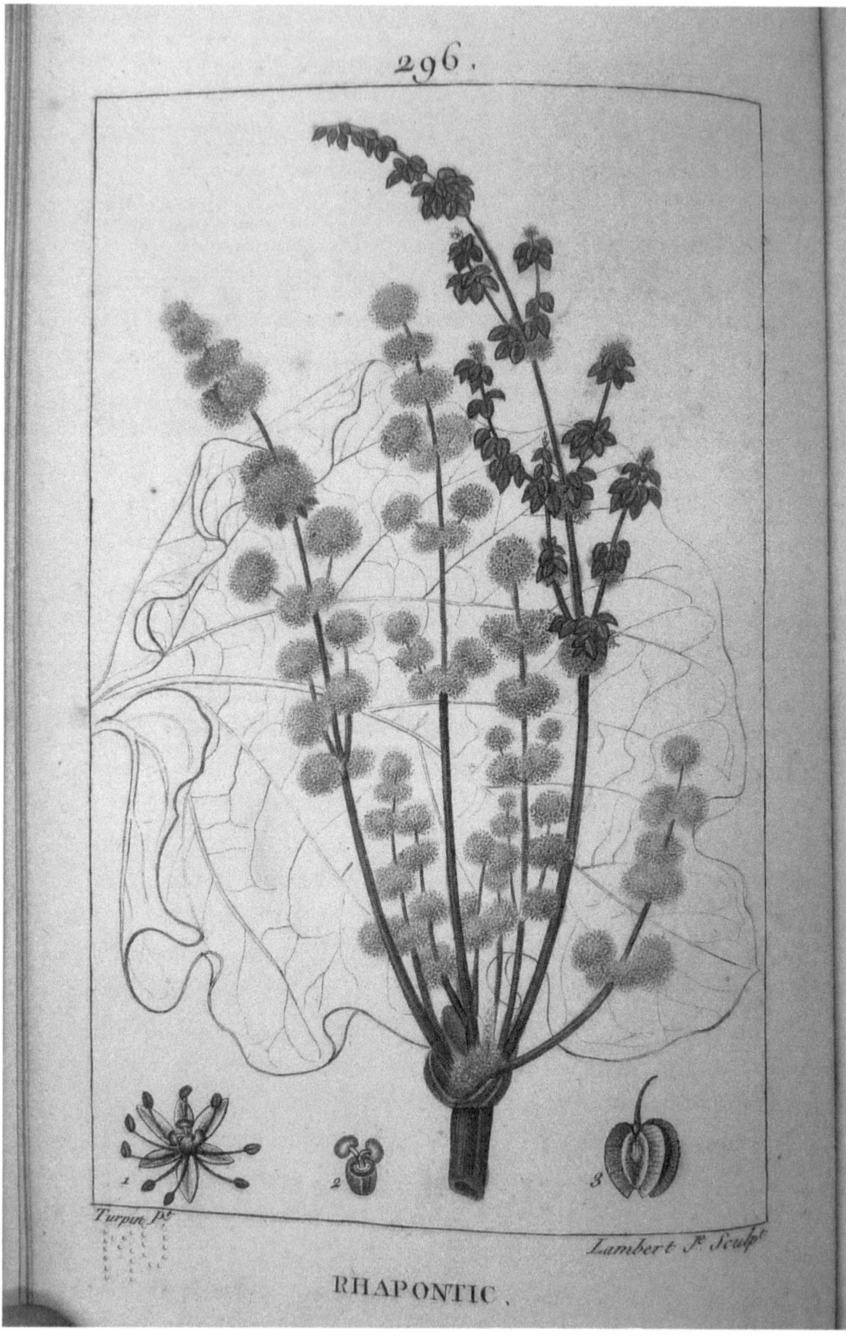

Figure 5.7 Pierre Jean François Turpin, *Rhapontic*, from François Pierre Chaumeton, *Flore Médicale*, 1814–1820, University of Michigan Library

a particular illustration of a plant. Yuichi, by contrast, possessed neither the intimate knowledge of *materia medica* nor the experience of debating and authenticating the names of plants. Without this fundamental questioning of knowledge, he did not think to interrogate how knowledge was constituted through the representational rhetoric of names and illustrations in the Western books.

Yuichi accordingly sees texts and *shashin* pictures together as one useful "technique" (*gigei*) for governance, and thus he implies that the *shashin* technique is necessarily always accompanied by captions or other texts. Yuichi believes that in the West textual and pictorial media were always effectively paired, and that their calculated use leads to something akin to newspapers, in which information from distant lands and people was relayed easily. In the atmosphere of apprehension about the unknown Western countries and the insecurity over Japan's preparedness that permeated the *bakufu* institutions, Yuichi not only appropriates the term *shashin* from Kōkan but also embraces it as a technique to govern and direct the people.

The contrastive analysis that I offer of two works by Yuichi—"Statements for the Office" vis-à-vis his illustrations included in "Miscellaneous Collections"—exemplifies the process through which Yuichi conceptualized *shashin*, as a technique that combines explanatory description and represented subjects (Figs. 5.8 and 5.9; see also color plates 11 and 12). By comparing and connecting the abstract concept of *shashin* to concrete pictorial representations, it is possible to perceive the contour of his interpretation of the concept.

Figures 5.8 and 5.9 come from the assembled picture album now titled "Miscellaneous Collections" mentioned earlier.[51] All of the thirty-nine illustrations by Yuichi are signed between 1863, after he started at the Bureau of Pictorial Studies in September 1862, and 1865, except for two that are undated. There is no historical document that explains the purpose of creating these illustrations, which were compiled collaboratively by the Office of the Study of Material Goods and the Bureau of Pictorial Studies. Yet the logic behind the establishment of the Office of the Study of Material Goods in 1861 as well as Keisuke's proposed projects at the institution in 1862 suggest that the Academy for Western Studies, the overarching organization overseeing these two offices, made an effort to gather illustrations of natural specimens available in Japan as part of a larger *bakufu* effort to update their knowledge of *bussangaku* for possible trading and commerce.[52]

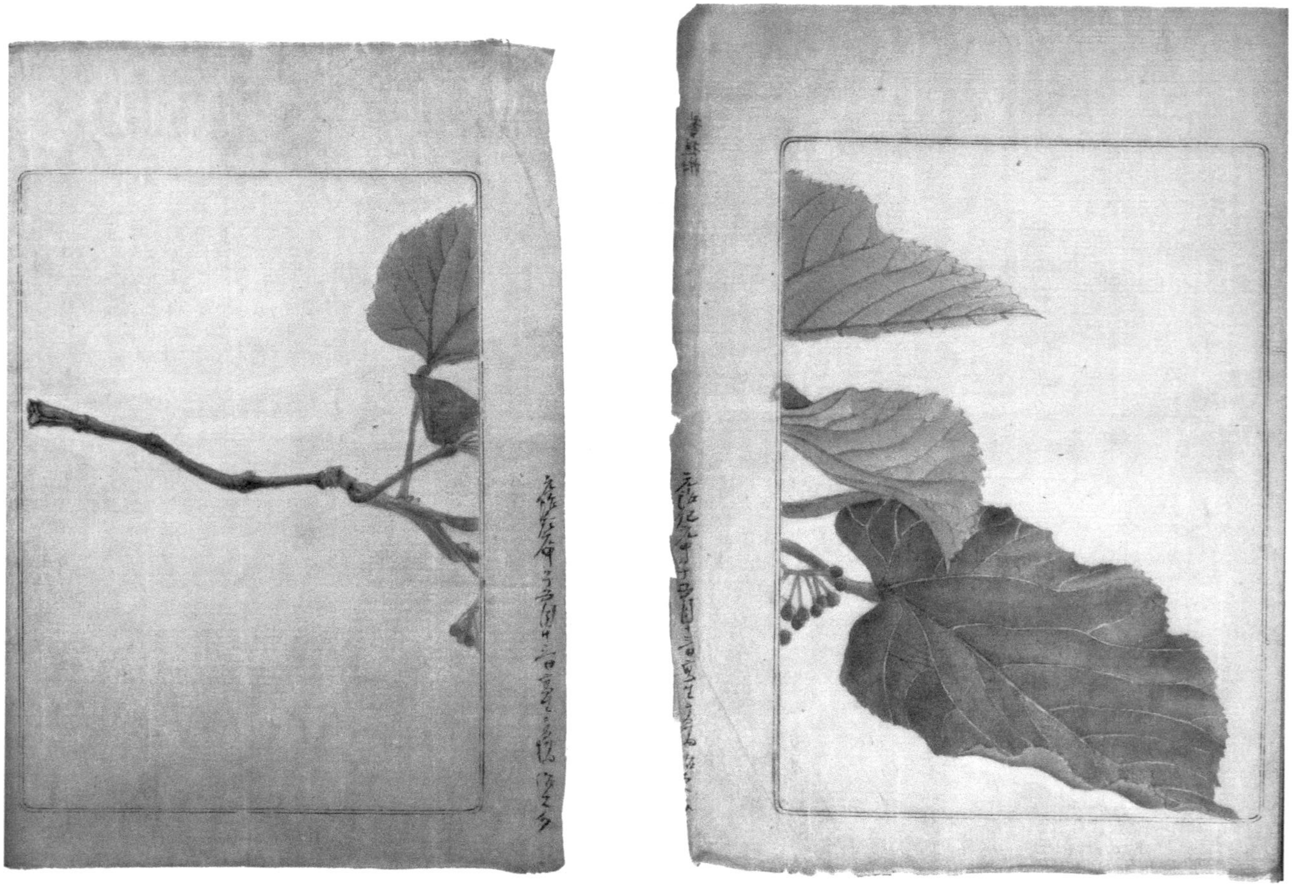

Figure 5.8 Takahashi Yuichi, *Bodaiju*, from "Kaiunrō zufu zassan," 1863–1865, ink on paper, Special Books, Tokyo Metropolitan Library

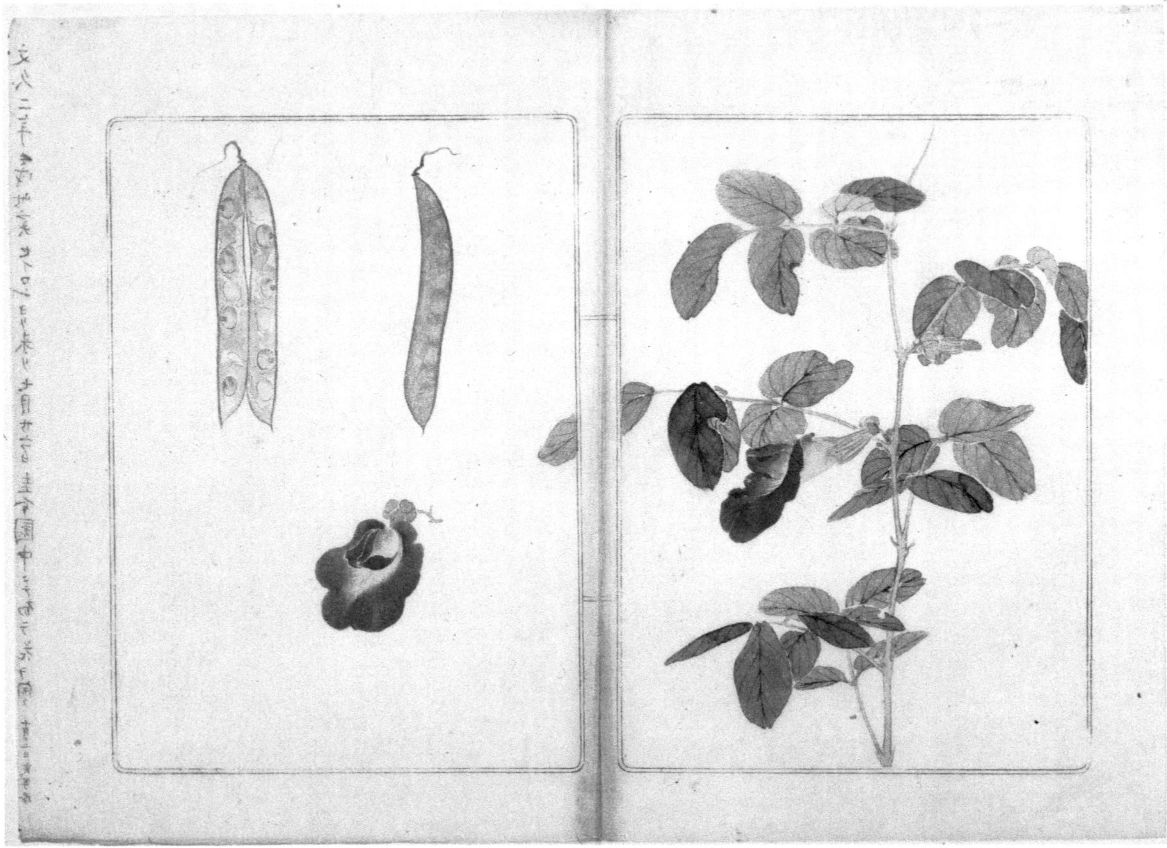

Figure 5.9 Itō Keisuke, *Snow pea*, from "Kaiunrō zufu zassan," 1863–1865, ink on paper, Special Books, Tokyo Metropolitan Library

A close look at Yuichi's illustration of *bodaiju* (*Tilia miqueliana*) indicates his abandonment of traditional outlining (*hitsu'i*) techniques for creating images in favor of a Western-inspired technique that employs shading (*butsu'i* in his words) to convey the three-dimensionality and surface quality of the leaf. Comparing this illustration with Yuichi's credo reveals how he tried to execute his ideas of *shashin*. He notes:

> In the pictorial methods of China and Japan, a painting starts with the concept of *hitsu'i* [a brush user's understanding of the attributes and conditions of his brush] and ends with the concept of *butsu'i* [the draftsman's understanding of the attributes and qualities of the object he is depicting], whereas in the Western pictorial method, pictures begin with the concept of *butsu'i* and end with *hitsu'i*.

In other words, according to Yuichi, the draftsman ought to start with looking at the subject of the illustration itself, before he contemplates his own tools. The connection has to be made first between the eyes of the draftsman and the subject, rather than between his eyes and the brush he is holding. In this illustration, the traces of brushstrokes that would be seen as expressions of *hitsu'i* are absent, and instead Yuichi carefully crafts the shape of the branch via darker shades. On the same folio Keisuke renders a sweet pea plant, which, the caption informs us, flowered for the first time at Keisuke's garden after its seed was brought from Ceylon in 1862 (Fig. 5.9). In basing his illustration on this specimen, Keisuke emphasizes his direct access to the physical flowering plant, an emphasis that parallels Yuichi's advocacy of *butsu'i* as the starting point of image making. This shared intellectual approach, moreover, is manifested in the illustrations, as Keisuke employs a representational technique resembling Yuichi's that minimizes the amount of outlining contour while shaping and modeling the plant with controlled brushwork. Yuichi saw *shashin* as a technique that combines descriptive captions and pictorial illustrations.[53]

Yanagawa Shunsan, the Concept of Shashin, *and Photographic Technology*

While Yuichi's manifesto embraced a concept of *shashin* aligned with how *bakufu* bureaucrats had evoked it a few years earlier, Yanagawa Shunsan, who also worked in Edo for the *bakufu*, applied the term *shashin* to photography. Shunsan's incorporation of *shashin* into his treatise on photography merges with that of Yuichi's to concretize the embodied meaning of *shashin* by pointing to both the existent subject and the textual description of the subject. Shunsan devoted much of his time on duty as one of the highest-ranking officials (*kyōju*) of the Academy of Developing Intelligence for Successful Enterprises selecting and translating articles that related to Japan from English language newspapers, an activity that would ultimately lead him to publish *Periodical on the West* (*Seiyō zasshi*), a magazine-format publication, in 1868, the same year that he published *Illustrated Book on Photography* (*Shashinkyō zusetsu*).[54]

In this publication, Shunsan identifies himself with a sinicized name, which is a combination of his various artistic names used for creative expression (*gigō*). Given that these *gigō* were used privately, it is unlikely that his general readers would be able to match his sinicized name with his official public name, Yanagawa Shunsan. Combining the Chinese characters that incorporate and play on the characters for his family name, Shunsan's sinicized name gives an impression that the book was translated by a scholar from the region of the Yangjiang River. What would be the purpose in 1868 of adding a layer of pretence and giving the impression that the book on photography was of Chinese origin? In Edo, the immediate association of photographic technology as "the Western" device, embodying Western pictorial values, as elaborated by Kōkan, was prevalent not only among the *bakufu* officials but also in the general public. By asserting through his sinicized name that this translated work on a Western device was mediated by a Chinese scholar, Shunsan situates photography in a triangulated relationship among the West, Japan, and China, replacing the dichotomous view of photography that prevailed at the time in Japan.

In the introduction to *Photography*, after noting how pictographs existed prior to letters in ancient societies, Shunsan claims:

> The pivotal role of pictures is to bridge over the time between thousands of years ago and today, and to shrink the hundred thousands of miles that lay between places. That is to say, if the shape of objects is transferred precisely and its spirit maintained, there is not even the slightest dissatisfaction in pictures. The purposes of pictures exist exactly in these aspects. A painter utters, "It is easier to depict the legendary monster *kami* than to draw [living] dogs and horses." I can only laugh at such remarks. Pictures happen by following *shashin* and *satsukei* [to capture a landscape or a scene].[55]

In his view, pictorial representations serve primarily as recordings, as a sort of eyewitness. Similar to the way in which ink-rubbing prints served as a visualized testimony to one's encounter with a plant and the opportunity to prove the existence of a specimen in the works of the Shōhyaku-sha, Shunsan also imagines pictures advantageously serving the role of testimony. In the chaotic environment of Edo, where the confrontation between *bakufu* and anti-*bakufu* forces was escalating, Shunsan's pressing concern to diminish temporal and spatial distance can be read as an expression of the rapidly changing reality that Shunsan himself witnessed.[56] Shunsan uses the concepts of *shashin* and *satsukei* as ones that generate such records. Tracing the photographic camera historically to the invention of camera obscura, Shunsan asserts the difference between camera obscura and photography in regard to the concepts of *shashin* and *satsukei*. He contends that whereas the former enabled man to capture landscape scenes, echoing the way in which Yōan endorsed the use of camera obscura in translating "Encyclopedia for Household Use" (1813–1846), the photographic camera made it possible "to *shashin*." His subtle differentiation between the two concepts hinges on the specific details of the techniques:

> The lens to *satsukei* [in the camera obscura] was invented about three hundred years ago by a man named Porta, and he had meticulously perfected its effect. That is to say, one would copy the shadow by using a brush. Further, there were those who believed that more could be improved on this lens. Around our period of *bunsei* [1819–1829], two men named [Joseph Nicephore] Niépce and [Louis] Daguerre invented a new lens to *shashin*. It involves chemicals to coat the silver plate, shining [images] on this plate by using sunlight, and adding more chemical liquid to this plate, thus never having to use the brush [to make a picture] again. Since its inception, this lens has worried various principles of nature and tormented creative logistics, while [people had concocted ways to] project them onto paper, stone, copper and glass. The speed of the sophistication of this projection technique is such that the lens to *satsukei* [camera obscura] carries the sounds of yesterday in the far distance. That is to say, the lens to *shashin* [photographic camera] gives almost no dissatisfaction or disappointment in cherishing the scenic landscapes of distant people and in imagining the earlier wise men. Therefore, among the historians that investigate facts, some could reproduce writing via this lens while others could take a picture using this lens, and doing so would be as if the historians are talking face to face with the historical figure. Indeed, given the superiority of this lens, its impact to social practices is significant, despite its seemingly minor branching of a pictorial tradition.[57]

Shunsan thus posits camera obscura as the device to *satsukei*, and photography as an improved version that allows one to *shashin*, hence his naming of the photographic camera as *shashin kyō* (*shashin* lens, or lens that *shashins*). Shunsan lists examples for the uses of photography that overlap with those proposed by Yuichi for the uses of pictures made in accord with *shashin* technique, in that both authors viewed obliteration of temporal and spatial distance as the key to the merit of *shashin*. *Shashin* images—whether illustrations or photographs—allowed viewers to come "face to face" with the temporally and geographically distant.

How *shashin* enables dissolution of temporal and spatial distance, moreover, depends on the combination of textual and pictorial representations. Like Yuichi, Shunsan reflects on the relationship between photographic technology and textual representation. In fact, Shunsan explicitly explores the application of photographic technology to recording textual representations for the future. Shunsan's uses of *shashin* thus expand beyond the pictorializing observed scenes or objects and actual people into two-dimensional representations that archive and preserve documents. His gesture toward employing photography for conservation, moreover, illuminates a particular understanding of the medium in Japan at this time, especially compared to popular European debates on photography in which questions about the new technology emerged from traditional Western pictorial conventions and were debated through the concepts and rhetoric of that discourse.[58]

Indeed, in Europe the stakes in arguing for or against photographic technology hinged upon the commentator's relationship to a particular pictorial tradition. In announcing the invention of daguerreotype in 1839, Daguerre himself described the medium through the language of aesthetic evaluation drawn from two-dimensional art, when he said "by this process, without any idea of drawing, without any knowledge of chemistry and physics, it will be possible to take in a few minutes the most detailed views, the most picturesque scenery."[59] One observer of the new medium responded to Daguerre's announcement by proclaiming that the discovery was "the most remarkable in the history of art."[60] William Henry Fox Talbot, in *Pencil of Nature* (1844–1846), narrates his invention of Talbot-type or calotype as an event that resulted from frustrated amateur painters' attempts to use camera lucida as a drawing aid.[61] He too estimated the potential of his invention by invoking specifically aesthetic terms; he evoked the "Dutch school of art," for instance, in describing the photographic image and described the photographic plate as "picturesque." Talbot's conceptual framing of photographic prints in *Pencil of Nature* through the lens of aesthetics "pitched art into the orbit of industrial knowledge."[62] By 1857, the particular relationship—both imagined and real—between pictorial traditions and photographic images had become serious enough in Europe that Lady Eastlake wrote that with "photography our chief object at present is to investigate the connection of photography with art."[63] This debate continued to circle around questions of the agency, mechanicity, and aesthetic aspect of photographic expression.[64] The language and concepts of cultivated pictorial and aesthetic traditions that European practitioners and consumers of photography employed to understand and explain the technology were thus distinctly different from the issue of textual captions that concerned Shunsan. In the case of the Academy of Developing Intelligence for Successful Enterprises in Japan, photography was grasped through the term *shashin* within an institutional educational context. The demands of this setting encouraged an approach to photography at the Academy that highlighted the medium's utility in responding to the dramatic cultural and political changes then facing Japan. In this light, Shunsan's claim that photography shortens the temporal and spatial distance between oneself and the past (the historian can talk face-to-face with his subject) or distant regions carries a historical import. Because the Academy was established as a repository of knowledge about and from the West, situating photography as a technology that can traverse the presumed distance between Japan and the West—both spatial and temporal—without "dissatisfaction" and "disappointment" allows Shunsan and his audience to approach photography as a technology that accelerates the process of gathering information from and understanding "the West" without hindrance.

Before publication of this book, Shunsan experimented with photography at the Academy, starting in 1864.[65] Proof of the competence in the collodion process that he acquired by the end of the year comes from a request by Ichikawa Kanenori, Shunsan's colleague at the Academy, to photograph his son, who was leaving for Russia as a member of a *bakufu* delegation.[66] As historian Miyachi Masato suggested in his essay on early modern imagery and photography, at the Academy the study of photography was embedded within the institutional goal of acquiring techniques and knowledge to better prepare the *bakufu* for the changing political and social environs.[67]

In this historical context, it becomes easy to imagine a scene at the Academy in which Shunsan was walking around and caught sight of Yuichi's "Statements for the Office" hanging on the wall of the Bureau of Pictorial Studies from time to time. Perhaps Yuichi's public declaration inspired Shunsan to think further about the concept of *shashin*, or even discuss the concept with Yuichi himself.[68] Shunsan also insisted that photographic imagery possessed real utility, an insistence that effectively incorporated the notion of *shashin* as a utilitarian tool promoted by both Kōkan and Yuichi.

As hinted at in the introduction to *Photography*, which was written by his friend Katsuragawa Hosaku, a physician and chemist working at the Academy, some regarded photography as magical, mysterious, and thus suspicious when Shunsan published this book.

> The more the strange and skillful art from the West becomes available, the more people find it uncommon and bizarre. I should characterize the photographic camera as the most uncommon device with sophisticated dexterity. There are only countable numbers of Japanese who lecture on the photographic camera, and thus they are so busy they cannot take a rest. They cannot prepare to learn how to mix and filter the chemicals, and thus depend completely on Western ships for the supply of chemicals in vain. That is to say, once a ship fails to come, all they can do is to hope for the arrival of the Western ships with their hands bundled. My friend [Shunsan] had deplored this situation, and began spending all his time and effort on searching knowledge [on the chemicals] by studying Western books by himself between his official duties. He planned to attain the thorough knowledge of its foundation and thus pass down this rare skill. However, as he began his research, he found the information to be unclear and the limit of knowledge seemed endless. It is like his investigation was executed in a thick fog. Once the clear way to attain the chemicals is established and the uncertainty of information is eliminated, however, it is no longer necessary to wait for the importation of chemicals via the Western ships. This is all thanks to the hard work of [Shunsan]. What I resent the most, moreover, is those who do not understand the circumstances that surround Western studies and render these efforts and goods as similar to mysterious lewd skills, giving us a look as if looking at kings of hell. Furthermore, these people also claim that photography is harmful to the society. How can they not understand a simple fact that photography is at the foundation of the opportunity to enrich and strengthen our country?[69]

Shunsan's *shashin* departs from the cultivated uses of the concept by the Shōhyaku-sha, where questions of fidelity and verifiability played a pivotal role in *shashin*; members of the group saw in the concept of *shashin* an assurance of the physical existence of the objects depicted in illustrations. If we return to Shunsan's introductory remark, discussing how laughable it is that a painter says a mythological monster is easier to depict than dogs and horses in actuality, we see that Shunsan foregrounds the assumption that it is far more difficult to depict the *shin* of existing entities than of fictional and imaginary creatures. And it is photography (*shashin kyō*) that facilitates this process of representing the real. At the same time, Shunsan also presumes that in using *shashin* the resulting image would necessarily represent an existing entity. Similarly in Yuichi's rendition, the imbricated existential certainty is taken for granted as a given element of *shashin*. Yuichi identified recording of facts for later generations or of distant places for those who would never visit them as the primary role of *shashin* pictures. In this respect, the relationship between the provided textual title and the pictorial representation became pivotal. Whereas *shashin* images capture what is existent, the captions assist in deciphering what one is looking at in the picture. For instance, *shashin* of a person must be accompanied with a text accurately naming the person in order for later generations to be able to identify the sitter. Moreover, understanding of the factual existence of a photographed figure, incident, or place was submerged, according to Yuichi's notion of *shashin*, within the larger project of constructing national wealth and strength.

Use of *shashin* by the Shōhyaku-sha was concerned primarily with establishing irrefutable fidelity between the physical availability of plants in their region and the names given to the plants; Yuichi and Shunsan focus on other concerns in their notions of *shashin*. On one hand, their use of the term extends the importance of the relationship between naming and the subject of representation, while on the other hand assuming that anything *shashin* must actually exist, thereby circumscribing *shashin* solely to the realm of reality. In other words, in Yuichi and Shunsan's notion of *shashin*, photographic prints *cannot* contain pictorial elements that are imaginary or fictional. Together Yuichi and Shunsan established *shashin* as a combination of visual and textual representations of the depicted subject, and their use of *shashin* solidified the premise of fidelity that had formerly engrossed the Shōhyaku-sha into an assumed quality of *shashin*. By integrating Kōkan's expansive uses for *shashin* pictures (in part instigated by the *bakufu* bureaucrats), Yuichi and Shunsan removed the tension that existed between the formulation of *materia medica* and the Shōhyaku-sha's particular process of vetting knowledge. Yuichi's appropriation of Kōkan's ideation of *shashin* as a useful tool for governance in particular reverberates with Yuichi's beliefs articulated in "Statements for the Office." (See Appendix.) In the closing three sentences of this text, Yuichi valorizes pursuit of pictorial studies at the Bureau of Pictorial Studies within the Academy of Developing Intelligence for Successful Enterprises particularly strongly:

> Indeed, pictures and letters serve the interest of a nation, and in that sense, are vital technologies [*saidai no gigei*] that should not be separated even for a brief moment. The reason why the Bureau of Pictorial Studies was established within the Academy rests precisely here. Because efficiency and urgency is demanded of the Bureau of Pictorial Studies, each of us at the Bureau must diligently pursue our study.

For Yuichi, pursuit of Western studies had to include both textual and pictorial representations. This necessity was conveyed through examples of illustrations that required a descriptive caption naming the subject of representation. Yuichi's statement shares directly with Keisuke's sentiment on the need for pictorial skills. Keisuke too took advantage of the teaching of *oranda-ryū* (Dutch-style) picture making within his capacity, working for the Office for the Study of Material Goods.

What took place at these *bakufu* institutions is a conflation of two strands of meaning and use of the term *shashin*. Kōkan's ideation of *shashin* as techniques of Western pictorial representation and Western apparatuses, merged with the import of *shashin* that Shunsan and Keisuke brought from their study of *materia medica* as a term that addressed and negotiated the relationship of fidelity among the three central components of their study: the name(s), the illustrations, and the verified existence of plants. In *Photography*, Shunsan modifies apprehension of *shashin* as an interlocutor of the three entities. Shunsan assumes that any photograph must represents something that already exists, and he emphasizes the seeming effect of shortening temporal and spatial distance as a result of photographic imagery. Indeed, the captions provided for photographic representations become pivotal in Shunsan's envisioned use of photography. Through a combination of caption and photographic image, Shunsan argues, one is able to depict events and figures in a distant land or time with no "disappointment" or "dissatisfaction."

I want to underscore that it is not the "nature" of the photographic medium per se that is highlighted by Shunsan's evocation of *shashin*. Rather, it is the specific uses of photographic representation combining pictorial and textual representations that motivate application of the term. The intimate and verified correspondence between the caption and the representation worked together to cement the notion of *shashin*. Appreciation and acceptance of photographic technology, expressed as *shashin*, therefore is indebted to the past practices of *materia medica* that the Shōhyaku-sha cultivated. The understanding of photographic technology, in this light, was not skewed by blind glorification of "the West" or "Western civilization" as some historians have suggested.[70] Rather, it was shaped by the specific changing social and political landscape particular to Japan at that time. Shunsan's expression of photography as *shashin* encompasses not only his ability to decipher foreign texts and technology but his own interpretation as well of the significance of the technology in the reality he faced.

What perpetuates contemporary confusion and conflation of the term *shashin* and the presumed realism of photography is an unexamined emphasis on the part of scholars on the meanings invested in the term by Chinese aesthetic discourses and Kōkan. They focus on these two strands of the genealogy of the term *shashin*—one developing out of aesthetic discourses from the Han dynasty and the other, out of Kōkan's late-eighteenth-century subjective use of *shashin* as a term specifically referring to a select group of representational techniques from the West. Contemporary scholars have effectively flattened out the epistemological significance that the term *shashin* carried for the Shōhyaku-sha and figures such as Shunsan, who lived after Kōkan's writings and before the spread of photographic equipment and knowledge in Japan. One can see the persuasiveness of drawing a neat causal link from Chinese aesthetics to Kōkan's discourse on Western pictures to photography for scholars seeking to understand how photography came to be understood through the term *shashin*. It is likely that Shunsan was familiar with the aesthetic use of the term *shashin*, and Kōkan's specific rendition of it, but what motivated his study of and practice with photographic technology was not his own aesthetic aspiration. Shunsan was keenly aware of the world outside of Owari and Edo; his fluency in languages such as Dutch, French, and English fueled interests to seek out and understand different cultural practices and histories. Indeed, his personal desire to produce representations that could shorten temporal and spatial distance attracted him to photography.

The teleological conflation of *shashin*, pictorial realism, and photographic technology thus not only fails to account for Shunsan's historical interests but also renders the historical development of photography in Japan laden with aesthetic concerns from the outset. As Joel Snyder has argued eloquently, what one perceives as a "realistic" representation is not a de facto effect of any medium, especially photography. Rather, what enables such a characterization intrinsically calls into question the "standardized, or characterized, or defined notion of vision itself."[71] In the case of Shunsan, his notion of vision hinged on extracting epistemological insight from the representations themselves, and as such it necessarily included legible texts to identify what is represented. Only if the represented subject, accompanying caption, and the representation itself were aligned was the premise of fidelity fulfilled. Only then did the photographic image achieve the status of representation without "disappointment" or "dissatisfaction" that Shunsan called *shashin*.

Rather than using photography as an exemplary medium with which a non-Western country's process of modernization can be measured (and thus its history is seen as a narrative of unfolding modernization), scholars have recently begun to identify photographic material and technology as a set of practices deeply embedded within specific historical, religious, and epistemological contexts. The conceptual grappling that took place surrounding *shashin* before photographic technology was available underscores the fact that as much as the technology was "imported" from the West, the very visual practices and environment to which photography was introduced shaped the reception and understanding of what photography was, or could do.

Identification of photography through the concept of *shashin* showcases the ways in which the very term *photography* was already marked with cultural and representational resonance. Yet *shashin* was not merely a theoretical concept operating within a limited discourse of aesthetics. For the Shōhyaku-sha, the term carried an epistemological import for their practice of *materia medica*. This asks us to move beyond the history of photography as one of technological adaptation and appropriation. We must then expand our trajectory of inquiry to take seriously the historicity of languages and the efficacy of pictorial representation. Such an expansion augments the existing narratives of photography to forge new paths in photographic history beyond national boundaries.

Reference Matter

APPENDIX

Takahashi Yuichi, "Yōgakyoku tekigen" (1865)[1]

Pictorial methods [*gahō*] in Western countries have prioritized *shashin* since the beginning. Because all that lies in front of one's eye is designed and created by principles of the heavens and nature, what the pen of a skilled man creates from this nature is an object that accords with those principles in reduced size. Although it is said that the uses of pictures and letters are similar, letters are limited only to inscription. In regard to semblance, there are many cases in which only pictures can delineate details adequately. For this reason, in publications from Western countries, many utilize pictorial illustrations to explain and inform readers. That is to say, pictures are understood to be superior to letters. This prioritization is rightly so. Relying on this logic of understanding the roles of pictures, it is not difficult to understand that the role that pictorial illustrations play on the national [*kokka*], everyday, and various other levels is not small. It is possible to demonstrate and insinuate the deeds and virtues of the sages and heroic historical figures [to people] by copying the portraits of them. In addition, during the time of peace, Western pictures can illustrate beautiful singing and dancing figures, while during a period of political unrest, pictures can simulate the events of military practice and actual battles. Despite the enormous geographical distance [between the viewer of the image and the location of the actual event that the picture depicts], pictures enable these events taking place at a distance to be depicted and known [to the viewers]. Every event and every object, there is none that does not result in pictures. In the West, the government of each country established an institution to show pictures and this facility is, indeed, a wonderful tool to assist governing [*chijutsu*].[2] In this respect, pictures are not merely a type of amusement. When one attempts to pass on the facts of one's life to later generations, it is impossible to convey them unless one uses letters to record things as much as possible and draws pictures of the situations. Without the pictures that represent the scenes, and without intermixing the textual recording of one's life into them, it is difficult to demonstrate and memorialize the facts of one's life. That is to say, the principal logic that flows between illustrations and sentences is of the sort that is quite difficult to evaluate in terms of superiority since both are equally excellent. The pictorial conventions of China and Japan originated from letters, whereas the conventions in the West started from pictures. In the pictorial methods of China and Japan, a painting starts with the concept of *hitsu'i* [a brush user's understanding of the attributes and conditions of his brush] and ends with the concept of *butsu'i* [the draftsman's understanding of the attributes and qualities of the object he is depicting], whereas in the Western pictorial method, pictures begin with the concept of *butsu'i* and end with *hitsu'i*. *Hitsu'i* interferes with objects, while *butsu'i* assists *hitsu'i*. *Hitsu'i* emerges in tracing the outlines, while *butsu'i* effuses from shades of colors when one assembles and uses these techniques of representation from the West. When illuminated by the light of the moon and the sun, various incomprehensible and uncertain elements of the world come to be shaped and [illegible] via expressions of shading, surface quality, distance, and depth. Discerning these [elements created by light], [someone in the West] has innovated artificial methods [to achieve the same effect as lighting], which is why there is a method to represent three dimensions [in the Western pictorial method]. In addition to this rule, there are additional separate rules to express distance in accord with the size of the picture. Those in the West have furthered the principle of this method from the basics out of their desire to approximate *shin* and to produce pictures so skillful that objects appear to be on the verge of motion. This is the process through which *shashin* become prioritized. Its basis, therefore, lies in the saturation of colors. Even the lines of letters of each country include various kinds of lines. The horizontal, vertical, slanting, reversed, wide and narrow lines are expressed with the natural saturation of colors. Because adequate saturations can only be obtained via the trained skill of a draftsman, the pursuit of the principle of saturation leads to the discovery of the hidden secret of the principle of the heavens and nature. Indeed, pictures and letters serve the interest of a nation, and in that sense, are vital technologies [*saidai no gigei*] that should not be separated even for a brief moment. The reason why the Bureau of Pictorial Studies was established within the Academy rests precisely here. Because efficiency and urgency is demanded of the Bureau of Pictorial Studies, each of us at the Bureau of Pictorial Studies must diligently pursue our study.

Early winter, 1865

Takahashi Yuichi, a student of Western pictures (*yōga*)
at the Bureau of Pictorial Studies, Academy of Developing
Intelligence for Successful Enterprises

Notes

Introduction

1. The term *Tokugawa* refers to the Tokugawa period, lasting from 1603 to 1868, during which the Tokugawa shogunate in the capital Edo was in power.

2. On the "conceptions" of photography in Europe, see Geoffrey Batchen, *Burning with Desire* (Cambridge, MA: MIT Press, 1997); and "The Naming of Photography 'A Mass of Metaphor.'" *History of Photography* 17, no. 1 (1993): 22–32. See also Joel Snyder's insightful argument against such impulse to identify "conceptions," in "Enabling Confusion," *History of Photography* 26, no. 2 (2002), 154–160.

3. To remain consistent with the practice of contemporary Romanization of Japanese terms, "*Honzō sjasin*" is rendered as "*Honzō shashin*" for the remainder of this book.

4. Sasaki Takami, "Kaisetsu," in *Nagoya sōsho sanpen*, ed. Nagoyashi hōsabunko (Nagoya: Nagoyashi kyōiku i'inkai, 1982), 8.

5. The wide dispersal and frequent inaccessibility of archival resources challenge efforts to grasp a fuller picture of the members of this group. For figures such as Itō Keisuke and Mizutani Hōbun, the material records are ample. For lesser known figures, such as Ōkubo Masa'aki and Yoshida Jakusōan, the records reveal only that they were *samurai* employed by the Owari domain.

6. Kitazawa Noriaki, for instance, traces the first appearance of the word *bijutsu*, which came to signify "visual art" in Japanese, to the translation of the official letter sent from Vienna to invite the Meiji government to participate in the world exposition in 1873. *Kyōkai no bijutsushi: 'bijutsu' keiseishi nōto* (Tokyo: Buryukke, 2000). Satō Dōshin explores the historical effort to standardize the concocted term in the records of world and domestic expositions. Satō Dōshin, *"Nihonbijutsu" tanjō: Kindainihon no "kotoba" to senryaku* (Tokyo: Kōdansha, 1996).

7. For instance, the major archives of the central member of the group, Itō Keisuke, are divided into three collections: Itō bunko at the National Diet Library includes approximately 2,000 books owned by Keisuke and his grandson Itō Tokutarō, which includes both published and unpublished manuscripts. Itō Keisuke bunko at the Nagoya University Library consists of about 180 volumes of published and unpublished manuscripts by Keisuke. Itō Keisuke Kinen Toshoshitsu at the Higashiyama Botanical Garden houses books, manuscripts, and

medical tools that belonged to Keisuke, as well as his abundant correspondence. In addition, I have consulted these archives for the research of this book: Gotō Museum of Art, Tokyo Metropolitan Library, Waseda University Library, Kyōu Library of the Takeda Science Foundation, Nagoya Municipal Museum, the University of Tokyo Library, and Gifu Historical Museum. The collections in the Netherlands—in the Rijksmuseum voor Volkenkunde, the University of Leiden Library, and the National Herbarium, Leiden—contain the largest archival holdings related to the Shōhyaku-sha in Europe. Approximately two hundred drawings by the members of the group, along with several hundred herbaria, are now in Russia's Royal Scientific Academy in St. Petersburg. Russian botanist Carl Maximovitch purchased the collection during his stay in Japan between 1860 and 1864. The British Library also has a few manuscripts by the members of the group that probably came as part of the Siebold collection.

8. *Compendium of Materia Medica* was imported to Japan by 1604. As I describe in more detail in Chapter Three, there were several translations of this Chinese text published after the text was imported. Isono Naohide, *Nihon hakubutsushi nenpyō* (Tokyo: Heibonsha, 2002), 102–103. See also Li Shizen, *Chinese Medicinal Herbs: A Modern Edition of a Classic Sixteenth-Century Manual*, F. Porter Smith, and G. A. Stuart, trans. (Mineola, NY: Dover, 2003).

9. The full title is *Naeukeurige beschryving van de natuur der viervoetige dieren, vissen en bloedloze waterdieren, vogelen, kronkel-dieren, slangen en draken.*

10. For a detailed study of the social and intellectual influences of these imported books, see, for instance, Timon Screech, *The Western Scientific Gaze and Popular Imagery in Later Edo Japan: The Lens Within the Heart* (Cambridge: Cambridge University Press, 1996); and W. F. Vande Walle and Kazuhiko Kasaya, eds., *Dodonaeus in Japan: Translation and the Scientific Mind in the Tokugawa Period* (Leuven, Belgium: Leuven University Press, 2001).

11. Peter Dear, *The Intelligibility of Nature* (Chicago: University of Chicago Press, 2006), particularly 173–195.

12. See Doris Croissant, "Meiji shoki yōga no riarisumu ni tsuite," *Jinbun gakuhō* 53 (1982), 157–187; Satō Dōshin, *Meiji kokka to kindai bijutsu* (Tokyo: Yoshikawa kōbunkan, 1999); Imahashi Riko, *Edo no kachōga* (Birds and flowers: The representation of natural history during the Edo period, Tokyo: Sukaidoa, 1995); Naruse Fujio, "Edojidai no seiyōgaron ni tsuite-sono toyōshisō tono kankei," *Bijutsushi* 85 (1972), 1–17; Kōno Motoaki "Edojidai 'shasei' kō," in *Nihon kaigashi no kenkyū*, ed. Yamamoto Yunezō sensei kokikinenkai (Tokyo: Yoshikawa kōbunkan, 1989), 388–417; "'Shasei' no gensen-chūgoku," in *Akiyama Terukazu hakushi koki kinen bijutsu ronbunshū*, ed. Akiyama Terukazu hakushi koki kinen bijutsu ronbunshū kankōkai (Tokyo: Benridō, 1991), 481–514; and "Edo kaiga to kyakkan shugi," in *Edo no nakano kindai: Akita ranga to kaitai shinsho*, ed. Takashina Shūji (Tokyo: Chikuma shobō, 1996), 233–288.

13. Croissant, 156.

14. Shiba Kōkan, *Seiyō gadan*, in *Yōgaku*, vol. 2 , ed. Numata Jirō, Matsumura Akira, and Satō Shōsuke (Tokyo: Iwanami shoten, 1976), 493.

15. Shiba, *Seiyō gadan*, 492.

16. Shiba Kōkan, "Oranda tsūhaku," in *Yōgaku*, eds. Numata Jirō, Matsumura Akira, and Satō Shōsuke, Tokyo: Iwanami shoten, 1976), 497–540. The layered meaning of *kyō* as referring to both lens and mirror is similar to the case of "glass" in English of the period.

17. Yokoe Fuminori, "Kamera obusukyura no denrai kara 'shashinjutsu' no toraimade" (Between the Arrival of the Camera Obscura and the Daguerreotype in Japan), in *The Advent of Photography in Japan*, ed. Tokyo Metropolitan Museum of Photography and Hakodate Museum of Art (Tokyo: Tokyo Metropolitan Foundation for History and Culture, Tokyo Metropolitan Museum of Photography, 1997), 14–15, 163–165.

18. Peter Galassi's claim that photography "was a legitimate child of the Western pictorial tradition" has played a significant role in understanding the reception and uses of photographic media in Japan. The inherent and embedded "Western" aspects are often overblown by historians in Japan, who adhere to a dichotomy between Japan and the West in writings on the history of photography in Japan. Although Galassi's work focuses on the invention of photographic technology and thus is unconcerned with the reception of the technology, the writings on photographic history in Japan uncritically absorb Galassi's claim, by decontextualizing and dehistoricizing it as if a similar narrative arc ought to be drawn. See Peter Galassi, *Before Photography: Painting and the Invention of Photography* (New York: Museum of Modern Art, 1981); Ozawa Kenji, *Nihon no shashinshi--Bakumatsu no denhan kara meijiki made* (Tokyo: Nikkor club, 1986); Anne Tucker, ed., *The History of Japanese Photography* (New Haven: Yale University Press, in association with Museum of Fine Arts, Houston, 2003).

19. It is no surprise then that accounts of the history of science in Japan often begin with the Meiji restoration of 1868, when "Western science" was institutionalized and validated as part of the government's program to "Westernize" the country. See Masao Watanabe, *The Japanese and Western Science*, trans. Otto Theodor Benfey (Philadelphia: University of Pennsylvania Press, 1976); Nakayama Shigeru, David L. Swain, and Yagi Eri, eds., *Science and Society in Modern Japan* (Tokyo: University of Tokyo Press, 1974); Morris Low, *Building a Modern Japan* (New York: Palgrave Macmillan, 2005); Murakami Yōjirō, *Bunmei no naka no kagaku* (Tokyo: Seidosha, 1994).

20. The engagements with both Western science and scientific fields that existed during the Tokugawa period serves as a critical issue for the field of Practical Learning (Jp: *jitsugaku*; Ch: *shih hsueh*) that derived from ideas of Chu Hsi, a significant intellectual in the system of Neo-Confucian thought. As I will explore in Chapter One, the centrality of philosophically grounded practice of *jitsugaku* and its relation to the prevailing intellectual concerns for natural ontology in the eighteenth and nineteenth centuries have produced a large body of exegesis and a variety of applications. Indeed, the Tokugawa intellectuals and their engagements with various aspects of Practical Learning have attracted many historians to this topic. See, for instance, Peter Nosco, *Remembering Paradise: Nativism and Nostalgia in Eighteenth-Century Japan* (Cambridge: Council on East Asian Studies, Harvard University,

1990); Maruyama Masao, *Studies in the Intellectual History of Tokugawa Japan,* trans. Mikiso Hane (Princeton: Princeton University Press, 1974); Harry Harootunian, *Toward Restoration: Growth in Political Consciousness in Tokugawa Japan* (Berkeley: University of California Press, 1970); Tetsuo Najita, *Visions of Virtue in Tokugawa Japan: The Kaitokudō Merchant Academy of Osaka* (Honolulu: University of Hawai'i Press, 1987). For a study that focuses on the explicit relationship between *materia medica* and Practical Learning, see Sugita Tsutomu, *Edo no hakubutsugakushatachi* (Tokyo: Kōdansha, 2006).

21. Even though the phrase "residue of the real" may evoke the concept of index for those who are familiar with photographic histories and theories, I am not invested in exploring the works of the Shōhyaku-sha through this concept. The Peircian notion of index operates *a part* of a complex tripartite system of signs (along with the notion of symbol and icon) and thus is not concerned with construction of epistemology and vetting of facts, which were the focus of inquiry for the Shōhyaku-sha. The universality of the system that Peirce assumes in his theorization also cannot account for the different systems of languages and scripts that faced the Shōhyaku-sha historically. The question of referentiality in the Peircian semiotic system from which the notion of index emerges falls short in historically contextualizing aspects of *shashin* that prove most useful and meaningful. I deliberately chose the concept of "fidelity" instead in order to convey *shashin*: the concept of index cannot account for the epistemological ramifications that various *shashin* pictures and commentaries carried. The discursive debate on the validity and usefulness of the concept of index for the study of photographic images has been coupled with Roland Barthes's notion of *punctum* expressed in his *Camera Lucida* and only amplified the mystical status of the concept without coming to a consensus as to *how* this theoretical concept operates meaningfully for varied historical and cultural expressions in photography. See Charles Sanders Peirce, "Logic as Semiotic: The Theory of Signs," in *Philosophical Writings of Peirce*, ed. Justus Buchler (New York: Dover, 1955), 98–119. For recent theoretical grappling on the concept, see James Elkins, "What Do We Want Photography to Be? A Response to Michael Fried" in *Critical Inquiry* 31:4 (2005), 938–956; Mary Ann Doane, "Indexicality: Trace and Sign—Introduction," and Thomas Gunning, "Moving Away from the Index: Cinema and the Impression of Reality" in the special issue of *differences* on indexicality, *differences: A Journal of Feminist Cultural Studies* 18:1(2007), 1–6, 29–52; Joel Snyder, "Pointless," in *Photography Theory*, ed. James Elkins (London: Routledge, 2007), 369–385.

22. Endō Shōji, *Honzōgaku to yōgaku* (Kyoto: Shibunkaku shuppan, 2003), 355–385. In this light, Kinoshita Naoyuki's assertion that "*shashin* was not applied to photography when it was first introduced to Japan" needs to be qualified further by asking what "first introduced" means. As is well documented, photographic devices were available to wealthy and curious *daimyō*, and thus introduction of the technology was uneven in society. Kinoshita Naoyuki, "The Early Years of Japanese Photography," in *The History of Japanese Photography*, ed. Anne Tucker (New Haven: Yale University Press, 2003), 26.

23. Ozawa, *Nihon no shashinshi*, 12–77.

24. See, for instance, Steve Edwards, *The Making of English Photography* (University Park: Pennsylvania State University, 2006).

25. In the anthology *Photography's Other Histories* (2005), Christopher Pinney, one of the volume's editors, argues for envisaging just such history. In a more recent edited volume, *Photographies East: The Camera and Its Histories in East and Southeast Asia* (2009), editor Rosalind Morris urges readers to reimagine the foreignness that the photographic camera symbolically embodied and physically exemplified in Asian contexts. In refusing to fall into the trap of either a universalizing teleological history of technology or an essentializing cultural concept of "Asia," Morris uses the question of "foreignness" to address the ambiguous dual temporality that cameras capture, what she calls "duration and a repeating drama of disappearance and persistence." Christopher Pinney and Nicolas Peterson, eds., *Photography's Other Histories* (Durham, NC: Duke University Press, 2003), 14; Rosalind C. Morris, ed., *Photographies East: The Camera and Its Histories in East and Southeast Asia* (Durham, NC: Duke University Press, 2009), 9. See also Karen Strassler, *Refracted Visions: Popular Photography and National Modernity in Java* (Durham, NC: Duke University Press, 2010).

26. Pinney and Peterson, *Photography's Other Histories*, 14.

Chapter One

1. The Dutch made obligatory visits to the shogun in Edo from the southern port of Dejima, to which they were restricted as residents. These Dutch ambassadorial groups were led by the chief of the factory, who was accompanied by a physician and a scriber. Between 1633 and 1850, the Dutch made 166 trips. The Dutch East India Company (Vereenigde Oost-Indische Compagnie, VOC) went bankrupt in 1799. In their homeland, as a result of the Napoleonic war, King Louis Napoleon would rule the country from 1806 to 1810, renaming it the Kingdom of Holland. In 1811, moreover, the British would take over Dutch East Indies, and twice (in 1808 and 1813) they attempted to seize the factory in Dejima. The changes in the broader political landscape of empires left the national identity of the Dutch factory in Dejima in hanging. However, the entity would continue to interact with the *bakufu* officials in Japan as "Dutch" factory until 1858, when the Dutch-Japan Treaty of Amity and Commerce was signed. For an account of an earlier journey to the Edo castle by Blomhoff, see Matthi Forrer, *The Court Journey to the Shogun of Japan: From a Private Account by Jan Cock Blomhoff* (Leiden: Hotei, 2000). For the cultural impact of these trips, see Timon Screech, *The Western Scientific Gaze and Popular Imagery in Later Edo Japan* (Cambridge: Cambridge University Press, 1996), 15–20. It must be noted that other Asian countries also visited the shogun numerous times during the Tokugawa period. Ronald Toby, *State and Diplomacy in Early Modern Japan: Asia in the Development of the Tokugawa Bakufu* (Princeton: Princeton University Press, 1984).

2. Since 1690, when Engelbert Kaempfer came as a physician for the Dutch East India Company, the Dutch envoys assiduously collected and preserved various

types of fauna and flora while in Japan. Confinement of the Dutch to the artificial island of Dejima restricted their access to a broader variety of specimens. Kaempfer's successor, Carl Thunberg, as well as Jan Cock Blomhoff, who came as the captain of VOC, accumulated more specimens before returning to Holland: see *A Very Unique Collection of Historical Significance: The Kaptain (*The Dutch Chief*) Collection from the Edo Period—The Dutch Fascination with Japan*, exhibition catalogue, ed. Nagasaki City Museum (Nagasaki: Nagasaki City Museum, 2000).

3. This is included in his *Nippon. Archiv zur Beshreibung von Japan und dessen Neben- und Schutzländern: Jezo mit den Südlichen Kurilen, Sachalin, Korea und den Liukiu-Inslen* (Würzberg and Leipzig: Verlag der K.U.K. Hofbuchhandlung von Leo Woerl, 1897).

4. Phillip Franz von Siebold, *Edo sanpu kikō*, trans. Saito Makoto (Tokyo: Heibonsha, 1967), 169–171. Because a foreigner's travel was strictly monitored and restricted by the shogun, Siebold had not been able to travel widely outside the artificial island of Dejima widely since his arrival in 1823. Unless otherwise noted, all translations from Japanese and classical Chinese are my own.

5. J. Hoffmann, comp., *Catalogus librorum et manuscriptorum Japonicorum a Ph. Fr. De Siebold collectorum, annexa enumeratione illorum, qui in Museo Regio Hagano servantur* (Lyon: J. Hoffmann, 1845), 16, in the collection of Rijksmuseum voor Volkenkunde.

6. Commonly known as the Siebold Incident, Siebold was found in possession of detailed maps of Japan as well as an outer coat bearing the family crest of the shogun when he was preparing to leave Japan after his five-year assignment in 1828. The shogunate was suspicious of a foreigner possessing Japanese maps, as well as an official garment of the shogunate, so Siebold was put under house arrest immediately. Takahashi Kageyasu, the chief astronomer working for the shogun, had given the maps to Siebold; he was put in prison. In addition to Kageyasu, ten others were also charged by the shogunate for treason and imprisoned. Siebold was expelled from Japan. However, with the signing in 1858 of the Japan-Holland Treaty of Amity and Commerce, the prohibition was lifted and Siebold returned to Japan in 1859. Although Kageyasu died in prison before the sentence was delivered, the shogun preserved Kageyasu's dead body in salted water, and later publicly decapitated the corpse in compliance with the sentence.

7. These are two of the seven herbs used in making autumnal porridge (*aki no nanakusa*).

8. Each illustration is painted in ink on medium-weight Japanese paper measuring approximately 32 x 24 cm. Each illustration was later numbered and stamped by the Rijksmuseum on recto. The serialized numbers given to the illustrations in Leiden show that the cataloguers viewed the illustrations as individual, distinct images, rather than as components of a manuscript.

9. In correspondence with the author, Kato Nobushige and Yamaguchi Takao, professors of botany and biology who work with Dutch botanical collections of Japanese specimens, state that to their knowledge this manuscript is the only one that employs this method of picture making.

10. Siebold, *Edo sanpu kikō*, 170.

11. Yoshikawa Yoshiaki, *Ikgaku yōgaku honzōgakusha no kenkyū* (Tokyo: Yasaka shobō, 1993), 330.

12. Hōbun himself identified *rōtō* as *hashiri dokoro* in his 1809 *Clarification on the Names of Things* (*Buppin shikimei*). Mizutani Hōbun, *Buppin shikimei*, in *Nagoya sōsho sanpen*, ed. Nagoyashi hōsabunko (Nagoya: Nagoya kyōiku i'inkai, 1982), 32.

13. Hayashi Razan made this insertion in his 1630 *Tashikihen*. Quoted in Sōda Hajime, *Zusetsu nihon iryō bunkashi* (Kyoto: Shibunkaku shuppan, 1989), 258. Ono Ranzan, in his *Honzō kōmoku keimō*, alerted readers that identifying *rōtō* as tobacco was "completely wrong." Ono Ranzan, *Honzō kōmoku keimō*, ed. Sugita Tsutomu (Tokyo: Waseda University Press, 1974), 233.

14. In his diary entry in 1838, a decade after his meeting with Siebold, Keisuke noted that he gave belladonna to a local physician, Taguchi Yōsei. Itō Keisuke, *Kinka'ō nikki* (Nagoya: Higashiyama shokubutsuen, 1996), vol. 2, 4. In 1892, ninety-year-old Keisuke put together illustrations by Hōbun of *hashiri dokoro* and tobacco plants that grew from a seed given by Siebold, with Keisuke's own descriptions of their differences and characteristics, on a hanging scroll. In the collection of Itō Keisuke kinenshitsu at Higashiyama Botanical Garden.

15. In *Dictated Compendium of Materia Medica* (*Honzō kōmoku keimō*, 1803), Ono Ranzan, Hōbun's teacher of *materia medica*, in fact identified *hashiri dokoro* as belladonna. Sōda, *Zusetsu nihon iryō bunkashi*, 259. Although Siebold had only two years' experience of practice after graduation, he successfully operated on a number of patients with cataract or infected cornea in Japan. Harold E. Henkes, "The Influence of the Dutch on Japanese Ophthalmology in the 19th century," in *Red-Hair Medicine: Dutch-Japanese Medical Relations*, ed. H. Beukers, A. M. Luyendijk-Elshout, M. E. van Opstall, and F. Vos (Amsterdam: Rodopi, 1991), 79–82.

16. Sōda, *Zusetsu nihon iryō bunkashi*, 258–261. Both of these assessments of the plant's property were correct: it can be used as a remedy so long as it is prepared by first soaking the plant overnight and then cooking until it forms a congealed body. Sakai Shizu, *Nihon no iryō-shi* (Tokyo: Tokyo shoseki, 1982), 358. Direct consumption of the plant, moreover, leads to hallucinatory reactions, often causing one to run frantically, whence derived the Japanese name *hashiri dokoro*, literally "running around wild."

17. Indeed, Keisuke's work has been examined through the lens of Practical Learning. Sugimoto Isao, "Itō Keiuke no jitsugaku no seikaku," *Nihondaigaku jinbunkagaku kenkyūjo kenkyū kiyō* 2 (1960), 39–70. See also Minamoto Ryōen, *Jitsugaku shisō no keifu* (Tokyo: Kōdansha, 1986).

18. Ryōen, *Jitsugaku shisō no keifu* (Tokyo: Kōdansha, 1986).

19. Ryōen, "'Jitsugaku' and Empirical Rationalism," in *Principle and Practicality: Essays on Neo-Confucianism and Practical Learning*, eds. William Theodore de Bary and Irene Bloom (New York: Columbia University Press, 1979), 378. See also Ellen Gardner Nakamura, *Practical Pursuits: Takano Chōei, Takahashi Keisaku, and Western Medicine in Nineteenth-Century Japan* (Cambridge, MA: Harvard University Press, 2005). For the importance of Practical Learning throughout Asia, see the other essays in the volume. Within this broad category, in fact, contemporary

historians have placed and interpreted the works of diverse Tokugawa thinkers, including Kaibara Ekken, Miura Baien, Andō Shōeki, and Kiho Seiryū, as sharing a pragmatic approach to the reality they faced.

20. Tetsuo Najita, *Visions of Virtue in Tokugawa Japan: The Kaitokudō Merchant Academy of Osaka* (Honolulu: University of Hawai'i Press, 1987), 10–13; and "History and Nature in Eighteenth-Century Tokugawa Thought," in *The Cambridge History of Japan*, vol. 4, *Early Modern Japan*, ed. John Whitney Hall (Cambridge: Cambridge University Press, 1991), 596–659. See also John A. Tucker, ed. and trans., *Ogyū Sorai's Philosophical Masterworks: The Bendō and Benmei* (Honolulu: University of Hawai'i Press, 2006).

21. See Sasaki Takami, "Kaisetsu," in *Nagoya sōsho sanpen* (Nagoya: Nagoyashi kyōiku i'inkai, 1982), 12–14. It is worth noting that although every member in 1835 is verified by name, nothing else is known about some of the members.

22. Moreover, we must be mindful of categories such as Chinese and Western medicine here. What I refer to as Chinese medicine in fact encompasses two traditions—the school of "old way" medicine (*ko'ihō*) and the school of "latter-day" medicine (*goseihō*)—both diverging from the medical tradition imported through Chinese texts and practiced in Japan. What came to be known as the school of "old way" medicine in Japan does not exist in China. Rather, in China the school of "latter-day" medicine tradition continued to thrive during the Ming (1368–1644 C.E.) and Qing dynasties (1636–1912 C.E.), thereby accentuating the differences between Chinese medicine in China and that in Japan. For detailed accounts of the history of Chinese medicine during the Tokugawa period, see Nihon gakushi'in and Nihon kagakushi kankōkai, eds., *Meijizen nihon igakushi*, 5 vols. (Tokyo: Nihon gakujutsu shinkōkai, 1955–1964); and Sakai, *Nihon no iryō-shi*. The fact that the school of "old way" medicine does not exist in China—in other words, it is a branch of medicine practiced only in Japan despite the foundational texts being from China—indicates the problem with identifying national categories with specific medical teachings. For the problems with such identification, see Joseph Alter, "Introduction: The Politics of Culture and Medicine," in *Asian Medicine and Globalization*, ed. Joseph Alter (Philadelphia: University of Pennsylvania Press, 2005), 1–20.

23. For instance, John Bowers writes, "By the middle of the nineteenth century, Western medicine had gained a firm foothold in Japan. A core of enlightened physicians understood its supremacy over *Kampo* [Chinese methods], and were eager to propagate its benefits." The assumption of incommensurability of Western and Chinese medicines aside, Bowers also ascribes a hierarchical order to them. John Bowers, *When the Twain Meet* (Baltimore: Johns Hopkins University Press, 1980), 7. For a history of Western medical knowledge in Japan during the seventeenth century, see Harold J. Cook, *Matters of Exchange: Commerce, Medicine, and Science in the Dutch Golden Age* (New Haven: Yale University Press, 2007), especially chapter 9, "Translating What Works: The Medicine in East Asia," 339–377.

24. In this respect, the "confrontational" encounter of the two medical traditions as described in John Bowers in *When the Twain Meet: The Rise of Western Medicine in Japan* seems to be grounded in a universalist assumption of their differences.

25. Itō Keisuke, draft of "Asai Tokuin ni atafuru sho" (A speech for Asai Tokuin), reprinted in Endō Shōji, *Honzōgaku to yōgaku* (Kyoto: Shibunkaku shuppan, 2003), 284.

26. Ibid., 283–284.

27. Ibid., 284.

28. Owari occupied the highest rank in the hierarchy of domains during the Tokugawa period; it was one of the *shinpan*, the domains whose lord was a blood relation of the shogun's family in Edo. Situated on the *Tōkaidō*, one of the main "highways" leading from Edo to the Imperial city of Kyoto, Owari's castle town flourished as a stronghold of commerce and Tokugawa power. The alternate attendance system (*sankin kōtai*) mandated by the Tokugawa shogun for more than two hundred lords (*daimyō*) across the country brought official travelers on their duty to Owari and, along with them, ideas and goods from far regions of the country. For detailed regional histories of the Owari domain, see Kishino Toshihiko, ed., *Owarihan shakai no sōgō kenkyū*, 3 vols. (Osaka: Seibundō shuppan, 2007); and *Owarihan shakai no bunka jōhō gakumon* (Osaka: Seibundō shuppan, 2002). The twenty-five-volume series *Nagoya sōsho*, published by Nagoya kyōiku i'inkai, contains reprints of publications from the Tokugawa period that provide rich material for the regional culture of Owari.

29. The castle town of Nagoya had five hundred physicians at the beginning of the century, to an estimated population of a little less than a hundred thousand, meaning the town would have had a few physicians on every city block. By mid-century this ratio of physicians to the general population was halved. See Sugimoto Isao, *Itō Keisuke* (Tokyo: Yoshikawa kōbunkan, 1960), 71.

30. The domain also looked for opportunities to recruit physicians from private practice (*machi'i*) to domainal positions (*han'i*) in accordance with their reputation and skill. Keisuke was himself hired by the domain in this fashion. He came from a family of physicians and was initially trained in *gosei-hō* tradition. After his six-month apprenticeship with Siebold in 1828, the twenty-six-year-old Keisuke returned to Owari and opened a clinic as one of the three *ranpō* physicians in a *machi'i* position. According to Yoshikawa, in 1835 a little more than half of the 120 *machi'i* physicians in Owari were products of the hereditary system. The other half came from merchant backgrounds. Yoshikawa, *Igaku yōgaku honzōgakusha no kenkyū*, 3–29. There is very little material documenting Keisuke's medical practice from this period. In his diary dated June and July 1838, for instance, Keisuke details the breadth of his interpersonal communications with various scholars of *materia medica* but says very little of his medical profession. During the period of the diary, Keisuke was accompanying a group of domainal officers on their travel to the *Kiso* forests. The group was sent there to select trees to cut and present to the shogun's castle in Edo, which burned in the third month of the year. Itō Keisuke, *Kinkaō nikki* (Nagoya: Keisuke monjo kenkyūkai, 1996), vol. 2, 3–75.

31. See *Oranda naigai yōhō*, n.p., in Waseda University Waseda University Japanese and Chinese Classics Database, http://archive.wul.waseda.ac.jp/kosho/ya04/ya04_01443/ (accessed May 13, 2010).

32. Shizan's thought on "Western medicine" (*ranpō*) is reproduced partially in Sugimoto, *Itō Keisuke*, 74.

33. Shibukawa Rokuzō, an astronomer at the official astronomical and calendrical bureau, pleaded in 1840 for granting of a monopoly governing publications on Western medicine. The shogunate eventually accepted a plea from the Hayashis to control publications on teachings on Western medicine in 1849.

34. In *Information for Nonspecialists in Regard to Cholera* (*Bakushabyō teate shirōto kokoroe sho*), Keisuke translated selected sections on the disease from *Volksgeneeskunde* by T. F. Osiander (1854) into Japanese.

35. The practice of placing *kunten* marks in classic Chinese texts was frequently employed during the Tokugawa period. It allows those with lesser knowledge of classical Chinese (the *lingua franca* of the time) to read by guiding rearrangement of Chinese texts into Japanese grammatical order.

36. Stauton's friend, the physician Alexander Pearson, gave the lecture in Guangdong in 1805. See Sōda, *Zusetsu nihon iryō bunkashi Zusetsu nihon iryō bunkashi*, 280. Angela Ki Che Leung offers a historical analysis of the reception and promotion of vaccination in Canton in "The Business of Vaccination in Nineteenth-Century Canton," *Late Imperial China* 29, no. 1 (2008), 7–39. For a detailed study on the various ways in which information related to vaccination was disseminated in Japan in the late Tokugawa Japan, see Ann Jannetta, *The Vaccinators: Smallpox, Medical Knowledge, and the "Opening" of Japan* (Stanford: Stanford University Press, 2007), particularly 78–101.

37. Keisuke's intellectual and professional investment in smallpox vaccination before vaccination itself was physically available places him within the national network of physicians articulated in *The Vaccinators* by Ann Jannetta.

38. For a range of translations on methods of vaccinations available around this time, see ibid., 149–154.

39. This announcement was made after the head of the shogunal medical academy, Taki Motokata, appealed to the shogunate; it was followed by another edict that centralized publications on *ranpō* medicine through the shogunal medical academy. See Sōda, *Zusetsu nihon iryō bunkashi*, 278.

40. Quoted in ibid., 75.

41. After this statement, Shizan is said to have hit Ishii Takuan, one of the domainal physicians, on the head. Quoted in Kishino, *Owarihan shakai no bunka jōhō gakumon*, 324.

42. Sugimoto, *Itō Keisuke*, 100.

43. Ibid., 103.

44. The absence of theorized erudition on smallpox in the archives of Keisuke suggests that his particular concern was limited to providing effective prevention rather than increased understanding of how vaccination works within the body.

45. Itō Keisuke, *Taisei honzō meiso*, in *Nagoya sōsho sanpen*, ed. Nagoyashi hōsa bunko (Nagoya: Nagoyashi kyōiku i'inkai, 1982), 311. Keisuke's intensive effort to disseminate knowledge of the Linnaean system is reflected in the various debts he shouldered in order to gather financial resources for the publication. For instance,

Keisuke placed a total of 1,061 *tsubo* (1 *tsubo* is about 3.3 square meters) for a credit of 328 *ryō*, including his home in Nagoya. To give a comparative picture of the magnitude of this debt, Keisuke paid 184 *mon*, 0.046 of 1 *ryō*, to stay at an inn, and 6 *mon*, 0.0015 of 1 *ryō*, for a sticky rice confection in 1827. These figures are taken from his "Accounting Book" ("*Kinsen shutsunō oboe*"), included in *Keihō yūki*, Keisuke's diary. See Itō, Keisuke, "Keihō yūki," In *Itō Keisuke nikki daiisshū*, edited by Keisuke monjo kenkyūkai (Nagoya: Nagoyashi higashiyama shokubutsuen, 1995), 78.

46. In this respect, Keisuke's actions echo those of other physicians of Western medicine (*ranpō*) from the same period. In his erudite essay "Ambiguous Encounters: Ogata Kōan and International Studies in Late Tokugawa Osaka," Tetsuo Najita investigates the role of the private academy in Osaka, *Tekijuku*, during the turbulent years of the Tokugawa period from 1838 to 1863. Najita illuminates how Ogata Kōan's commitment to teach and practice compassionate healing (*jinjutsu*) is in accordance with the teachings of ethics propounded by Itō Jinsai, who described compassion as a "never stable reference within the self, but rather manifested itself as an activity." Najita argues that Ogata Kōan, the founder of *Tekijuku*, "believed in the basic action ethics related to this idea, that to save others was a human imperative." Tetsuo Najita, "Ambiguous Encounters: Ogata Kōan and International Studies in Late Tokugawa Osaka," in *Osaka: The Merchants' Capital of Early Modern Japan*, ed. James McClain and Wakita Osamu (Ithaca, NY: Cornell University Press, 1999), 221, 213–240.

47. Endō, *Honzōgaku to yōgaku*, 278–283.

48. To this day history textbooks in Japanese high schools highlight *New Book on Anatomy* (*Kaitai shinsho*), Genpaku's translation of the Dutch anatomical book, as an innovative result of Western learning during the Tokugawa period. For a detailed chronology of dissections in the history of Japan before the Meiji period, see Ogawa Kenzō, "Meiji mae nihon kaibōgaku shi," *Meijizen nihon igakushi*, ed. Nihon gakushiin nihon kagakushi kankōkai (Tokyo: Shuppan kagaku sogō kenkyusho, 1978), vol. 1, 47–249. Timon Screech argues that "the actual human interior was accessible to hand and eye by means only of anatomy or surgery. Those two fields appeared for the time as serious clinical concerns in Japan in the late eighteenth century, and largely as a result of interventions by *Rangaku*-inspired doctors (Ran'i).... Neither anatomy nor surgery were much practiced in Japan, for cadavers were scarce and few wished to subject themselves to the agonies of surgery, which was still conducted without anesthetics. Western medicine was a discourse." Timon Screech, *The Western Scientific Gaze and Popular Imagery in Later Edo Japan* (Cambridge, UK: Cambridge University Press, 1996), 88–89. Although Western medicine played a pivotal role in inviting Japanese physicians such as Tōyō to see the interior of a human body for themselves, it is important not to overlook that these dissections took place in order to verify and criticize the school of Chinese medicine in which they were trained.

49. Sugita Genpaku, *Rangaku jishi*, in *Bunmei genryū sōsho*, ed. Hayakawa Junsaburō (Tokyo: Kokusho kankōkai, 1913), vol. 1, 13. For an English translation of Genpaku's book, see Sugita Genpaku, *Dawn of Western Science in Japan: Rangaku Kotohajime*, trans. Ryōzō Matsumoto and Eiichi Kiyooka (Tokyo:

Hokuseidō, 1969). Sugimoto Tsutomu, a noted historian of Western learning (*rangaku*), cautions that Genpaku's dramatic accounts of the beginnings of Western learning is biased in that Genpaku gives a misleading impression of the Tokugawa cultural and intellectual activities as "secluded." This perception was uncritically accepted and perpetuated by the later influential philosopher and intellectual historian Watsuji Tetsurō in his theory of Tokugawa Japan in *Sakoku-Nihon no higeki* (Sakoku—the tragedy of Japan). Sugimoto Tsutomu, "Sakoku ga souzō shita 'kindai'," and "'Rangaku jishi' fusetsu no koshō," in Sugimoto Tsutomu, *Edo yōgaku jijō* (Tokyo: Yasaka shobō, 1990), 16–26 and 157–179. Yōrō Takeshi, an anatomist and one of the most popular writers on issues of medicine and society in Japan today, argues that the engagements with anatomical study that began in eighteenth-century Japan stem from the articulation of differences between what is "artificial" (society) and "natural" (human body), a differentiation corresponding to Ogyū Sorai's concepts of "man" as having artificial principles that govern society and "nature" as having its own internal logic that does not correspond to rules created by men. Yōrō Takeshi, "Edo no kaibōzu," in *Edo no naka no kindai*, ed. Takashina Shūji (Tokyo: Chikuma shobō, 1996), 189–228.

50. The dissection took place at the *Rokkaku gokusha* prison. In fact, there is a commemorative stone at the site today, celebrating (ironically) "the advent of modern Japanese medicine." The body was of a man named Kutsuyoshi, a criminal who repeatedly swindled money. The actual handling of the body was done by a professional *fuwake*, and the dissection was attended by four others, among them Asanuma Sukemitsu, a sketch artist employed for this occasion. Yamawaki Tōyō, *Zōshi*, in *Nihon kagaku koten zensho*, ed. Saegusa Hiroto (Tokyo: Asahi shinbunsha, 1948), vol. 8, 146–203.

51. Ibid., 152–153.

52. He was able, however, to distinguish them at the second dissection he attended in 1762.

53. Ibid., 153. Sugita Genpaku identifies the imported book Tōyō owned as a book on anatomy by Johann Vesling (1598–1649), in his *Kaitai shinsho*.

54. Quoted in Ogawa Kenzō, "Meiji mae nihon kaibōgaku shi," in *Meijizen nihon igakushi*, vol. 1, 92–95.

55. Yamawaki, *Zōshi*, 156.

56. Sugita, *Rangaku jishi*, 13–14.

57. The original Japanese for the word "dissection" is *fuwake* in this quote. The custom set forth by the shogunate was to have *eta*, the outcast group, perform *fuwake*, or dissection. On the day of this particular dissection, an *eta* named Toramatsu was assigned the task, but Genpaku recalls that Toramatsu was unable to come due to illness, and instead Toramatsu's ninety-year-old grandfather took his place. For historical and political establishment of the outcast during the Tokugawa period, see Daniel V. Botsman, *Punishment and Power in the Making of Modern Japan* (Princeton: Princeton University Press, 2005), especially 50–58.

58. The Dutch translation of Kulmus's book is titled *Ontleedkundige Tafelen*, translated from German by G. Dicten.

59. Sugita Genpaku, Nakagawa Jun'an, Ishikawa Genjō, and Katsuragawa Hoshū, eds., *Kaitai shinsho* in *Nihon kagaku koten zensho*, ed. Saegusa Hiroto (Tokyo: Asahi shinbunsha, 1948), vol. 8,236–241.

60. The most efficient way to learn Western medicine was to study directly with the few Western physicians themselves, and Dejima, where the Dutch were confined to live, attracted many Japanese who sought such opportunities. Naritakejuku, a private academy of Western medicine that Siebold founded in Dejima in 1824, is the prime example. The fact that Siebold himself required his students to write a "dissertation" also suggests that even at his own academy of Western medicine the textual format was prioritized over actual practice. Sakai, *Nihon no iryōshi*, 284–312; Jannetta, *The Vaccinators*, 90–92. For the interaction and influences of Western physicians before Siebold, see Sakai, *Nihon no iryōshi*, 251–261, 285–298; and Sōda, *Zuhan nihon iryō bunka shi*, 255–256.

61. Yoshikawa Yoshiaki, *Igaku yōgaku honzō gakusha no kenkyū* (Tokyo: Yasaka shobō, 1993), 154.

62. The *bakufu* purchased the eight-volume encyclopedia in Dutch from Captain Hendrik Doeff in 1810. Other copies of the book, moreover, had been circulating among the Western learning (*rangaku*) scholars since the late 1780s. For a detailed historical background of the inception of the translation project, see Sugimoto Tsutomu, *Edo jidai seiyō hyakkajiten: "kōseishinpen" no kenkyū* (Tokyo: Yūzankaku shuppan, 1999), 40–51.

63. For a brief overview, see Jirō Numata, "The Introduction of Dutch Language to Japan" *Monumenta Nipponica* 19, no. 3, 243–253.

64. This includes the arrival of a Russian ship in 1792, when the Russian naval officer Adam Laxman returned a Japanese castaway Daikokuya Kōdayū and demanded a commercial treaty from the *bakufu* in return. In 1796, a British naval officer named Broughton arrived at what is today Muroran in Hokkaidō, in an attempt to gather more information on their coastal geography. In 1797, the Dutch-chartered American ship *Eliza* arrived in Dejima, followed by the 1799 arrival of *Franklin*. The American whaling ships also asked for fair treatment of their shipmen in case of shipwreck. Another Russian, Nicolai Petrovich Rezanov, led a fleet to Dejima in 1804. See Ronald Toby, *State and Diplomacy in Early Modern Japan: Asia in the Development of the Tokugawa Bakufu* (Princeton: Princeton University Press, 1984), 231–246. For a detailed analysis of the state of military forces that were placed by the *bakufu* around Dejima from 1789 to 1818, see Kajiwara Yoshinori, "Kanseibunka ki no Nagasaki keibi to fētongōjiken," *Fukuokadaigaku jinbun ronsō* 37 no. 1 (2005), 339–377.

65. Udagawa Genshin was taken in once by Sugita Genpaku, one of the founders of Western learning (*rangaku*). However, he was turned away from the family for tardiness and an inappropriate attitude. He was adopted into the Udagawa family in 1798. The Udagawas joined the translation team when Baba left to facilitate negotiations with the Russians in Ezo (present-day Hokkaidō). This office was established as part of the *bakufu* Office of Maps and Calendars (*tenmon gata*). Sugimoto Tsutomu speculates that the translation group slowly dissolved around 1845, and

Yōan's death in 1846 marked the decisive loss of an energetic member. Sugimoto, *Edo jidai seiyō hyakkajiten*, 65. Yōan's adopted son, Kōsai became involved in 1846. Kōsai was a biological son of Iinuma Yokusai, an affiliate of the Shōhyaku-sha.

66. Ibid., 52–61, 65. Of course, as Sugimoto asserts, these numbers are not absolute, since the translators and editors of some articles are unknown and the length and depth of the articles vary. But they do serve as an indication of the extent of the involvement of the Udagawa family in this project.

67. *Bangosen* (http://archive.wul.waseda.ac.jp/kosho/bunko08/bunko08_c0514/bunko08_c0514_p0039.jpg, accessed June 1, 2010); *Ransetsu benwaku* (http://archive.wul.waseda.ac.jp/kosho/ni01/ni01_00511/ni01_00511_p0057.jpg, accessed June 1, 2010); Sugita Genpaku, *Rangaku jishi*, 9. Shiba Kōkan, *Shunparō hikki* in *Nihon zuihitsu taisei*, ed. Nihon zuihitsu taisei henshubu (Tokyo: Yoshikawa kōbunkan, 1975), 22. Sugita Genpaku referred to camera obscura as *anshitsu shashinkyō* (literally, "darkened room *shashin* lens") in his *Rangaku jishi*. Morishima Chūryō's *Lexicon* translated *daguerreotype* as *inshō kyō* (impressed-shape lens).

68. Yabe Ichirō, "Honzō kara kindai shokubutsugaku e—'botanika kyō' to 'shokugaku dokugo,'" *Shokugaku keigen*, ed. Yabe Ichirō (Tokyo: Kōdansha, 1980), 253. In the draft, the Western words are rendered in transliteration using *katakana* script.

69. Sugimoto, *Edo jidai seiyō hyakka jiten*, 491. I am using the spelling as it appeared in this Dutch translation of Chomel's encyclopedia to maintain the historical spelling and am aware of the alternative spelling "Cruydeboek" from the eighteenth century and a contemporary Dutch spelling "Kruidenboek."

70. Noel Chomel, *Algemeen huishoudelijk-, natur-, zedekundig-en konst-/woordenboek* (Leiden: Joh le Meir, 1778), 1651.

71. Uchiyama Jun'ichi, *Edo no kōkishin: bijutsu to kagaku no de'a'i* (Tokyo: Kōdansha, 1996).

72. Naruse Fujio, "Edojidai no seiyōgaron ni tsuite-sono toyōshisō tono kankei," *Bijutsushi* 85 (1972), 1–17.

73. Ibid., 10.

74. Ibid., 11.

75. These concatenations are used extensively in discussion of early Meiji paintings as well. Employing these two pillars of "utilitarianism" and "prioritizing the actual" put forth by Naruse analytical key terms in examining imported illustrations as well as Western-style pictures from the Tokugawa period, many scholars have relied on and fitted these concepts to their analysis in navigating historical materials. The 1985 exhibition *Shajitsu no keifu* at the National Museum of Modern Art typifies my point. Haga Tōru, "Takahashi Yuichi to riarisumu," *Mizue* 713 (1979), 58–64. While trying to clarify the process and the degree of their interactions, Imahashi Riko thoughtfully historicizes the concepts of *shashin* and realism in her rigorous work *Edo no kachōga: Hakubutsugaku wo meguru bunka to sono hyōshō* (Birds and flowers: The representation of natural history in the Edo period) and equates the illustrations in imported text on natural history to the concept of *shashin* along with the concepts of *shasei* and *seisha* circulating in artistic discourse.

She reasons that "because illustrations of *hakubutsugaku* [a neologism of *materia medica*] also originate from the desire to transcribe pictorially a being as it exists in nature," these concepts can be understood under the same umbrella of *shashin*, *shasei*, and *seisha* of the Rimpa school of painting. Imahashi's thorough analysis of the process of copying and use of the copied images as sketches illuminates the absence of differentiation between "copies" and "sketches." Imahashi Riko, *Edo no kachōga: Hakubutsugaku wo meguru bunka to sono hyōshō* (Tokyo: Sukaidoa, 1995), 106–107. Imahashi's interpretive analysis emphasizes the subjectivity of each painter discussed, and in that sense it responds directly to Suzuki Susumu's call to examine the interaction between the Japanese artistic work and the imported Western illustrations through the lens of active subjective selections, rather than that of the passive "epigone." Suzuki Susumu, "Kinsei shaseiga no keifu—nihonga no shasei shugi," in *Kinsei shaseiga no keifu*, ed. Kanagawa Prefectural Museum (Yokohama: Kanagawa Prefectural Museum, 1971), 5–8.

76. Kōno classifies the meanings of *shasei* in the Tokugawa period into four major (but not absolute) conceptual categories: (1) *shōi shasei* refers to the attempt to grasp and sketch the interiority of the object of depiction, but it does not necessarily need to occur simultaneously with the act of observation; (2) *kyakkan shasei* signifies depiction of the objective accuracy of the object; (3) *seimitsu shasei* refers to detailed and fine depictions; and (4) *taikan shasei* is a concept equivalent to "sketch" in our contemporary usage that signifies the act and work of representation created while looking at the object of depiction. Kōno asserts that although these four categories of the meaning of *shasei* overlap with embedded meanings in Chinese *xiesheng*, the employment of this term in Japan presents more ambiguity than in China, and thus its applicability reaches wider genres of painting. Furthermore, Kōno argues that an imported Chinese concept even more widely applied than *shasei* is *shashin*. Kōno's insightful observation signals the elasticity of the concept of *shashin*, suggesting that its employment in Tokugawa artistic discourse is not sufficiently explained by following the set formula of East and West; rather, it should be studied by considering the multilevel interactions of concepts, traditions, and uses that pictorial representations affected. Kōno Motoaki, "Edojidai 'shasei' kō," in *Nihon kaigashi no kenkyū*, ed. Yamamoto Yunezō sensei kokikinenkai (Tokyo: Yoshikawa kōbunkan, 1989), 388–427; idem, "'Shasei' no Gensen—Chūgoku," in *Akiyama Mitsukazu hakushi koki kinen bijutsu ronbunshū* (Tokyo: San'ichi shobō, 1991), 481–514; idem, "Edo kaiga to kyakkan shugi," in *Edo no nakano kindai—Akita ranga to 'kaitai shinsho,'* ed. Takashina Shūji (Tokyo: Chikuma shobō, 1996), 231–288.

77. In this light, I am equally wary of applying the notion of objectivity formulated and structured through Western scientific studies to the cases I examine in this book. Articulation of "subject" or "subjectivity" in the context of Tokugawa Japan requires thorough and careful consideration of writings in fields of intellectual, social, and religious history that lie outside the scope of the present work. Murakami Yōjirō, one of the pioneers of history and philosophy of science in Japan, has written extensively on the entrenchment of the concepts of "subject" and "object" embedded in modern scientific and social theories of Japan. See, for instance, Murakami

Yōichirō, *Nihonjin to kindaikagaku* (Tokyo: Shinyōsha, 1979). In his study of how the notion of the "West" (*ran*) was articulated and represented in the popular culture of late Tokugawa Japan, particularly through the faculty of sight, Timon Screech argues: "Objectivity was strenuously held to be possible in Europe, and consequently also argued for by *Rangaku* (Western learnings) commentators, and yet in the Japanese context such a notion might carry rather little metaphysical ballast, and even seem quaint. The very sense of objectivity had few resonances in Japanese thought." Screech, *The Western Scientific Gaze*, 172. Lorraine Daston and Peter Galison offer a detailed analysis of the process through which Western scientific discourse has embraced, adjusted, and identified the notion of "objectivity" since the seventeenth century; Daston and Galison, *Objectivity* (New York: Zone Books, 2007).

78. Satō Dōshin, *Meiji kokka to kindai bijutsu* (Tokyo: Yoshikawa kōbunkan, 1999).

79. Ibid., 216–217.

80. Excavating a pictographical history of the ideograms, Satō contends that although the concepts of *shin*, *sei*, and *i* overlap in what he terms "perceptual realism," the concept of *shin* differs from others in that it is historically impregnated with the connotations of universality extended beyond spatial and temporal specificities. On the other hand, the concept of *sei* points to the livelihood of the living entity, and *jitsu* to the objecthood of the represented, which is confined to the spatially and temporary determined specificity of the depicted object. He further considers the religious connotations carried by the concept of *shin* and argues that the concept of *shashin* encompasses inclinations toward sacralization; ibid., 218–222. Satō's point here is particularly insightful and provocative in regard to implementation of *goshin'ei*, photographic reproductions of a copper-etching portrait of the Meiji emperor in the early Meiji period.

81. The notion of principle is a product of Yōan's interpretive translation; it does not appear in the Dutch translation. I thank Christian de Pee for his help in translating the Dutch.

82. Ibid., 384.

83. See *Algemeen huishoudelijk-, natur-, zedekundig-en konst-/woordenboek, A-D* (Leiden: Joh le Meir, 1778), 331.

84. The literal translation of *matō* is "magic light." Sugimoto, *Edo jidai seiyō hyakka jiten*, 528.

85. For resources on magic lantern shows during the Tokugawa period, see Iwamoto Kenji, *Gentō no seiki—eiga zen'ya no shikaku bunkashi* (Tokyo: Shinwasha, 2002), 85–111; and the exhibition catalog, Tokyo Metropolitan Museum of Photography, *The Advent of Photography in Japan* (Tokyo: Tokyo Metropolitan Museum of Photography, 1997), 43–45.

Chapter Two

1. Lydia H. Liu, *Translingual Practice* (Stanford: Stanford University Press, 1995), 16.

2. Michel Foucault, *The Order of Things: An Archaeology of the Human Sciences* (New York: Vintage Books, 1994), 132.

3. Ibid., 129–130.
4. Ibid., 157.
5. I am fully aware that translation of *shin* as "the real" conjures a variety of associations, especially in relation to photographic history and theory. I therefore want to clarify that my use of the "real" in this project refers, first and foremost, to tangible objects. Whereas Lacan's "touché" originates from psychoanalytical and subjective, and thus intangible responses, the discourse of *materia medica* grapples with a dichotomy between the real and the imposter or forgery, both of which are physically extant. This sense of the real as antithetical to the fake differs from the Lacanian notion of "touché," and thus consequentially from Roland Barthes's application of the concept to photographic theory explored in his *Camera Lucida*. See Jacques Lacan, "Touché and Automaton," in *The Four Fundamental Concepts of Psychoanalysis*, trans. Alan Sheridan (New York: Norton, 1978); and Roland Barthes, *Camera Lucida: Reflections on Photography*, trans. Richard Howard (New York: Hill and Wong, 1981).
6. Liu, *Translingual Practice*, 35.
7. The first shogun of the Tokugawa, Tokugawa Ieyasu, is said to have regarded *Compendium* as one of the essential texts in his collection. Moreover, throughout the Tokugawa period, a total of fourteen versions were published in Japan. Isono Naohide classifies these copies into three major types according to the year of publication: (1) a Kanei version, which includes six types, published in 1634; (2) a Manji version, published in 1660, which includes three types; and (3) a Kanbun version, published in 1672, which includes five types. Each of these versions supplements information lacking in the original *Compendium*. Isono, *Nihon hakubutsugakushi nenpyō*, 121. On the process of importation of *Compendium* from China, see Yabe Ichirō, *Edo no honzō* (Tokyo: Saiensusha, 1984); Sugita Tsutomu, *Edo no hakubutsugakushatachi* (Tokyo: Seidosha, 1985); Ueno Masuzō, *Nihon hakubutsushi* (Tokyo: Heibonsha, 1986); Kimura Yōjirō, *Edoki no nachurarisuto* (Tokyo: Asahi shinbunsha, 1988); Shimonaka Hiroshi, ed., *Saishoku edo hakubutsugaku shūsei* (Tokyo: Heibonsha, 1994); and Doi Yasuhiro, *Honzōgaku sha Hiraga Gennai* (Tokyo: Kōdan sha, 2008), 12–31.
8. For a detailed study of Ekken's thought and intellectual influences, see Mary Evelyn Tucker, *Moral and Spiritual Cultivation in Japanese Neo-Confucianism: The Life and Thought of Kaibara Ekken, 1630–1740* (Albany: State University of New York Press, 1989).
9. Carla Nappi's exploration of Li Shizhen's *Compendium* exposes a fascinating cultural history of Li's own experiments with "textual, physical, and perceptual manifestations" in compiling this book. By avoiding comparison to more familiar European works on natural history, Nappi's study of *Compendium* underscores the intellectual merit of approaching this canonical work as a product deeply embedded in the cultural and historical context of early modern China. Carla Nappi, *The Monkey and the Inkpot: Natural History and Its Transformations in Early Modern China* (Cambridge: Harvard University Press, 2009).
10. Mary Evelyn Tucker's analysis of Ekken's oeuvre focuses on the underpinning ethical currents in his thought. My interest in Ekken's work here is limited to making intellectual connections between the concerns expressed by Ekken and

those of the Shōhyaku-sha, particularly in regard to the paramount importance of exploration and investigation of locally available natural specimens; see Tucker, *Moral and Spiritual Cultivation.*

11. Of the 1,362 items included in *Materia Medica in Japan*, 358 do not have Chinese names, implying that they were not described in the imported Chinese texts. Ekken also listed the items whose Chinese names were uncertain. Kaibara Ekken, "Honzō no sho wo ronzu," in *Yamato honzō*, vol. 1, annotated by Shirai Kōtarō (Tokyo: Ariake shobō, 1983), 5.

12. The inadequacy of *Compendium* in the Japanese context that Ekken points out is also reflected in the categories he used in *Materia Medica in Japan*. Ekken divided objects into thirty-four main categories, as opposed to the sixty-one used in *Compendium*. Ibid., 5. For a detailed study of categories used in *Compendium* and other Chinese *honzō* books, see Yamada Keiji, "Honzō ni okeru bunrui no shisō," in *Higashiajia no honzō to honzōgaku no sekai*, ed. Yamada Keiji (Kyoto: Shibunkaku shuppan, 1995), vol. 1, 3–42.

13. The emphasis that Li Shizen placed on the ever-changing life force in *Compendium*, as well as Ekken's absorption of Chu Hsi's Neo-Confucian thought, especially the identification of nature as an ever-changing, all-encompassing, and infinite resource of knowledge, underlies Ekken's supposition. Nappi, *The Monkey and the Inkpot*; Tucker, *Moral and Spiritual Cultivation.*

14. Kaibara, "Honzō no sho wo ronzu," vol. 1, 5.

15. Mizutani Hōbun, *Buppin shikimei*, in *Nagoya sōsho sanpen*, ed. Nagoya hōsa bunko (Nagoya: Nagoyashi kyōiku iinkai, 1982), vol. 19, 25–26.

16. *Nominal Differentiations* consists of these sections: an introduction by a member of the Shōhyaku-sha, Ishiguro Sei'an; an introduction by Ōkōchi Zonshin, Keisuke's brother and a member of the group; an introductory remark by Keisuke asserting briefly the difference and importance of the Western theory of botany; a note that outlines the methodologies of translation and the structure of the listing section; a table of contents in the order of the Japanese style syllabary (*iroha*); a list of Linnaean nomenclature by category; the listings of plants in Roman alphabetical order; and closing remarks by Mizutani Hōbun. There is a supplementary volume to the text, and it consists of four more sections: an introduction by Keisuke; a list of common names for plants in the order of the Roman alphabet and their corresponding names in Chinese; an introductory guide to the principal structures of the Linnaean nomenclature; and a concluding remark by Yoshio Jō'san, a member of the group and the *ranpō* physician.

17. Yokusai's book, in fact, played a seminal role in modern botanical studies. It was republished, for instance, in 1875 by Keisuke with the help of Tanaka Yoshio and Ono Motoyoshi. In 1913, Makino Tomitarō, the pioneer of modern botany in Japan, re-edited the ten volumes on the section on trees that Yokusai had finished in the draft but not published, and he published them along with his commentary.

18. The historical Linnaean binominal scheme has been modified since its inception. The tier of "family" was added in 1800, and that of "phylum" much later.

19. Lisbet Koerner, *Linnaeus: Nature and Nation* (Cambridge: Harvard University Press, 1999), 15–16.

20. Itō, *Taisei honzō meiso*, 473.

21. For instance, although the Chinese character for chrysanthemum indicates that the plant is a grass (by the grass radical that makes up the character), not all names ascribed to chrysanthemum contain this character. Indeed, the regional names of the plant often give no indication of which categories an object belongs to. Take the case of the common *yomena* plant, an edible chrysanthemum available in Owari and identified in Chinese texts as *keijichō* (Ch: *jierchang*). The Chinese characters that convey *keijichō* signify chicken, child, and intestines. In other words, the semiotic components of the Chinese characters used in this name do not point to grass or chrysanthemum, either collectively or individually. If one did not know the plant beforehand and its name, it would be impossible to know from the characters on paper that it is a kind of chrysanthemum—or even a plant. In Keisuke's *Nominal Differentiations*, this plant is listed as "Aster indicus. Linn," belonging to the genus of "Aster" with "indicus" as its species name. Ibid., 362.

22. Ibid., 294–303.

23. Sugimoto Kaoru, *Itō Keisuke* (Tokyo: Yoshikawa Kōbunkan, 1960). On his way back from Dejima, Keisuke was robbed. However, Thunberg's *Flora Japonica* was not stolen because supposedly the robber did not see any monetary value in this book.

24. For Thunberg's activities in Japan and the significance of his botanical contribution, see Timon Screech, *Japan Extolled and Decried: Carl Peter Thunberg and the Shogun's Realm, 1775–1796* (London: Routledge, 2005); and Richard Rudolph, "Thunberg in Japan and His *Flora Japonica* in Japanese," *Monumenta Nipponica* 29, no. 2 (1974), 163–179.

25. Ōba Hideaki, ed., *Nihon shokubutsu kenkyū no rekishi—Koishikawa shokubutsu'en sanbyakunen no ayumi* (Tokyo: University of Tokyo Press, 1996), 69–72. Siebold also used Thunberg's *Flora Japonica* as the basic resource for his botanical study of Japan. Kimura Yōjirō, "Shokubutsu no shoku to shu ni tsuite," in *Higashi ajia no honzō to hakubutsugaku no sekai*, ed. Yamada Keiji (Kyoto: Shibunkaku shuppan, 1995), vol. 1, 62.

26. Itō, *Taisei honzō meiso*, 312.

27. These "sages" include Conrad Gesner, Robert Morison, and John Ray.

28. For instance, Yōan states that "the vein on the leaves are similar to nerves, the leaves are like lungs on the human body, the roots of the plants are like stomachs, and flowers are like genital areas." Udagawa, *Fukkō botanika kyō*, 584.

29. Keisuke sent a draft of *Nominal Differentiations* to Siebold in order to receive Siebold's opinion on some of the nomenclature in the book. Keisuke explains in the introduction that some of the identifications of Japanese names by Thunberg were wrong or not entirely certain. For Keisuke, *Flora Japonica* therefore served as the base from which he could draw a foundational structure and information, but not a document that could be entirely trusted. The draft of *Nominal Differentiations* with Siebold's markings is in the collection of the National Diet Library in Tokyo

and also at the library of University of Leiden. The copy at the National Diet Library is accessible online (http://www.ndl.go.jp/nature/thum/009.html; accessed May 13, 2010).

30. Keisuke assigned a number to each of the sixty-six possible combinations of twenty-four classes and their orders. The number fifty-three, for instance, corresponds to the sixteenth class and the seventh order, which groups plants that have more than nineteen stamens. Keisuke, *Taisei honzō meiso*, 343–352.

31. In fact, Keisuke compared *Flora Japonica* to *Clarification* during his stay in Nagasaki with Siebold in order to grasp the Linnaean system fully; see Kimura Yōjirō, *Edoki no nachurarisuto*, 211–212.

32. Keisuke uses the term *Hanburugu ben*, "Hamburg dialect," in characterizing the Roman alphabetical names in the appendix. Keisuke comments that he has read a book by a man named Hamburg in which names of plants were listed in dialects as well as several European languages. Itō, *Taisei honzō meiso*, 325–325, 453–455.

Chapter Three

1. Early displays took place at the *Igakukan*, the domainal medical school in Owari; later events were hosted at Keisuke's private garden, known as *Kyoku-en*. *Nagoya no hakurankai*, exhibition catalogue (Nagoya: Nagoya City Museum, 1982), 11–26.

2. The medical institute was renamed Igakukan in 1831, when it was administered by Asai Shizan. During the tenure of Asai Tei'an, the academy was known as Seikandō. Endo, *Honzōgaku to yōgaku* (Kyoto: Shibunkaku shuppan, 2003), 257.

3. *Honzō kai* hosted by *Igakukan* took place on the fifteenth day of the sixth month every year, beginning in 1832, and included approximately ten thousand items every time. Odagiri Shunkō, *Owari meisho zue* (Nagoya: Hishiya Kyubei, Katano Toshio, 1844). Kishi Masahiro discusses the relevance of the miscellaneous section for the nature of modern museum culture in "Owari igakukan yakuhine mokuroku—'zappin' tenji ni miru hakubutsukan no genten," *Nagoyashi hakubutsukan kiyō*, vol. 17 (1994), 19–26. It is uncertain which year of the *Igakukan honzō kai* this woodblock depicts. Kishi speculates the date is somewhere between 1836 and 1844. Furuta Ryō's essay on the origin and the development of art exhibitions chronicles and juxtaposes various types of gatherings during the Tokugawa and early Meiji periods. Furuta Ryō, "Nihon no bijututenrankai sono kigen to hattatsu," *Museum* 545 (December 1996), 29–56.

4. For a detailed account of the process of importing and learning to grow ginseng in Japan under the reign of the eighth shogun, Yoshimune, see Kazuhiko Kasaya, "The Tokugawa Bakufu's Policies for the National Production of Medicines and Dodonaeus' Cruijdeboeck," in *Dodonaeus in Japan*, eds. W. F. Vande Walle and Kazuhiko Kasaya (Leuven: Leuven University Press, 2001), 167–186.

5. The breadth of items displayed at these *honzō kai* occasions indicates that *materia medica* defies any facile categorizations that might be offered by modern disciplinary practices and demands. This historical material requires instead a more synthetic and inclusive approach that recognizes and accepts the historical reality

of study of *materia medica* attested by Figure 3.1, in which exhibited items included living animals, skeletal models, and minerals that were explored as part of *episteme* of *materia medica*. In other words, rather than approaching these occasions merely as physical manifestations of the Shōhyaku-sha's interests, we must posit *honzō kai* as a site of practice that allowed epistemological tensions to arise, dissolve, or manifest themselves. It is for this reason that throughout the project I privilege the term *materia medica* (Jp: *honzōgaku*) to convey its intertwined relation to *Compendium*. Even though some contemporary scholars have used *hakubutsugaku* (encyclopedic study) to characterize these gatherings and indeed the activities of the Shōhyaku-sha, recognition of the encyclopedic as a field of study began to appear only during the Meiji period in order to specifically indicate "Western" elements included in the *materia medica* materials from the Tokugawa period. Contemporary historian Kojima Kaoru argues that it was in Itō Keisuke's 1879 essay "General History of *hakubutsugaku* in Japan" ("*Honpō hakubutsu kigen enkakusetsu*") that the field of encyclopedic study is explicitly and decidedly projected retroactively to the study of *materia medica*. In other words, to apply the term *hakubutsugaku* retrospectively to historical activities such as those of the Shōhyaku-sha places a localized highlight on "Western" elements of their study, and it draws an arbitrary line within the multifarious fields of inquiry and historical materials. Kojima Kaoru, "Hattori Sessai, hakubutsu zufu no meishu," in *Bakumatsu Meiji no gakatachi*, ed. Tsuji Masao (Tokyo: Perikansha, 2008), 163–196. It is interesting to note that in China, too, the term *bowu xue* (Jp: *hakubutsugaku*) was a neologism that appeared in the midnineteenth century. See Fa-ti Fan, *British Naturalists in Qing China: Science, Empire, and Cultural Encounter* (Cambridge, MA: Harvard University Press, 2004).

6. From Plato's cave to Descartes's blind man with a stick to Locke's *tabula rasa*, many thinkers have proposed innumerable theories on the relationship between vision and cognition, between image and perception, and of vision itself in Western philosophy. Although I admit that the temptation exists to analyze the heavy reliance on perception by the Shōhyaku-sha as a critical tool of epistemological negotiation in light of Locke's notion of "human understanding," for instance, I choose to resist it for this project. The main reason for my reluctance is that the position of the "self" and human beings imagined, articulated, and practiced in Tokugawa Japan did not concur with the fixed (God-above and later Cartesian) and dichotomized (Nature versus Culture) framing of the world that underpins these Western theories. David Lindberg, *Theories of Vision from Al-kindi to Kepler* (Chicago: University of Chicago Press, 1976); Krzysztof Pomian, "Vision and Cognition," in *Picturing Science Producing Art*, eds. Caroline A. Jones and Peter Galison (New York: Routledge, 1998), 211–231.

7. Historically, various types of collective gatherings and exhibitions were practiced during the Tokugawa period, including banquets at which famous artists and writers would perform (*shoga kai*), showings of rare religious icons (*kaichō*), street spectacles (*misemono*), and *honzō kai*. Peter Kornicki, "Public Display and Changing Values: Early Meiji Exhibitions and Their Historical Precursors," *Monumenta*

Nipponica 49, no. 2 (1994), 167–196. Andrew Markus, "Shogakai: Celebrity Banquets of the Late Edo Period," *Harvard Journal of Asiatic Studies* 53, no. 1 (1993), 135–167. Especially in the Owari domain, these types of public display took place rather frequently, as the diaries of Kōriki Enkōan and Hosoe Yōsai indicate. Kōriki Enkōan was a samurai by origin. A Neo-Confucian scholar, Hosono Yōsai left a fifty-four-volume diary that dates from 1836 to 1879. See Nagoyashi Kyōiku i'inkai, ed., *Nagoya sōsho*, vol. 19, *Zuihitsuhen. Kankyō manhitsu* (Nagoya: Nagoya Kyōiku iinkai, 1960). The *honzō kai* differ from other collective gatherings specifically in its articulated purpose and intended outcome. This is not to say, however, that there is no overlap of items between these various types of exhibition practices. The case of the horseshoe crab showcases my point. Mizutani Hōbun and Ishikawa Hachita submitted horseshoe crabs to the *honzō kai* hosted by Ōkōchi even though just three months earlier a horseshoe crab was the highlight of a spectacle (*misemono*) on *Hirokōji* street at the center of the castle town. My point here is that the overarching epistemological framework that motivated and necessitated the *honzō kai* gatherings is distinctly different from that of spectacle shows. As this chapter explores, the *honzō kai* aimed to discriminate the real from the false; emphasis was placed on rarity and novelty in street spectacles.

8. For instance, in Owari in 1735 the domainal garden adjacent to the Owari Castle received seven samples of Korean ginseng from Shogun Yoshimune, and by 1745 nearly nine thousand ginseng plants were growing in the garden. One prime example of the impact of Yoshimune's support of the pursuit of *materia medica* is the loosening of regulations on importation of Dutch goods and ideas. Ueno Masuzō, *Kusawo tenishita shōzōga* (Tokyo: Yasaka shobō, 1986), 216–221; Endō, *Honzōgaku to yōgaku*, 19. For a chronological list of *honzō kai*, see Isono, *Nihon hakubutsushi nenpyō*, 761–770.

9. According to a Shōhyaku-sha account provided by Keisuke, moreover, the group's *honzō kai* took place on the seventh of every month. Itō Keisuke, "Owari hakubutsugaku shōhyaku-sha soshienkaku narabi shosentetsu rireki zakki," reprinted in Yoshikawa Yoshiaki, *Igaku yōgaku honzōgaku no kenkyū* (Tokyo: Yasaka shobō, 1993), 30–33.

10. Timon Screech characterizes these *honzō kai* as pharmacopoeial assemblies and discusses the extent to which the imported items from Holland and the West penetrated these gatherings. *Western Scientific Gaze in Later Edo Japan* (Cambridge, UK: Cambridge University Press, 1996), 137; Isono Naohide "Yakuhinkai, honzō kai," *Keio gijuku daigaku hiyoshi kiyō* (Shizenkagaku), no. 29 (2001), 6–14. For *honzō kai* organized by Yamamoto Reading Room Group (*Yamamoto Dokushoshitsu*), see the *Dokushoshitsu honzō kai mokuroku* in the collection of Kyōu Shoya, Takeda Science Foundation, Osaka. Kiyono Kenji's brief section on various *honzō kai* is also useful. Kiyono Kenji, *Nihon kōkogaku jinruigakushi* (Tokyo: Iwanami shoten, 1955), vol. 1, 400–406. Also see *Rūtsu Nihon no hakubutsukan* (Osaka: Osaka Municipal Museum, 1979).

11. Hiraga Gennai, *Kyokuzan sensei bunkairoku*, in *Hakubutsukan tanpenshū*, edited by Ueno Masuzō, vol. 45 of *Edo kagaku koten sōsho* (Tokyo: Kōwa shuppan, 1982).

12. Although Gennai's teacher, Tamura Ransui, is said to have organized the "first" *honzō kai* in 1757, recent scholarship has demonstrated the need to reexamine such a claim. See Endō, *Honzōgaku to yōgaku*, 57–69.

13. The list of regulations is reprinted in Kiyono, *Nihon kōkogaku jinruigaku zenshi*, vol. 1, 405–406.

14. In this respect, I disagree with W. F. Vande Walle's assessment that it was the West that introduced observation as an integral component of *honzōgaku*. Vande Walle claims that Dodonaeus's book on herbals, which was initially imported in 1659 but had not been studied until 1717 by the Shogun Yoshimune, who "introduced . . . scientific knowledge about nature based on observation." W. F. Vande Walle, introduction to *Dodonaeus in Japan*, eds. W. F. Vande Walle and Kazuhiko Kasaya (Leuven: Leuven University Press, 2001), 17. Characterizing Japanese *materia medica* (*honzōgaku*) as "the standard equivalent to the Western concept of 'natural history,'" Vande Walle also claims that Yoshimune's "rediscovery" of Dodonaeus's book and his subsequent order to study the book created a situation in which "the East and the West for the first time caught up with each other." The underlying logic presumes that "the East" had been "behind" "the West," and that history is best understood as unidirectional and sequential movement toward "progress."

15. It is worth noting here that the systematic categories established in *Compendium* went largely unquestioned from the late eighteenth century until after the Meiji Restoration of 1868. W.J.T. Mitchell's theorization of "media" is particularly useful for shedding light on the complex issues in addressing the question of media; W.J.T. Mitchell, *What Do Pictures Want?* (Chicago: University of Chicago Press, 2005), especially 201–221.

16. The decision to focus on *Compendium* visually is not unique to Shōhyakusha practice. Rather, it is a telling example of the depth and magnitude of the influence Ekken cast in formulating *honzōgaku*. For instance, the visual testimony of *Compendium* was also presented at several *honzō kai* hosted by the Yamamoto Reading Room Group in Kyoto, a private academy of *honzōgaku* scholars. For a more detailed discussion of Yamamoto Reading Room Group, see Yamamoto Bōyō, Endō Shōji, and Nanba Tsuneo, eds., *Hyappinkō* (Tokyo: Kagaku shoin, 1982).

17. The materials displayed at *honzō kai* share similarities with cabinets of curiosities that emerged during the Age of Enlightenment in respect to the types of objects included. However, it is also important to recognize that *honzō kai* were always temporary and were not considered to be physical repositories of objects. Rather than being an isolated and insulated space that symbolically expressed the wealth, knowledge, and power of a particular individual or institution, as was the case in Europe, the *honzō kai* was a public exhibition event that sought to bring together knowledge and objects for the purpose of developing knowledge of *materia medica*. Another element that differentiates the activities of the Shōhyaku-sha from the Renaissance cabinet of curiosity is the notion of "containment." The Shōhyaku-sha's lack of overt interest in putting themselves in the ratifying position (of "authority" of all knowledge associated with *materia medica*, or building a collection that would demonstrate the group's ability to address all items related to *materia medica*)

indicates rather that their intention emerged from acute awareness of their lack of knowledge. See Arthur MacGregor, *Curiosity and Enlightenment: Collectors and Collections from the Sixteenth to the Nineteenth Century* (New Haven: Yale University Press, 2007), particularly 11–44; Paula Findlen, *Possessing Nature: Museums, Collecting, and Scientific Culture in Early Modern Italy* (Berkeley: University of California Press, 1996).

18. Hacking carves out three fundamental elements that differentiate these categories: contingency, nominalism, and external explanations of stability. Ian Hacking, *Social Construction of What?* (Cambridge, MA: Harvard University Press, 1999), 36–62, 66–99.

19. Untitled flyer (1830). Private collection, Nagoya, Japan.

20. On the fifteenth day of the third month in 1827, Keisuke hosted a *honzō kai* at his home. Six months later, on the fifteenth day of the ninth month, Keisuke's older brother, Ōkōchi Zonshin, hosted another *honzō kai*, also in Owari. At the *honzō kai* hosted by Keisuke, more than 280 items were exhibited, submitted by eighteen people, including Mizutani Hōbun. Similarly, the *honzō kai* hosted by Zonshin included 320 items submitted by sixteen people. Ōkubo Masa'aki, another group member, listed the items displayed at these *honzō kai* in the order in which they were exhibited by copying what was on the tags attached to each object into a catalogue format. These records are kept under the title "Bunsei jūnen honzō kai mokuroku" in the collection of Daitōkyū Bunko, Gotō Art Museum, Tokyo, Japan.

21. Kaibara, *Yamato honzō*, "kusuri rui," vol. 1, 218.

22. Yamaoka Kyoan, *Honzō seiseigo*, reprinted in *Nagoya sōsho*, vol. 13, 198. It is noteworthy that one of the claims Kyoan recorded was Matsuoka Joan's assertion that there is a type of *hatsuyuri* with leaves that resembles the buckwheat plant, although Kyoan himself had not seen such a sample. In 1830, Hōbun walked in the mountains of Kiso and collected one *haruyuri* plant that Joan seemed to be referring to. See Mizutani Hōbun, *Kiso saiyakuki*, republished in *Nagoya sōsho*, vol. 13, 280.

23. These responses originally were delivered as lectures between 1799 and 1801 at the *Igakukan*, the official medical academy of the Tokugawa *bakufu*, in which he used both *Compendium* and *Materia Medica in Japan* as textbooks. Ranzan's pupil recorded the lectures and kept the notes as "Yamato honzō hisei," which then was circulated and copied by others. Endō Shōji has written an in-depth analysis of Ranzan's activity and influence, particularly in regard to his knowledge of European plants and animals. See Endō, *Honzōgaku to yōgaku*, 73–155.

24. Ono Ranzan, *Yamato honzō hisei*, in *Ekken zenshū*, ed. Ekken kai (Tokyo: Kokusho kankōkai, 1973), vol. 6, 566, emphasis added.

25. Mizutani, *Buppin shikimei*, vol. 19, 32.

26. Untitled flyer (1830). Private collection, Nagoya, Japan.

27. The communal and collective process that played such a significant role at the *honzō kai* is also reflected in the choice of the locations where they held the events. The *honzō kai* in 1835 took place at Ichigyō-in, located within a precinct of the Shōkō-in Temple in the Owari domain. This area, to the south of the Owari

Castle where a Tokugawa domainal lord resided, functioned as a transient entertainment area of the city. Famed for its constantly changing shows, be it a display of enormous bamboo-woven figurines of historical and mythological origins, statues of Kannon (the Buddhist bodhisattva of mercy) made of dried fish and shellfish, exotic birds made of seashells, impersonators of celebrated actors, a live camel, a demonstration of imported devices such as peep-boxes, or acrobatics by dwarves, this area of the town existed to satisfy the appetites for curiosity and the desires and imagination of the townspeople, samurai, and Buddhist monks alike. Although the reasons for partaking of the entertainment offered by this area varied, these shows offered a temporary experience of the "peculiar" and "unfamiliar" in an open public space. The decision to host a *honzō kai* here, in an area already codified as a space of overwhelming sensory excitement, amusement, and fascination, then, suggests anticipation of a certain crowd flowing in with a thirst for the novelty and sensory delight already associated with the space. Ichigyō-in, for instance, had hosted an acrobatic spectacle in 1821 and chrysanthemum viewing during the eleventh month in 1835, just eight months after holding a *honzō kai* organized by Shōhyaku-sha. During the *honzō kai* here, crowds overflowed the Ichigyō-in day and night. See Nagoya City Museum, *Sakariba: Matsuri, misemono, daidōgei Tokubetsuten* (Nagoya: Nagoya City Museum, 2002), 43–52.

28. Eiko Ikegami's exploration of various aesthetic circles from eighteenth- and nineteenth-century Japan offers a glimpse of how these amateur networks developed and flourished during the Tokugawa period. Eiko Ikegami, *Bonds of Civility: Aesthetic Networks and Political Origins of Japanese Culture* (Cambridge, UK: Cambridge University Press, 2005).

29. Kimura Kenkadō, a wealthy Osaka merchant, and his circle also practiced a nonhierarchical admission policy. The social networking of Kenkadō exemplifies the transgressive and communal effort that underscored *honzōgaku* gatherings. See *Kimura Kenkadō: Naniwa chi no kyōjin* (Osaka: Osaka rekishi hakubutsukan, 2003). For detailed discussion of the activities of the Shabenkai, see Hirano Mitsuru, "Tenpōki no honzōkenkyukai 'Shabenkai'—zenshi to seiritsu jijō oyobi katsudō no jittai," *Shundai shigaku* 98 (1996), 1–45.

30. Untitled flyer (1830). Private collection, Nagoya, Japan.

31. Aramata Hiroshi, *Hakubutsugaku no seiki* (Tokyo: NHK shuppan, 1998), 154.

32. Shōhyaku-sha, *Honzō kai buppin mokuroku*, in *Nagoya sōsho sanpen*, edited by Nagoyashi hōsa bunko (Nagoya: Nagoyashi kyōiku i'inkai), 1982, 230.

Chapter Four

1. Kōno Norihisa, "Wagakuni no inyō zufu ni tsuite" *Yokusai kenkyūkai dayori* 59 (1992), 2–8. Furthermore, there is one example of an ink-rubbing print in which the print was made from a completely dried specimen. Keisuke wrote that he had reapplied water to the specimen in order to make the impression. See Itō Keisuke, "Ogasawara-tō sanbutsu ki," Collection of Itō Keisuke kinenshitsu, Higashiyama Botanical Gardens.

2. Of the forty-nine albums of ink-rubbing prints that I was able to examine,

twenty-two titles include the word *shin'ei*. The dates of these albums vary from 1831 to 1881. Of the twenty-two, ten bear a title derived from two terms, *honzō* (*materia medica*) and *shin'ei*.

3. Udagawa Yōan "Shokugaku dokugo," in Kimura Yōjirō, ed., *Shokugaku keigen: Udagawa Yōan: fukkoku to yaku, chū* (Tokyo: Kodansha, 1980), 170. Historian Sugimoto Tsutomu speculatively dates the year of this manuscript to be between 1829 and 1831, based on Yōan mentioning an acquisition of a book by Siebold that took place in 1827. In this manuscript Yōan specifically uses the term *shokugaku* to translate *botanica*, stating that this field of study differs fundamentally from Chinese *materia medica* (*honzōgaku*) in its interest in all living plants, regardless of their medical efficacy. The contemporary Japanese term for botanical studies, *shokubutsugaku*, was introduced from China, read as *zhìwu xué* in Chinese, which was a translation of "botany" in English. See Sugimoto Tsutomu, *Edo yōgaku jijō*, 215–215.

4. Keisuke obsessively collected and accumulated information throughout his life. I am assuming that since the manuscript has a title that refers to a specific plant, it was compiled when Keisuke was still alive. After his death in 1901, his grandson Tokutarō, who was also a botanist, spent years reorganizing the pieces of information that Keisuke had gathered and gave it a title: "Plants Miscellaneous" ("*Shokubutsu zusetsu zassan*"). There are more than 100 volumes of "Plants Miscellaneous" at the National Diet Library in Tokyo, and more than 140 volumes under the same title at the Nagoya University.

5. The first illustration in the pamphlet is a pasted picture from *Zhíwù míngshí túkao*, published in China in 1848 by Wu Qijun. Keisuke added Japanese names to this book and republished it in 1885 with the assistance of Ono Yoshitsune.

6. Noted in the manuscript "Ōkubo sensei sōmoku fu," Collection of Kyōu Library, Takeda Science Foundation, Osaka, Japan.

7. The catalogue is called "Jakusōan hakurankai mokuroku" and is located in the collection of Nagoya City Museum. According to the preface of the catalogue, Jakusōan hosted a *honzō kai* annually on the last day of spring. Yoshikawa notes that Jakusōan arranged the gathering for the group to last for one day and opened his gate to the general public for two days. Yoshikawa, Yoshiaki, *Igaku yōgaku honzōgaku sha* (Tokyo: Yasaka shobō, 1993), 12.

8. This assertion was first made by Ueno Masuzō, a prominent and pioneering historian of *materia medica* in Japan in his canonical *Nihon hakubutsugakushi* (Tokyo: Kōdansha, 1989), 175. Also see Kawamura, "Wagakuni no inyō-zu fu ni tsuite," 2–8. The full title of Kniphof's book is *Botanica in Originali seu Herbarium vivum*.

9. None of these copies are dated. Copies by Niwa and Yoshio are in the collection of Kyōu Library, Takeda Science Foundation. Ono Ransai's copy is at Nishio City Library. Iinuma's copy is housed in a private collection. It cannot, however, be assumed that all are copied from Keisuke's "original." Endō Shōji speculates on how Ema Ransai and Keisuke both came to own "original" Kniphof in his article "Iinuma Yokusai to hito: Sōki zusetsu shippitsu chū no Iinuma

Yokusai," in *Iinuma Yokusai*, ed. Iinuma Yokusai seitan nihyakunen kinenshi henshū i'inkai (Gifu: Iinuma Yokusai seitan nihyakunen kinen jigyōkai, 1984), 277. In fact, Yokusai and Tanba included Kniphof's name in the titles given to their facsimiles. Yokusai titled his facsimile "Kniphof Botanical *Inyō-zu*" ("*Kinihofu shokubutsu inyō-zufu*"), while Tanba referred to his as "Kniphof Materia Medica in Ten Volumes" ("*Kinihofu honzō jukkan*"). Kawamura Norihisa, "Itō Keisuke to inyō-zu-hō," in *Seitan nihyakunen kinen Itō Keisuke no shōgai to sono jisseki* (Nagoya: Higashiyama Botanical Garden, 2003), 104–106.

10. Ueno, *Nihon hakubutsugakushi*, 175–176.

11. The technical definitions of *nature printing* vary, mostly depending on where this printing method is located within the genealogy of printing technology. Wilfrid Blunt explains that nature printing was "discovered in the fifteenth century and improved circa 1730 by D. Bruckmann." An online exhibit at the University of Delaware Library narrows the scope of its definition by asserting that nature printing is "the process involved in passing the object to be reproduced between a steel plate and a lead plate, through two rollers closely screwed together. The high pressure imbeds the object—for example a leaf—into the lead plate. When colored ink is applied to this stamped lead plate, a copy can be produced." Wilfrid Blunt, *The Art of Botanical Illustration* (New York: Dover, 1994), 138; and http:www.lib.udel.edu/ud/spec/exhibits/color/natures.htm (accessed December 18, 2004).

12. Quoted in Blunt, *The Art of Botanical Illustration*, 139.

13. Ibid., 139.

14. Maruzen acquired the copies at the time of Tokutarō's death in 1941. Kimura Yōjirō, "Nachurarisuto Iinuma Yokusai," in *Iinuma Yokusai*, 98. The National Diet Library owns the second edition of Kniphof's *Botanical in Originali seue*, published between 1757 and 1764. All the images included in the volumes are hand-colored.

15. In the period before the Shōhyaku-sha, the only example of someone using this rubbing method I could find was the painter Maruyama Ōkyo, an artist working in the imperial capitol of Kyōto, who incorporated ink-rubbing pictures into his sketchbook dated in 1771. This too signals the possibility that Kniphof's book was not the sole inspiration behind development of the ink-rubbing technique. See Sakakibara Yoshirō, "Juhasseiki no shokubutsu shasei," in *Higashi ajia no honzō to hakubutsugaku no sekai*, ed. Yamada Keigo (Kyoto: Shibunkaku shuppan, 1995) vol. 1, 198–273.

16. Imahashi Riko, *Edo no kachōga: Hakubutsugaku wo meguru bunka to sono hyōshō* (Birds and flowers: The representation of natural history during the Edo period, Tokyo: Sukaidoa, 1995), 96–130. See also Isono Naohide, "Edojidai dōbutsuzufu ni okeru tensha," in *Higashi ajia no honzōto hakubutsugaku no sekai*, vol. jō, ed. Yamada Keiji (Kyoto: Shibunkan shuppan, 1995), 299–333.

17. One way of understanding this historical shift from the *ranga* group to the Shōhyaku-sha lies in the social roles of each group. The core members of the Shōhyaku-sha worked as physicians, whereas the members of the Akita *ranga* school occupied high-ranking domainal positions. The latter were far removed from the daily reality that many members of the Shōhyaku-sha faced, seeing and

curing the sick. The contrast between the two groups is best summarized by the characterizations of the groups on the part of contemporary historians: Imahashi refers to the Akita group as a "salon," whereas Endō describes the Shōhyaku-sha as "sober and stoic." The urgency with which the Shōhyaku-sha sought to verify the existence of actual plants, as we saw in the previous chapters, was fueled by the present situations causing them great distress. Indeed, during the Great Tenpō famine (1833–1839) that affected much of the mainland of the Japanese archipelago, Keisuke responded by publishing *Listing of Edible Plants in the Times of Poor Crops* (*Kyūkō shokumotsu benran*, 1837), a one-page flyer that compiled a wide range of edible plants found within the region. Keisuke notes at the beginning of the flyer that the included plants are drawn neither from Chinese nor Dutch textual sources; rather, the selection is based on the commonly shared knowledge drawn from direct experiences (*tashikani kenbun seri*), thus reassuring the readers that the plants in question should be easily located in their own region. The Owari domainal lord bought the woodblock from Keisuke and printed several thousand copies to be distributed within the domain. The list uses the common Japanese name and also lists corresponding Chinese names used in *materia medica*. Keisuke calls attention as well to the fact there are various regional names for each plant and that should one have trouble identifying the plants using the Japanese names, he should consult a scholar of *materia medica* nearby. Itō Keisuke, *Kyūkō shokumotsu benran*, reprinted in *Kōnō sōshi*, ed. Kōnōshoin (Tokyo: Kōnō shoin, 1882), 6–12; Yoshikawa, *Igaku yōgaku honzōgakusha no kenkyū*, 25.

18. The letter is partly reprinted in Endō Shōji, "Kaho yūki go no Keisuke," in *Kinkaō nikki* (Nagoya: Higashiyama Botanical Garden, 1996), vol. 2, 87.

19. Zonshin noted this in the 1835 *honzō kai* catalogue. *Honzō kai buppin mokuroku*, in *Nagoya sōsho sanpen*, ed. Nagoyashi hōsa bunko (Nagoya: Nagoyashi kyōiku iinkai, 1982), 273.

20. Sugano Yō, *Edo no dōbanga* (Tokyo: Shinchōsha, 1983), 12; Nishimura Tei, *Nihon dōbangashi* (Tokyo: Shomotsu tenbōsha, 1941).

21. *Shiba Kōkan hyakka jiten*, ed., Tsukahara Akira (Kōbe: Kōbe shiritsu hakubutsukan, 1996), 56; Timon Screech, *The Western Scientific Gaze and Popular Imagery in Late Edo Japan* (Cambridge: Cambridge University Press, 1999), 94–107; *The Shogun's Painted Culture: Fear and Creativity in the Japanese States 1760–1829* (London: Reaktion Books, 2000), 43–48.

22. The 1774 *New Book on Anatomy* (*Kaitai shinsho*) by Sugita Genpaku and Maeno Ryōtaku reproduced the etching prints in the Dutch anatomical book in woodblock prints. For the historical inquiry into how the relatively unknown and unskilled Odano Naotake came to produce the woodblock prints in this book, see Haga Tōru, "Akita ranga no fushigi," in *Edo no naka no kindai*, ed. Takashina Shūji (Tokyo: Chikuma shobō, 1996), 22–96.

23. The person who came to be known as Aōdō Denzen was originally named Nagata Zenkichi. Matsudaira Sadanobu awarded Zenkichi a new "studio label" Aōdō, whose characters signified "Euro-Asian." Screech, *Shogun's Painted Culture*, 44.

24. Maki Bokusen was born in 1775 to an Owari samurai family and lived in Edo

from about 1791 to 1804. He studied *ukiyo-e* with Kitagawa Utamaro and took the name of Utamasa. Bokusen is believed to have begun studying with Hokusai after Utamaro's death in 1806 and took the *gō* names (artistic names) of Hokutei, Hokusen, Gekkotei, Hyakusai, and Gakyojin. As for how Bokusen learned the copper-etching technique, there is, as of yet, no historical resource that directly answers the question. However, there are three possibilities put forward by historians. Nishimura Tei suggests that the knowledge might have been passed down originally from Katsuragawa Hoshū, who translated *New Book on Anatomy* with Maeno Ryōtaku and Sugita Genpaku. Through Hoshū's apprentice, the knowledge was transmitted to Udagawa Bōsai, who then told his apprentice Ogawa Morinaka, a physician in Owari domain. On the other hand, Yoshida Susugu argues that Bokusen's friend, the Owari pharmacist Fukutokuya Ichibei, introduced Bokusen to his patron physicians, Nomura Ritsuei and his circle of scholars and intellectuals interested in Western learning (*rangaku*). Yoshida Toshihide conjectures that Bokusen could have learned the technique while he lived in Edo, between approximately 1791 and 1801, especially because one of his copper-etching prints dates from 1801. See Nishimura Tei, "Maki Bokusen to chūkyō no dōbanga," in *Nihon dōbangashi*, 117–147; Yoshida Susugu, "Chūkyō no senkusha Maki Bokusen," *Hokusai kenkyū* 9 (April 1975), 16–35; and Yoshida Toshihide, "Maki Bokusen saikenshō," *Nagoyashi bijutsukan kenkyū kiyō* 1 (1991), 1–30. For a detailed study of the relationship between Bokusen and Hokusai, see Kamiya Hiroshi, "Hokusai to Nagoya—kenkyūjosetsu," *Nagoyashi hakubutsukan kenkyūkiyō* 18 (1994), 1–34.

25. Yoshida Toshihide claims that Kondō Shūen was "an important member of the Shōhyaku-sha." However, I have been unable to locate materials to demonstrate this assertion. See Yoshida, "Maki Bokusen saikenshō," 21.

26. Sasaki Takami, "Kaisetsu," in *Nagoya sōsho sanpen*, ed. Nagoyashi hōsabunko (Nagoya: Nagoyashi kyōikuiinkai, 1982), 15; Sugimoto Isao, *Itō Keisuke* (Tokyo: Yoshikawa kōbunkan, 1960), 132. Sugimoto notes that the vine flower that frames Thunberg is black-eyed Susan vine (*Thunbergia alata*; Jp: *yabane kazura*), which Thunberg helped identify.

27. This combination of characters would be recognizable as a transliteration of "Thunberg" to readers as Keisuke provides Thunberg's full name in Chinese characters in the explanatory note to *Nominal Differentiations*. Itō Keisuke, *Taisei honzō meiso*, republished in *Nagoya sōsho sanpen*, ed. Nagoyashi hōsa bunko (Nagoya: Nagoya kyōiku i'inkai, 1982), 317.

28. Stephen Addiss, "The Role of Calligraphy in Japanese Society," in *Words in Motion: Modern Japanese Calligraphy*, ed. Yomiuri shinbun (Tokyo: Yomiuri shinbun, 1984), 31.

29. Maki Bokusen, *Yōka seisen zukai* (Nagoya: Eirakuya, 1820), Collection of the University of Tokyo Library. There is an unresolved issue regarding Bokusen's advertisement for *Taisei shin'ei honzō* in 1815. The earliest mention of the Linnaean binominal system in Japan is believed to be 1817, when Yōan translated a section of "Encyclopedia," so it is unclear as to how Bokusen would have been aware of the Linnaean system before Yōan and the members of the Shōhyaku-sha.

30. Maki Bokusen, *Shashin gakuhitsu*, collection of Nagoya Municipal Museum (Kyoto: Hishiya Magobei, 1815). Bokusen passed away before completing either title.

31. Nishimura, *Nihon dōbangashi*, 125.

32. Although outside the scope of this investigation, the popularity and pervasiveness of *shingyōsō* styles in Japan since their introduction via calligraphy is noteworthy. Beyond calligraphic discourse, practices of the tea ceremony, architecture, theater, and floral arrangements have adopted these styles as a method to teach and demarcate lineage.

33. Addiss, "The Role of Calligraphy in Japanese Society," 31.

34. As illustrations for an instructional book on picture making, these botanical representations and their appreciation stand outside the field of *materia medica*. However, the distinction between the discourses on illustration and *materia medica* should not be overstressed. Keisuke referenced *Primer on Eight Varieties of Paintings* (*Hasshu gafu*) in his unpublished pamphlet on oleander; varying kinds of books and illustrations served as resources for studying *materia medica*. The overlap, interaction, and mutual relevance of these two seemingly distinct discourses is more elusive and yet common than we might assume today.

35. In fact, Nagata Seiji mentions that even though *Manual* was published in 1816, the sleeve in which the first printing of the book was sold indicates it was intended for publication in 1815, the same year the advertisement for "Picture Book" appeared. Nagata Seiji, "Sakuhin Kaisetsu," in *Hokusai no edehon*, ed. Nagata Seiji (Tokyo: Iwanami bijutsu shuppan, 1994), vol. 4, 272. The fact that the same pictorial idea behind the *shingyōsō* style occupied both Bokusen and Hokusai around the same time is not surprising, given that in 1812 fifty-five-year-old Hokusai finished three hundred sketches for the first volume of a series known as *Hokusai's Sketches* (*Hokusai manga*) while staying with Bokusen in Owari. Katsushika Hokusai, *Hokusai manga* (Tokyo: Shogakkan, 2005), 14–16. The first printing of *Sketches* was published through a local Owari publisher, Eirakuya Tōshirō, in 1814. Matthi Forrer, *Eiryakuya Tōshirō: A Publisher at Nagoya* (Amsterdam: J. C. Gieben, 1985), 195–197.

36. Between 1810 and 1849, the year of his death, Hokusai oversaw publication of twenty *etehon* titles, and according to an account given by Nagata Seiji forty-nine more titles were advertised but never published. These *etehon* series served to respond to the increasing popularity and demand of the Katsushika school and of Hokusai himself, who was in great demand as a teacher.

37. Maki, *Yōka seisen zukai*.

38. Originally, the leopard was brought to the Owari castle by a delegation from Chōsen (present-day North and South Korea) in the third month of 1830. Yamamoto Bai'itsu drew the leopard on a silk scroll at the request of the Owari domainal Lord Tokugawa. Tabei Ryūtarō, *Kokon chūkyōgadan* (Nagoya: Kōfū shoin, 1911), 158. Later, the live leopard was displayed (with the label "tiger") at a local street show (*misemono*) with an entry fee of 50 mon, almost three times more than cost of a regular street show. At the show, the leopard was kept inside a double bamboo cage, and the people could watch the leopard hunt by purchasing a live pigeon for the additional cost of 32 mon. Keisuke and the Shōhyaku-sha are directly associated

with the copper etching of a seal that Shūen produced in 1833. The seal appeared on the shore of Atsuta on the third day of the seventh month in 1833 during the reclamation works, and immediately a large crowed gathered to watch the rare mammal. A local fisherman captured the seal the next day in order to display it as a part of a street show, which incited a lawsuit between him and the owner of the land on which the seal was captured; the landowner insisted on setting the seal free into the ocean. During the dispute, the seal was kept at Seiju-in Temple in the central castle town area, where the animal is said to have learned to do some tricks. The seal was then put up for display from the fourteenth day of the eighth month, attracting a large crowd until its death three days later. There were a variety of goods sold associated with the popularity of the seal. See Nagoya City Museum, ed., *Tokubetsuten sakariba: Matsuri, misemono, daidōgei Tokubetsuten* (Nagoya: Nagoya City Museum, 2002), 39–40.

39. The relationship between copying and original in Kōkan's oeuvre is well documented in Tsukahara Akira, ed., *Shiba Kōkan hyakka jiten* (Kobe: Kobe Municipal Museum, 1996), 99–121. Denzen applied this technology to print on cotton and silk, producing tobacco cases and baskets. See Fuchū Art Museum, ed., *The Age of Aōdō Denzen*, exhibition catalogue (Fuchū: Fuchū Art Museum, 2006).

40. Timon Screech, *The Western Scientific Gaze and Popular Imagery in Later Edo Japan* (Cambridge: Cambridge University Press, 1996), especially 52–60 and 94–106.

41. The key terms that inform Screech's notion of "precisionism" are the Chinese character *sei*, which is read *kuwashiki* in Japanese, and the compound *saiku*, which Screech translates as "precision or close-wrought devices." Ibid., 47–60.

42. Kawahara Keiga was born in 1786 into a family of artisans in Dejima. In 1811 he acquired the status of *dejima deiri eshi* (picture maker allowed to work in Dejima, the restricted residential and commercial area for the Dutch), a position that led him to meet and work with Siebold. As a privileged *eshi*, Keiga produced pictures for foreigners staying in Dejima, ranging from whales to insects to people and daily scenes of Dejima. Satisfied with the quality and the level of Keiga's picture making, Siebold gave him a microscope to produce more illustrations for his *Flora Japonica* and *Fauna Japonica*. The woodblock illustrations of this book were printed in black and white ink in the initial publication of 1836 with this title, and his nephew Ryo'a republished it in color in 1868, changing the title to *Shashin Illustrated Book on Plants, Trees, Flowers, and Fruits* (*Sōmoku kajitsu shashin zufu*). In the two-volume 1836 publication, Keiga included detailed instructions for coloring the images. See Sugimoto Tsutomu, "Kawahara Keiga to shokubutsu zukan 'Keiga shashin sō'," in *Edo yōgaku jijō* (Tokyo: Yasaka shobō, 1990), 308–317; Ueda Kyōichirō, "Kawahara Keiga," in *Saishoku edo hakubutsugaku shūsei*, ed. Shimonaka Hiroshi (Tokyo: Heibon sha, 1994), 293–308; and Matthi Forrer, "Kawahara Keiga-oranda shōkan wo meguru kaiga," and Ken Vos, "Kirokueshi toshiteno Keiga," in the exhibition catalogue *Kawahara Keiga ten* (Tokyo: Seibu Museum of Art, 1987).

43. Kawahara Keiga, *Sōmoku kajitsu shashin zufu* (1868), republication of *Keiga shashin sō* (1836), in the collection of Waseda University Library.

44. See Melinda Takeuchi's description of how the Japanese literati painters absorbed and understood the concept of *shin* in these treatises. Melinda Takeuchi, *Taiga's True Views: The Languages of Landscape Painting in Eighteenth-Century Japan* (Stanford: Stanford University Press, 1992), 82–88. Takeda Koichi provides a detailed account of the relationship between painting manuals from China and the traditions of Southern School painting in "Chūgoku gafu to nihon nanga no kankei," in *Kinsei Nihonkaiga to gafu, e-tehon ten*, ed. Machida shiritsu kokusai hanga bijutsukan (Machida: Machida shiritsu kokusai hanga bijutsukan, 1990), 147–158.

45. See Kōno Motoaki, "Edojidai 'shasei' kō," in *Nihon kaigashi no kenkyū* (Tokyo: Yoshikawa Kōbunkan, 1989), 387–427.

46. Sketchbooks of Maruyama Ōkyo, Watanabe Shikō, Ogata Kōrin, and Kano Tanyū attest to the prevalence of not only the subjects but also the interest in attempting to depict these living subjects as a part of pictorial training.

47. In fact, there are cases where an illustration from a Western book was incorporated as part of a painting manual published in Japan.

48. Keiga demonstrates some familiarity with the Linnaean system and includes the Latin names for certain of the illustrations. In the explanatory notes (*bonrei*), Keiga briefly mentions that the "wicks" of flowers can be divided into two types, *yin* and *yang*, as well as mentioning Linné and Thunberg as reference resources for the Latin names. However, I want to emphasize that the Latin names are provided inconsistently and that the numbers of stamen and pistils are absent in the publication. The dried bud of magnolia has been used traditionally in *materia medica* as a medicine that warms the body, in particular to alleviate headaches and cold symptoms.

49. Keisuke, in fact, was familiar with Keiga's work through the German Felix Meijlan's 1829 publication *Japan*. Keisuke characterized Keiga's illustration in Meijlan's book as "extremely impressive" but did not speak to its quality as accurate or detailed. See Ema monjo kenkyūkai, ed., *Emake raikan shū* (Kyoto: Shimonkaku Shuppan, 1984), 100.

50. Indeed, Kan'en's endeavor was supported enthusiastically by Hotta Atsuhiro, one of the junior elders of the *bakufu* (*waka doshiyori*), who himself had published illustrated encyclopaedic books on the topic of birds.

51. Yabe Ichirō, "Iwasaki Kan'en," in Shimonaka, ed., *Saishoku edo hakubutsugaku shūsei*, 285.

52. *Erya* is one of the first encyclopedic and thesaurus dictionaries compiled in China around the third century. *Illustrated Compendium of the Three Powers* (*Sancai tuhui*) by Wang Qi was another encyclopedic work published in China in 1607. Inspired by this title, Terashima Ryōan published *Japanese and Chinese Illustrated Compendium of the Three Powers* (*Wakan sansaizue*) in 1712. Originally from China, *Hiden kakyō* was edited by Hiraga Gennai and published in Kyoto in 1829 in six volumes. More appropriately characterized as a how-to book, it is a guide on how to grow particular plants.

53. Iwasaki Kan'en, *Honzō zufu*, republished in *Honzō zufu* (Kyoto: Dōhōsha, 1980), vol. 1, 8. This publication is based on the copy of *Honzō zufu* at Jingū bunko, Ise Shrine. The original owner of the copy was Nishimura Hiroyasu, a wealthy

merchant of Ise who also pursued *materia medica* under the guidance of Yamamoto Bōyō in Kyoto.

54. Iwasaki Kan'en, *Honzō zufu*, vol. 1, 3–4.

55. Ibid., 5.

Chapter Five

1. Although there are earlier texts written about photographic media, such as Kawamoto Kōmin's *Odd and Rare Devices form the West* (*Ensei kikijutsu*, 1854), which summarized the technical how-to of the daguerreotype, or the circulated manuscript "potokarahī" by Udagawa Kōsai, the extent to which the information on these texts was actually useful was fairly limited compared to Shunsan's work, for two reasons: (1) the materials necessary for making daguerreotypes were not readily available or cheap, and (2) they were direct translations of one Western source and thus did not offer the kind of synthesized and experimented account that Shunsan's writing conveyed.

2. Before moving to Edo, Shunsan and Keisuke collaborated on translating a book on niter (potassium nitrate), titled *Banpō sōsho shōseki hen* (1854). The Owari domain ordered Keisuke to translate and revise a book on the subject that had become increasingly potent as the political climate grew unstable: the *bakufu* had become visibly weakened, and foreign attacks were felt to be imminent. Shunsan's comical personality and his relationships with Fukuzawa Yukichi and Utsunomiya Saburō, the "father" of chemistry in Japan, are documented in a memoir by Imaizumi Mine, who was a thirteen-year-old girl when Shunsan was living with the Katsuragawa family. See Imaizumi Mine, *Nagori no yume—ran'i Katsuragawake ni umarete* (Tokyo: Heibonsha, 1963).

3. In the introduction, Shunsan mentions that he had access to French and English writings by Dagron, but he does not give the titles of the works he consulted. In the body of the text, Shunsan quotes a note written by Dagron on June 19, 1864, but the quote did not come from Dagron's *Traité de Photographie Microscopique*, published that year, or any other source that I have been able to identify. The 1864 *Traité* was a small booklet that explained the process of making microfilm positives. Dagron was the first person to receive a patent for microfilm. Shunsan additionally lists seven books on photography in Dutch, French, English, and Russian in his introduction but offers only the transliterated form of the authors' last names in *katakana* script. From his list, I have been able to identify only the Dutch translation of Désiré van Monckhoven's *Traité général de photographie* (1862).

4. The capital Edo was renamed in July 1868 as Tokyo by the Imperial Rescript following transfer of ruling power from the Tokugawa shogunate to the imperial order in November 1867.

5. Ishii Kendō "Shishi to shashin to kanzō," in *Zōtei Meiji jibutsu kigen* (Tokyo: Shunyōdō, 1936), 463–464.

6. Historiographical approaches to Iinuma Yokusai's work are indicative of the compartmentalizing tendencies of modern academic practice. Hasegawa Hitoshi, scholar of natural history, begins an essay "Iinuma Yokusai" with the sentence

"Iinuma Yokusai was a physician, but he was also a great naturalist [*nachura-risuto*]," implying that there is little in common with being a physician and being a "naturalist." In the official history of Ōgaki-city, where Yokusai resided, his name appears in the section of "chemistry and botany," but not in "Chinese medicine" (*kanpō gata*) or "Western medicine" (*yōhō gata*), both disciplines in which Yokusai was trained. Shimonaka Hiroshi, ed., *Saishoku edo hakubutsugaku shūsei* (Tokyo: Heibonsha, 1994), 252; *Ōgaki shishi*, Ōgaki-city, ed. (Ōgaki: Ōgaki-city, 1930), vol. 2, 410–450.

7. Yokusai intended to publish this body of work in several parts and forty-nine volumes. Between 1856 and 1862, only the twenty-volume section on plants was published. Yokusai published them as four sets of five volumes. The first set was published in 1856, the second and third in 1861, and the last set in 1862. In 1874, Tanaka Yoshio and Ono Motoyoshi re-edited the volumes on plants with the assistance of Paul Amèdèe Ludovic Savatier, a French doctor who worked in Yokosuka from 1865. The re-edited version was published as *Shintei sōmoku zusetsu* by Tanaka Yoshio in 1874. In 1913, a leading Japanese botanist, Makino Tomitarō, revised *Shintei sōmoku zusetsu* and published *Illustrated Explanations of Plants and Trees of Japan, Additionally Revised* (*Zōtei sōmoku zusetsu*). The ten-volume section on trees, moreover, was not published until 1977, when Kitamura Shirō, professor of botany at Kyoto University, annotated Yokusai's original and published them as *Illustrated Explanations of Plants and Trees of Japan, Section on Trees* (*Sōmoku zusetsu mokubu*).

8. Kawamoto Kōmin's lectures based on Peiter van der Berg's *Eerste Grondbeginselen der Natuurkunde* (1844) were transcribed by his apprentice Tanaka Kōki and published in *Odd and Rare Devices from the West* (*Ensei kikijutsu*, 1854). In this book, Kōmin speaks about daguerreotype photography as *choku sha'ei kyō*, while referring to camera lucida as device of *shashin* (*shashin ki*). For a thorough study of terms applied to photography and daguerreotype in Tokugawa Japan, and the wet collodion process in early Meiji, see Koizumi Sadahiro, "Nihon shashin yōgoshi no kenkyū," *Nihon daigaku geijutugakubu kiyō* (1984), 3–13 (1985), 11–17.

9. Ozawa Kenji, *Bakumatsu Meiji no shashin* (Tokyo: Chikuma shobō, 1997), 24–26. Ozawa reads Ueno's first name as "Shin'nojō." I am following the reading provided in the National Diet Library catalogue and read it "Toshinojō."

10. The case of the Kaga domain stands out in that it was not the domainal lord who took initial interest in the technology. There Ōno Benkichi (1801–1870), a trained craftsman of toys and automata, took several wet collodion portraits of himself and his wife and friends. Motoyasu Hiroshi, "Shoki shashinshi no chiikiteki shozentei," in *Ishikawa kenritsu rekishi hakubutsukan kiyō*, (1997), vol. 10, 23–48.

11. See Kinoshita Naoyuki, "The Early Years of Japanese Photography," in *The History of Japanese Photography*, ed. Anne Tucker (New Haven: Yale University Press in association with Museum of Fine Arts, Houston, 2003), 14–99.

12. Ozawa Kenji, *Nihon no shashinshi—bakumatsu no denhan kara meijiki made* (Tokyo: Nikkor club, 1986), 28–34 and 69–74.

13. For a detailed study of Ueno Hikoma and Shimo'oka Renjō, see Ozawa Kenji, *Bakumatsu Meiji no shashin* (Tokyo: Chikuma shobō, 1997), 86–125; Kinoshita

Naoyuki, "Shashin no hajimari" and "Shashin wo meguru kotoba," in *Ueno Hikoma to Bakumatsu no shashinkatachi*, vol. 1, *Nihon no shashinka*, eds. Nagano Shigeichi, Iizawa Kōtarō, and Kinoshita Naoyuki (Tokyo: Iwanami shoten, 1999), 3–8, 57–65. For Ugai Gyokusen's work, see Ichikawa Jinzō, "Tōto ryōgoku yagenbori shashinshi Ugai Gyokusen shoki," *Risshō daigaku kyōyō gakubu kiyō* 22 (1988), 27–39. An exhibition catalogue *Nihon no dandizumu* (Tokyo: Nigensha, 2003) provides interesting accounts of the Japanese bureaucrats who were photographed both in Japan and abroad. What these conflicting records demonstrate, moreover, is that acceptance of photography did not come overnight. One response that highlights the general curiosity and fascination with the medium is in *The Chronicle of Edo* (*Bukō nenpyō*, 1868). Saitō Gesshin noted in his entry from 1854 that "the subjects of photographic depiction range from the beautiful mountains and rivers, the majestic buildings of temples and shrines, the crowded housings of rich and poor, to humble tiny shacks. Also, there are portrait photographs, subjects of which include respectable persons, *samurai* and ordinary people, sensual images of women of brothels, made-up faces of *kabuki* actors, as well as other various groups. It's possible to say that there is nothing that has not been made into a photograph. In addition, because even plants, trees, animals are depicted to come to life, everyone regardless of gender and age is desperately seeking photographs." Saitō's account illuminates the gradual shift in popular familiarity with photographic practice in the capital Edo. Moreover, in Shunsan's *Illustrated Book on Photography* (*Shashinkyō zusetsu*, 1868–69), we also find claims that some people find photography to be magical and fearsome. Saitō Gesshin, *Zōtei bukō nenpyō*, ed. Asakura Kamezō (Tokyo: Kokusho kankōkai, 1913), 267.

14. In November 1843 the Tsuyama domain officially accepted Udagawa Yōan's request to adopt Kōsai in order to continue the Udagawa family line and its high reputation as a family of scientists. He worked as an official physician of the Tsuyama domain, a job passed down from his adopted father Yōan, and as a translator.

15. Endo Shōji suggests the possible interpersonal exchange of photographic information between Kōsai and Kawamoto Kōmin because they were close friends and collaborators in Edo. Endo Shōji, "Gifu no bakumatsu meijiki no shashin jijō," in *Kanzō koshashin shū*, ed. Gifuken rekishi shiryōkan (Gifu: Gifuken rekishi shiryōkan, 1999), 36. Kōsai's correspondence with Taketani Ryōtei, the chief vassal of the Fukuoka domain, also indicates the extent of Kōsai's knowledge of photography, and by extension that of Yokusai. Ryōtei sent three wet collodion images to Kōsai in 1860 as examples of a successful experiment with photography in the Fukuoka domain. Endo Shōji, "Owari mino no shashinjutsu," *Kagakushi kenkyū* 28, no. 1, 20.

16. This translation is the only work in Japanese that introduced albumen printing. It is now in the collection of Waseda University Library. For the result of Kōsai's instruction, see Miyagawa Toshio, Shirai Yasuo, Danki Reiji, Kashimoto Akihiko, and Hiraki Osamu, "Udagawa Kōsai no 'potokarahī' ni tsuite—komonjo no kaidoku to ranpakushippan no shashin tokusei," *Tōkyo kōgei daigaku kōgakubu kiyō* 11, no. 1, 14–25.

17. The wet collodion process, perfected by Frederick Archer in the 1850s, involves covering a piece of glass with wet solutions before an exposure is made. The photographer then has about ten minutes to fix the exposed plate, which is typically done in a portable darkroom. Unlike the earlier daguerreotype, this process allows multiple prints when a glass negative is used, and it greatly shortens the exposure time to seconds. Collodion glass is also much less expensive than plates for daguerreotype, which contributed to the proliferation of portrait photographs in mid-nineteenth-century Europe and America. The wet collodion process dominated production of photographic images in Japan until 1881, when Ezaki Reiji successfully made an image using a dry collodion process. The tintype process, popular in the United States, never spread to Japan.

18. The poem reads *kuku tada sue kyūnari touen shissyō mata kini yoru*, and roughly translates, "I have already passed my eighty-first year, have only hastened feelings of the end, but I only return to my armchair and laugh at myself."

19. Iinuma Yokusai, *Sōmoku zusetsu*, republished *Sōmoku zusetsu* (Gifu: Heirinsō, 1874), vol. 1.

20. Itō Keisuke, "jo" (preface), included in *Sōmoku zusetsu*, vol. 1.

21. The manuscript, now housed at the Tokyo Metropolitan Library, also includes pictorial works by Kawakami Tōgai, an instructor at the Bureau of Pictorial Studies, and Nakajima Gyōzan, a fellow student of Yuichi.

22. Various writings by Yuichi, which include drafts of letters, published commentaries, and rules for his art school, were collected as "Takahashi Yuichi abura'e shiryō," a five-volume set of albums. The volumes were donated to the Tokyo University of the Arts, which was then called Tokyo Bijutsu Gakkō, in 1914 by Yuichi's daughter-in-law. The entire manuscript is republished as *Takahashi Yuichi abura'e shiryō*, ed. Aoki Shigeru (Tokyo: Chūō kōron bijutsu shuppan, 1984).

23. The name changes between 1855 and 1863 tellingly reflect the shifting identities of the institution: in 1855 the name changed from *Tenmon dai bansho wage goyō gata* (Branch for Translating Barbaric Books) to *Yōgaku sho* (Institute of Western Study); in 1856, it changed to *Bansho shirabejo* (Academy for Western Studies); in 1862, it was renamed to *Yōsho shirabejo* (Institute of Studying Western Books); and in 1863, it changed to *Kaiseijo* (Academy of Developing Intelligence for Successful Enterprises). In September 1868 *Kaiseijo* was amalgamated into the new Meiji government.

24. This is the former body that eventually became the University of Tokyo in 1877. The "origins" of the Academy for Western Studies can be traced back to the Bureau of Western Books of the Tenmon gata, a *bakufu* bureaucracy that controlled and distributed calendars, maps, and astronomical illustrations. I am relying on an exhibition catalogue, *The Archaeology of Science*, and Kurasawa Takashi's exhaustive research on the educational institutions and their roles toward the end of Tokugawa period for the reproductions of primary texts and insightful observations. See Kurasawa Takashi, *Bakumatsu kyōikushi no kenkyū* (Tokyo: Yoshikawa kōbunkan, 1983); and *Gakumon no arukeoroji*, ed. University of Tokyo (Tokyo: University of Tokyo Press, 1997).

25. Reprinted in its entirety in Doi Yasuhiro, "Bansho shirabejo no bussan kenkyū to Itō Keisuke no kankei," *Hōseidaigaku daigakuin kiyō* 36 (1996), 165–167.

26. Suzuki Hiroyuki, *Kōkokatachi no jūkyūseiki* (Tokyo: Yoshikawa kōbunkan, 2003), 11.

27. Doi, "Bansho shirabejo shutsueki," 215.

28. Kawakami Tōgai was appointed as a *kyōju* of *kudoku* (reading Dutch) earlier in the year and then was appointed *ezu shirabe shutsu yaku* of the Office of Pictorial Illustrations (*Ezu shirabe gata*). In addition, after his appointment to Ezu shirabe gata in June, he was employed at the naval school in Nagasaki. "Kawakami Tōgai nenpyō," in *Kawakami Tōgai to sono shūhen: Bakumatsu kara Meiji e* (Nagano: Nagano ken shinano bijutsu kan, 1990), 51.

29. Reprinted in Kurasawa, *Bakumatsu kyōikushi no kenkyū*, 293.

30. Ibid., 294

31. Quoted in Kurasawa, 294.

32. Shiba Kōkan, "Oranda tensetsu," in *Shiba Kōkan zenshū*, vol. 3 (Tokyo: Yasaka shobō); also available in the Waseda University Japanese and Chinese Classics Database, http://archive.wul.waseda.ac.jp/kosho/ni05/ni05_00776/ni05_00776_p0007.jpg (accessed June 1, 2010).

33. Shiba Kōkan, *Seiyō Gadan*, in *Nihon shisō taikei* (*Yōgaku*), vol. 64, 491–492. Several scholars have noted the significance of Kōkan's understanding of *shashin*. In translating this essay into English, Calvin L. French conveyed the term *shashin* as "true representation of reality," "represent reality faithfully," and "realistic painting." Tom Looser renders the same concept as "imagining reality" and *shin* as "truth" in his translation. I find the translation *shashin* as "imagining" properly reflects both physical and illusionary aspects; the concept of "truth" is by comparison too elusive and overloaded with wide-ranging intellectual assumptions. Calvin L. French, *Shiba Kōkan*, 81–84, 171–174. Tom Looser, trans., "Seiyōga dan," in *Readings in Tokugawa Thought*, 3rd ed., ed. Tetsuo Najita (Chicago: Center for East Asian Studies, 1998), 175–181. Doris Croissant consistently translates the word "*ga*" as "paintings" in her translation of Kōkan's writings. Given the types of representations with which Shiba was undoubtedly most familiar—copper etching prints, *vues d'optique*, and illustrations in Western natural history books—I find it more appropriate to translate "*ga*" as "pictures," rather than the more specific category of "paintings." See Doris Croissant, "In Quest of the Real," in *Challenging Past and Present: The Metamorphosis of Nineteenth-Century Japanese Art*, ed. Ellen P. Conant (Honolulu: University of Hawai'i Press, 2006), 153–176.

34. Ibid., 493.

35. The entire manuscript is republished in Yoshikawa Yoshiaki, *Igaku yōgaku honzōgakusha no kenkyū* (Tokyo: Yasaka shobō, 1993), 115–118. The report offers an intriguing description of the interior of the Ijinkan Inn where Siebold was staying, as well as the social landscape surrounding the delegation.

36. Doi Yasuhiro speculates that these were probably four volumes taken from Siebold's *Flora Japonica* and *Fauna Japonica* (1833–1850). However, I have not found historical material that would support this conjecture. Doi, "Bansho shirabejo shutsu'eki," 78.

37. Yoshikawa, *Igaku yōgaku honzōgakusha no kenkyū*, 116. Carl Hubert de Villeneuve (1800–1874) came with Siebold during his first mission between 1825 and 1829 as an artisan.

38. Ibid., 117.

39. Wirgman began to publish the satirical journal *Japan Punch* in 1862 and continued it monthly until 1887. Wirgman also illustrated Sir Rutherford Alcook's *The Capital of Tycoon* (London, 1863) and taught sketching to Takahashi Yuichi.

40. In the 1862 *honzō kai* held by the Shōhyaku-sha in Owari, Keisuke submitted this *shashin* picture along with that of Siebold.

41. Tanaka Yoshio asserts, however, that Keisuke's decision was based on "the complicating and difficult social situation in Edo" (*yononaka nakanaka muzukashi'i sawagi ni natta kara*), and that Keisuke "ended up going back to his home town because he got tired of the situation in Edo." In this interpretation of Keisuke's decision to return to Nagoya, Yoshio depicts his teacher as someone unable to cope with the current situation. See Tanaka Yoshio, "Tanaka Yoshio kun no keireki dan," in *Tokyo kokuritsu hakubutsukan hyakunenshi, shiryō hen*, edited by Tokyo National Museum (Tokyo: Dai'ichi hōki shuppan, 1973), 565. Historian Doi Yasuhiro speculates that Keisuke might have become tired of the conservative attitude of the *bakufu* institution as well as the low status of the institution itself. See Doi, "Bansho shirabejo shutsueki," 190.

42. Haga Tōru considers Yuichi's pursuit of pictorial studies the result of a firm resolution that was forcefully developed in him by his grandfather, who was a master of archery and *kendō* for the Sano domain. Haga Tōru, *Kaiga no ryōbun* (Tokyo: Asahi shinbunsha, 1990), 47.

43. "Takahashi Yuichi rirekisho," reprinted in exhibition catalogue, Kanagawa Prefectural Museum of Modern Art, ed., *Botsugo Hyakunen Takahashi Yuichi ten* (Kamakura: Kanagawa Prefectural Museum of Modern Art, 1994), 195.

44. Haga Tōru extends the speculation proposed by Takumi Hideo that what Yuichi refers to in this text could be the stone-etching prints included in illustrated books on the Mexican War and the American Revolutionary War gifted by Matthew Perry to Abe Masahiro, the chief senior councilor of *bakufu* in 1854 Tokugawa when he revisited Japan. Haga, *Kaiga no ryōbun*, 49–53. See also Hijikata Tei'ichi, *Takahashi Yuichi gashū* (Tokyo: Kōdansha, 1972); Takashina Shūji, *Nihon kindai bijutsushiron* (Tokyo: Kōdansha, 1977); Furuta Ryō, *Kanō Hōgai Takahashi Yuichi* (Tokyo: Mineruba shobō, 2006), especially 87–95.

45. Considering the frequency and the centrality of the concept of *shashin* within the oeuvre of Yuichi, moreover, it would be misleading to assume that his understanding of the concept of *shashin* as articulated in this manifesto remained unchanged throughout his long and complex career. The concept of *shashin* and *shinsha* (reversing the order of characters from *shashin*) surfaces frequently in Yuichi's writings, particularly when discussing the significance of portrait oil paintings of historical and political figures. In February 1871 Yuichi expounded the necessity of *shinsha* in petitioning the new government to establish a "pictorial parliament" (*zugaku in*), an institution that Yuichi envisioned would gather old paintings from across the

country and display them chronologically to teach a general survey of Japan's history, while also exhibiting pictures of recent disasters, such as earthquakes, to assist later generations in imagining historical events. Also, in October 1867 Yuichi used both *shinsha* and *shashin* in a letter to ask for the loan of a portrait painting. The terms occur again in his diary of a trip to the Nara and Kyoto area in 1872, "Shinsha goyō junkaiki," on which Yuichi served as a draftsman. It is easy to imagine how widely the meanings of this concept fluctuated for Yuichi as he engaged in a variety of concerns and activities ranging from pedagogy to political lobbying over a long career. I want to emphasize that my interest in analyzing his "Statements for the Office" here is not in using the concept of *shashin* as a methodological tool to better investigate Yuichi's artistic development per se. Rather, I examine his manifesto as an important historical text from which to extricate the subtle bifurcations of the concept of *shashin* within the *bakufu* institutions. Kōno Motoaki's examination of Yuichi's work historicizes Yuichi's life within the shifting conditions of the late Tokugawa period. Kōno takes Yuichi's insistent pursuit of Western pictorial techniques to reveal a fixation on realistic depiction and states that "this absolute trust in realism always existed in Yuichi." See Kōno Motoaki, "Takahashi Yuichi," in *Bakumatsu Meijiki no gakatachi*, ed. Tsuji Nobuo (Tokyo: Perikansha, 2002), 91–129.

46. Kōkan writes in *Western Pictures* that "in Western countries, pictures are regarded higher than written texts, and thus both pictures and texts are used for the nation, and are not devised for amusement." Similar sentiment echoes in Yuichi's statement, yet Yuichi does not introduce the comparison between "serving the nation (*kokuyō wo nasu*)" and "amusement (*ganbutsu*)."

47. Yuichi's title for the manifesto, in which he calls the office *yōgakyoku*, also highlights the extent to which Yuichi identified with "*yōga*" practices, rather than the much broader and general designation of Pictorial Studies (*gagaku*) that was used in the official title of the bureau.

48. Takahashi Yuichi, "Takahashi Yuichi rireki Yanagi Genkichi hen," in *Nihon kindai shisō taikei*, ed. Aoki Shigeru and Sakai Tadayasu (Tokyo: Iwanami shoten, 1989), vol. 17, 171.

49. Kōkan uses the verb *benjiru*, meaning to demarcate, distinguish, and separate.

50. Shiba Kōkan, *Seiyō Gadan*, 491.

51. The name *Kai'unrō* is a literary name (*gō*) of a Meiji bureaucrat Shishido Sakari (1841–1900) who worked for the Ministry of Agriculture. Shishido came to work closely with Yoshio. Together they organized an exhibit commemorating Keisuke's eighty-eighth birthday (*beiju*) in 1890.

52. Harada Noriko concurs that it was probably Yoshio who asked the students at the Bureau of Pictorial Studies to make sketches of material objects. Harada Noriko, "Hakubutsu zufu de tadoru futatsu no hakubutsukan no rekishi," in *Nihon no hakubutsu zufu jūkyūseiki kara gendai made*, ed. National Science Museum (Tokyo: Tōkai daigaku shuppan, 2001), 52.

53. After spending just over one year at the Bureau of Pictorial Studies, Yuichi became frustrated with the teachings there and left the institution. He then began to study with Charles Wirgman, who sketched Keisuke during the 1861 meeting with

Siebold, and another Yokohama resident, an American named Raphael Schoyer, in 1870. Yuichi's self-designed career to master *shashin* would later take him to photographer Felix Beato and the painter Antonio Fontanesi, who came to work as an instructor at the "first" "art" school in Japan, *Kōbu bijutsu gakkō*. For the legacy of Fontanesi and *Kōbu bijutsu gakkō*, see the exhibition catalogue *Antonio Fontanesi and Japanese Modern Art: Revolutionary Artists* (Tokyo: Tokyo tei'en bijutsukan, 1997). For consideration of Fontanesi's teaching influence on Yuichi, see Sakamoto Kazumichi, "'Sake' wo yomu," in *Takahashi Yuichi* (Tokyo: Shinchōsha, 1998), 73–85.

54. Osatake Takeki, "Yanagawa Shunsan den," in *Meiji bunka zenshū*, ed. Meijibunka kenkyūkai (Tokyo: Nihon hyōron shinsha, 1955), vol. 4, 21–26.

55. Yanagawa Shunsan, *Shashinkyō zusetsu*, republished in *Nagoya sōsho*, ed. Nagoyashi kyōiku i'inkai (Nagoya: Aichiken kyōdoshiryō hakkōkai, 1983), 469.

56. In *Zōtei bukō nenpyō*, for instance, Saitō Gesshin notes the publication of Shunsan's *Photography* in his entry of 1868. The same entry for the winter of 1868 is filled with descriptions of fights, arsons, and new *bakufu* patrolling troops; it provides a vivid image of the unsettled and uncertain atmosphere that covered much of Edo. Saitō, *Zōtei bukō nenpyō*, 333–334.

57. Yanagawa, *Shashinkyō zusetsu*, 469.

58. Talbot was aware of the reproductive capacity of his process for textual materials. In his *Pencil of Nature*, he included one plate to demonstrate the potential application of photography to reproduce texts: "fac-simile of an old printed page (Plate IX)."

59. Louis Jacques Mandé Daguerre, "Daguerreotype," reprinted in *Classic Essays on Photography*, ed. Alan Trachtenberg (New Haven: Leete's Island Books, 1980), 12.

60. Quoted in Joel Snyder, "Res Ipsa Loquitur," in *Things That Talk: Object Lessons from Art and Science*, ed. Lorraine Daston (New York: Zone Books, 2004), 198.

61. The camera lucida is a drawing device that, with the aid of a mirror, allowed artists to trace the outlines of subjects on a flat surface. Compared to the camera obscura, it is much more portable and was used widely among artists and amateur draftsmen in Europe.

62. Steve Edwards, *The Making of English Photography* (University Park: Pennsylvania State University, 2006), 29.

63. Lady Elizabeth Eastlake, "Photography," *Quarterly Review* (London) 1, no. 101 (April 1857), reprinted in Alan Trachtenberg, ed., *Classic Essays on Photography* (New Haven: Leete's Island Books, 1980), 43.

64. Debate on the "artistic" status of photographic expression continued into the twentieth century. Publication of *The History of Photography* by Beaumont Newhall in 1937 solidified the status of photography as a unique artistic medium. In 1940, Newhall established the Department of Photography at the Museum of Modern Art in New York, the first department dedicated to photography in any museum in the world.

65. Ichikawa Kanenori, a senior lecturer at the institution, documents Shunsan's experimentations in his diary. See the excerpts of the diary in Miyachi Masao, "*Kinsei gazō no shokinō to shashin no shutsugen*," in the exhibition catalogue *Baku-*

matsu maboroshi no abura'eshi Shima Kakoku, ed. Matsudoshi tojō rekishikan (Matsudo: Matsudoshi tojō rekishikan, 1996), 172.

66. Miyanaga Takashi, "Bakumatsu roshia ryūgakusei Ichikawa Bunkichi ni kansuru ichishiryō," *Shakai rōdō kenkyū* 39, no. 4 (1991), 129–188.

67. Ibid., 172.

68. In 1870 Shunsan wrote a letter to the government on behalf of Yuichi, who wanted his new teacher, Charles Wirgman, to live with him in Tokyo, outside of the designated foreign residential area where Wirgman stayed. The request was not granted. See Kumamoto Kenjirō, "Takahashi Yuichi no shōgai to sakuhin," *Bijutsu kenkyū* 59 (1936), 20.

69. Yanagawa, *Shashinkyō zusetsu*, 371–372.

70. Kaneko Ryuichi, "Japan," *Early Works of Photography* (Tokyo: Tokyo Metropolitan Museum of Photography, 1992).

71. Joel Snyder, "Picturing Vision," in *The Language of Images*, ed. W.J.T. Mitchell (Chicago: University of Chicago Press, 1980), 223.

Appendix

1. This translation is based on the transcription of "Yōgakyoku tekigen" included in Aoki Shigeru, ed., *Takahashi Yuichi abura'e shiryō* (Tokyo: Chū'ō kōron bijutsu shuppan, 1984), 215–216.

2. Kitazawa Noriaki warns that Yuichi's conceptualization of an institution here should not be equated to a "museum" in the contemporary sense, noting that "Yuichi's imagination stems more from the thinking process of a technocrat, and not from that of an artist." Kitazawa Noriaki, "Kōgyō nashionarisumu, bijutsu, 'bijutsu' gainen keiseishi sobyō," in *Gakumon no arukeorojī*, ed. University of Tokyo (Tokyo: University of Tokyo Press, 1997), 272–292.

Glossary

Individuals and Collectives

Aōdō Denzen (1748–1822)	亜欧堂田善
Hiraga Gennai (1728–1780)	平賀源内
Iinuma Yokusai (1782–1865)	飯沼慾斎
Ishiguro Sei'an (1787–1836)	石黒済庵
Ishizaki Yūshi (1768–1846)	石崎融思
Itō Keisuke (1803–1901)	伊藤圭介
Iwasaki Kan'en (1786–1842)	岩崎灌園
Kaibara Ekken (1630–1714)	貝原益軒
Kawahara Keiga (1786–1860)	川原慶賀
Koga Kin'ichirō (1816–1884)	古賀謹一郎
Kondō Shūen	近藤集延
Maeno Ryōtaku (1723–1803)	前野良沢
Maki Bokusen (1775-1824)	牧墨僊
Mizutani Hōbun(1779–1833)	水谷豊文
Ono Ranzan (1729–1810)	小野蘭山
Ōkōchi Zonshin (1796-1883)	大河内存真
Shōhyaku-sha	嘗百社
Sugita Genpaku (1733–1817)	杉田玄白
Takahashi Yuichi (1828–1894)	高橋由一
Tanaka Yoshio (1838–1912)	田中芳男
Tanba Shūji (1828–1908)	丹波修治
Udagawa Yōan (1798–1846)	宇田川榕庵
Yamawaki Tōyō (1706–1762)	山脇東洋
Yanagawa Shunsan (1832–1870)	柳河春三

Yoshida Jakusōan (1805–1859)	吉田雀巣庵
Yoshio Jōsan (1787–1843)	吉雄常三

Disciplinary Categories

Goseihō	後世方
Honzōgaku	本草学
Honzō kai	本草会
Jitsugaku	実学
Ko'ihō	古医方
Rangaku	蘭学
Ranpō	蘭方

Titles of Books and Manuscripts

Bencao gangmu	本草綱目
Buppin shikimei	物品識名
"*Honzō shashin*"	本草写真
Kaitai shinsho	解体新書
"Kōsei shinpen"	厚生新編
Taisei honzō meiso	泰西本草名疏
Yamato honzō	大和本草
Zōshi	蔵志

Concepts

Inyō-zu-hō	印葉図法
Shasei	写生
Shashin	写真
Shingyōsō	真行草

Works Cited

Primary Sources

Chomel, Noel. *Algemeen huishoudelijk-, natur-, zedekundig-en konst-/woordenboek.* Leiden: Joh le Meir, 1778.

Daguerre, Louis Jacques Mandé. "Daguerreotype." In *Classic Essays on Photography*, edited by Alan Trachtenberg, 11–13. New Haven: Leete's Island Books, 1980.

Eastlake, Lady Elizabeth. "Photography." *Quarterly Review* (London) 1, no. 101 (April 1857). Reprinted in *Classic Essays on Photography*, edited by Alan Trachtenberg, 39–68. New Haven: Leete's Island Books, 1980.

Hiraga Gennai. *Kyokuzan sensei bunkairoku* (1760). In *Edo kagaku koten sōsho*, vol. 45, *Hakubutsukan tanpenshū*, edited by Ueno Masuzō. Tokyo: Kōwa shuppan, 1982.

Hoffmann, J., comp. *Catalogus librorum et manuscriptorum Japonicorum a Ph. Fr. De Siebold collectorum, annexa enumeratione illorum, qui in Museo Regio Hagano servantur.* Lyon: J. Hoffmann, 1845.

Iinuma Yokusai. *Sōmoku zusetsu*, vol. 1. Gifu: Heirinsō, 1874.

Itō Keisuke. *Kyūkō shokumotsu benran.* In *Kōnō sōshi*, edited by Kōnō shoin. Tokyo: Kōnō shoin, 1882.

———. *Taisei honzō meiso.* Reprint. In *Nagoya sōsho sanpen*, edited by Nagoyashi hōsa bunko. Nagoya: Nagoya kyō'iku i'inkai, 1982.

———. *Keihō yūki.* In *Itō Keisuke nikki daiisshū*, edited by Keisuke monjo kenkyūkai. Nagoya: Nagoyashi higashiyama shokubutsuen, 1995.

———. *Kinkaō nikki*, vol. 2, edited by Keisuke monjo kenkyūkai. Nagoya: Nagoyashi higashiyama shokubutsuen, 1996.

———. "Ogasawara-tō sanbutsu ki." Collection of Itō Keisuke kinenshitsu, Higashiyama Botanical Gardens (n.d.).

Iwasaki Kan'en. *Honzō zufu.* Kyoto: Dōhōsha, 1980.

Kaibara Ekken. *Yamato honzō*, annotated by Shirai Kōtarō. Tokyo: Ariake shobō, 1983.

Katsushika Hokusai. *Santei gafu.* Edo: Takekawa Fujibei, 1816.

———. *Hokusai manga.* Reprint. Tokyo: Shōgakkan, 2005.

Kawahara Keiga. *Sōmoku kajitsu shashin zufu* (1868). Republication of *Keiga shashin sō* (1836) in collection of Waseda University Library, 1868.

Maki Bokusen. *Shashin gakuhitsu*. Kyoto: Hishiya Magobei, 1815.

———. *Yōka seisen zukai*. Ise: Yamagataya, 1820.

Mizutani Hōbun. *Buppin shikimei*. Reprint. In *Nagoya sōsho sanpen*, edited by Nagoyashi hōsabunko. Nagoya: Nagoya kyōiku i'inkai, 1982.

———. *Kiso saiyakuki*. In *Nagoya sōsho*, edited by Nagoyashi kyōiku i'inkai, vol. 13. Nagoya: Nagoyashi kyōiku i'inkai, 1963.

Morishima Chūryō. *Bangosen*. Osaka: Gungyodō, 1798. Waseda University Japanese and Chinese Classics Database. http://www.wul.waseda.ac.jp/kotenseki/. Accessed June 1, 2010.

Odagiri Shunkō. *Owari meisho zue*. Nagoya: Hishiya Kyubei, Katano Toshio, 1844.

Ōkubo Masa'aki. "Ōkubo sensei sōmoku fu." Collection of Kyōu Library, Takeda Science Foundation, Osaka, Japan, n.d.

Ono Ranzan. *Yamato honzō hisei*. In *Ekken zenshū*, vol. 6, edited by Ekken kai. Tokyo: Kokusho kankōkai, 1973.

———. *Honzō kōmoku keimō*. Reprint. ed. Sugita Tsutomu. Tokyo: Waseda University Press, 1974.

Ōtsuki Gentaku. *Ransetsu benwaku*. Ise: Yamagataya, 1799. Waseda University Japanese and Chinese Classics Database. http://www.wul.waseda.ac.jp/kotenseki/. Accessed June 1, 2010.

Saitō Gesshin. *Zōtei bukō nenpyō*. Edited by Asakura Kamezō. Tokyo: Kokusho kankōkai, 1913.

Siebold, Phillip Franz von. Translated by Saito Makoto as *Edo sanpu kikō*. Tokyo: Heibonsha, 1967.

Shiba Kōkan. *Shunparō hikki*. In *Nihon zuihitsu taisei*, edited by Nihon zuihitsu taisei henshūbu. Tokyo: Yoshikawa kōbunkan, 1975.

———. *Oranda tensetsu*. In *Yōgaku*, edited by Numata Jirō, Matsumura Akira, and Satō Akisukeō. Tokyo: Iwanami shoten, 1976.

———. *Oranda tsūhaku*. In *Nihon shisō taikei* (Yōgaku), vol. 64. edited by Numata Jirō, Matsumura Akira, and Satō Shōsuke. Tokyo: Iwanami shoten, 1976.

———. *Seiyō gadan*. In *Nihon shisō taikei* (*Yōgaku*), vol. 64. edited by Numata Jirō, Matsumura Akira, and Satō Shōsuke. Tokyo: Iwanami shoten, 1976.

Shōhyaku-sha. *Honzō kai buppin mokuroku*. Reprint. In *Nagoya sōsho sanpen*, edited by Nagoyashi hōsa bunko. Nagoya: Nagoyashi kyōiku i'inkai, 1982.

Sugita Genpaku. *Rangaku jishi*. In *Bunmei genryū sōsho*, vol.1, edited by Hayakawa Junsaburō. Tokyo: Kokusho kankōkai, 1913.

———, Nakagawa Jun'an, Ishikawa Genjō, and Katsuragawa Hoshū. *Kaitai shinsho*. In *Nihon kagaku koten zensho*, vol. 8, edited by Saegusa Hiroto. Tokyo: Asahi shinbunsha, 1948.

Takahashi Yuichi. "Yōgakyoku tekigen." In *Takahashi Yuichi abura'e shiryō*, edited by Aoki Shigeru. Tokyo: Chū'ō kōron bijutsu shuppan, 1984.

———. "Takahashi Yuichi rireki Yanagi Genkichi hen." In *Nihon kindai shisō taikei* (*bijutsu*), vol. 17, edited by Aoki Shigeru and Sakai Tadayasu. Tokyo: Iwanami shoten, 1989.

Tanaka Yoshio. "Tanaka Yoshio kun no keireki dan." In *Tokyo kokuritsu hakubutsu-*

kan hyakunenshi, shiryō hen, edited by Tokyo National Museum. Tokyo: Dai'ichi hōki shuppan, 1973.

Udagawa Yōan. *Fukkō botanika kyō*. Reprint. Tokyo: Ino'ue shoten, 1965.

———. *Shokugaku keigen: Udagawa Yōan- fukkoku to yaku, chū*, edited by Yabe Ichirō. Tokyo: Kōdansha, 1980.

Yamamoto Bōyō. *Hyappinkō*, edited by Endō Shōji and Nanba Tsuneo. Tokyo: Kagaku shoin, 1982.

Yamawaki Tōyō. *Zōshi*. In *Nihon kagaku koten zensho*, vol. 8, edited by Saegusa Hiroto. Tokyo: Asahi shinbunsha, 1948.

Yanagawa Shunsan. *Shashinkyō zusetsu*, In *Nagoya sōsho*, vo. 13, edited by Nagoyashi kyō'iku i'inkai. Nagoya: Aichiken kyōdoshiryō hakkōkai, 1983.

Yoshida Jakusōan. "Jakusōan hakurankai mokuroku." (Catalogue.) Collection of Nagoya City Museum.

Yoshio Shunzō. *Oranda naigai yōhō*. Owari: Eiryakuya Tōshirō, 1820. Waseda University Japanese and Chinese Classics Database. http://archive.wul.waseda.ac.jp/kosho/ya04/ya04_01443/. Accessed on May 13, 2010.

Secondary Sources

Addiss, Stephen. "The Role of Calligraphy in Japanese Society." In *Words in Motion: Modern Japanese Calligraphy*, edited by Yomiuri shinbun, 30–37. Tokyo: Yomiuri shinbun, 1984.

Alter, Joseph. "Introduction." In *Asian Medicine and Globalization*, edited by Joseph Alter, 1–20. Philadelphia: University of Pennsylvania Press, 2005.

Aramata Hiroshi. *Hakubutsugaku no seiki*. Tokyo: NHK shuppan, 1998.

Barthes, Roland. *Camera Lucida: Reflections on Photography*. New York: Hill and Wong, 1981.

Batchen, Geoffrey. "The Naming of Photography 'A Mass of Metaphor.'" *History of Photography* 17, no. 1 (1993): 22–32.

———. *Burning with Desire*. Cambridge, MA: MIT Press, 1997.

Blunt, Wilfrid. *The Art of Botanical Illustration*. New York: Dover, 1994.

Bowers, John Z. *When the Twain Meet: The Rise of Western Medicine in Japan*. Baltimore: Johns Hopkins University Press, 1980.

Botsman, Daniel V. *Punishment and Power in the Making of Modern Japan*. Princeton: Princeton University Press, 2005.

Cook, Harold J. *Matters of Exchange: Commerce, Medicine, and Science in the Dutch Golden Age*. New Haven: Yale University Press, 2007.

Croissant, Doris. "Meiji shoki yōga no riarisumu ni tsuite." *Jinbun gakuhō* 53 (1982): 157–187.

———. "In Quest of the Real." In *Challenging Past and Present: The Metamorphosis of Nineteenth-Century Japanese Art*, edited by Ellen P. Conant, 153–176. Honolulu: University of Hawai'i Press, 2006.

Daston, Lorraine, and Peter Galison. *Objectivity*. New York: Zone Books, 2007.

Dear, Peter. *The Intelligibility of Nature*. Chicago: University of Chicago Press, 2006.

de Bary, William Theodore, and Irene Bloom. *Principle and Practicality: Essays on Neo-Confucianism and Practical Learning*. New York: Columbia University Press, 1979.
Doane, Mary Ann. "Indexicality: Trace and Sign: Introduction." *differences: A Journal of Feminist Cultural Studies* 18, no.1 (2007): 1–6.
Doi Yasuhiro. "Bansho shirabejo no bussan kenkyū to Itō Keisuke no kankei." *Hōseidaigaku daigakuin kiyō* 36 (1996): 174–49.
———. "Bansho shirabejo shutsueki ki no Itō Keisuke ni zuikō shita monjin no dōkō." In *Kinka'ō nikki*, vol. 3, edited by Keisuke monjo kenkyūkai, 209–226. Nagoya: Higashiyama Botanical Garden, 1996.
———. *Honzōgaku sha Hiraga Gennai*. Tokyo: Kōdansha, 2008.
Edwards, Steve. *The Making of English Photography*. University Park: Pennsylvania State University, 2006.
Elkins, James. "What Do We Want Photography to Be? A Response to Michael Fried." *Critical Inquiry* 31, no. 4 (2005): 938–956.
Ema monjo kenkyūkai, ed. *Emake raikan shū*. Kyoto: Shimonkaku shuppan, 1984.
Endō Shōji. "Iinuma Yokusai to hito: Sōki zusetsu shippitsu chū no Iinuma Yokusai." In *Iinuma Yokusai*, edited by Iinuma Yokusai seitan nihyakunen kinenshi henshū i'inkai, 260–282. Gifu: Iinuma Yokusai seitan nihyakunen kinen jigyōkai, 1984.
———. "Kaho yūki go no Keisuke." In *Kinkaō nikki*, 75–90. Nagoya: Higashiyama Botanical Garden, 1996.
———. "Gifu no Bakumatsu Meijiki no shashin Jijō." In *Kanzō koshashin shū*, edited by Gifuken rekishi shiryōkan, 33–57. Gifu: Gifuken rekishi shiryōkan, 1999.
———. *Honzōgaku to yōgaku*. Kyoto: Shibunkaku shuppan, 2003.
———. "Owari Mino no shashinjutsu." *Kagakushi kenkyū* 28, no. 1 (2000): 112–114.
Fan, Fa-ti. *British Naturalists in Qing China: Science, Empire, and Cultural Encounter*. Cambridge: Harvard University Press, 2004.
Findlen, Paula. *Possessing Nature: Museums, Collecting, and Scientific Culture in Early Modern Italy*. Berkeley: University of California Press, 1996.
Forrer, Matthi. *Eiryakuya Tōshirō: A Publisher at Nagoya*. Amsterdam: J. C. Gieben, 1985.
———. *The Court Journey to the Shogun of Japan: From a Private Account by Jan Cock Blomhoff*. Leiden: Hotei, 2000.
Foucault, Michel. *The Order of Things: An Archeology of the Human Sciences*. New York: Vintage Books, 1994.
French, Calvin L. *Shiba Kōkan: Artist, Innovator, and Pioneer in the Westernization of Japan*. New York: Weatherhill, 1974.
Fuchū Art Museum, ed. *The Age of Aōdō Denzen*. Fuchū: Fuchū Art Museum, 2006.
Furuta Ryō. "Nihon no bijututenrankai sono kigen to hattatsu." *Museum* 545 (December 1996): 29–56.
———. *Kanō Hōgai Takahashi Yuichi*. Tokyo: Mineruba shobō, 2006.
Galassi, Peter. *Before Photography: Painting and the Invention of Photography*. New York: Museum of Modern Art, 1981.

Gunning, Thomas. "Moving Away from the Index: Cinema and the Impression of Reality." *differences: A Journal of Feminist Cultural Studies* 18, no. 1 (2007): 29–52.
Hacking, Ian. *Social Construction of What?* Cambridge, MA: Harvard University Press, 1999.
Haga Tōru. "Takahashi Yuichi to riarisumu." *Mizue* 713 (1979): 58–64.
———. *Kaiga no ryōbun*. Tokyo: Asahi shinbunsha, 1990.
———. "Akita ranga no fushigi." In *Edo no naka no kindai*, edited by Takashina Shūji, 22–96. Tokyo: Chikuma shobō, 1996.
Harada Noriko. "Hakubutsu zufu de tadoru futatsu no hakubutsukan no rekishi." In *Nihon no hakubutsu zufu jūkyūseiki kara gendai made*, edited by National Science Museum, 52–55. Tokyo: Tōkai daigaku shuppan, 2001.
Harootunian, Harry. *Toward Restoration: Growth in Political Consciousness in Tokugawa Japan*. Berkeley: University of California Press, 1970.
Henkes, Harold E. "The Influence of the Dutch on Japanese Ophthalmology in the 19th Century." In *Red-Hair Medicine: Dutch-Japanese Medical Relations*, edited by H. Beukers, A. M. Luyendijk-Elshout, M. E. van Opstall, and F. Vos, 79–82. Amsterdam: Rodopi, 1991.
Hijikata Tei'ichi. *Takahashi Yuichi gashū*. Tokyo: Kōdansha, 1972.
Hirano Mitsuru. "Tenpōki no honzōkenkyukai 'Shabenkai'- zenshi to seiritsu jijō oyobi katsudō no jittai." *Shundai shigaku* 98 (1996): 1–45.
Ichikawa Jinzō. "Tōto ryōgoku yagenbori shashinshi Ugai Gyokusen shoki." *Risshō daigaku kyōyō gakubu kiyō* 22 (1988): 27–39.
Ikegami Eiko. *Bonds of Civility: Aesthetic Networks and Political Origins of Japanese Culture*. Cambridge, UK: Cambridge University Press, 2005.
Imahashi Riko. *Edo no kachōga: Hakubutsugaku wo meguru bunka to sono hyōshō* (Birds and Flowers: The Representation of Natural History During the Edo Period). Tokyo: Sukaidoa, 1995.
Imaizumi Mine. *Nagori no yume—ran'i Katsuragawake ni umarete*. Tokyo: Heibonsha, 1963.
Ishii Kendō. *Zōtei Meiji jibutsu kigen*. Tokyo: Shunyōdō, 1936.
Isono Naohide. "Edojidai dōbutsuzufu ni okeru tensha." In *Higashi ajia no honzōto hakubutsugaku no sekai*, vol. 1, edited by Yamada Keiji, 299–333. Kyoto: Shibunkan shuppan, 1995.
———. "Yakuhinkai, honzō kai." *Keio gijuku daigaku hiyoshi kiyō* (Shizenkagaku), no. 29 (2001): 6–14.
———. *Nihon hakubutsugakushi nenpyō*. Tokyo: Heibonsha, 2002.
Iwamoto Kenji. *Gentō no seiki—eiga zen'ya no shikaku bunkashi*. Tokyo: Shinwasha, 2002.
Jannetta, Ann. *The Vaccinators: Smallpox, Medical Knowledge, and the "Opening" of Japan*. Stanford: Stanford University Press, 2007.
Kajiwara Yoshinori. "Kansei-bunka ki no Nagasaki keibi to fēntongōjiken." *Fukuoka-daigaku jinbun ronsō* 37, no.1 (2005): 339–377.
Kamiya Hiroshi. "Hokusai to Nagoya—kenkyūjosetsu." *Nagoyashi hakubutsukan kenkyūkiyō* 18 (1994): 1–34.

Kanagawa Prefectural Museum of Modern Art, ed. *Botsugo hyakunen Takahashi Yuichi ten*. Kamakura: Kanagawa Prefectural Museum of Modern Art, 1994.

Kaneko Ryūichi. "Japan." In *Shashin no reimei*, edited by Tokyo Metropolitan Museum of Photography. Tokyo: Tokyo Metropolitan Museum of Photography, 1992.

Kasaya Kazuhiko. "The Tokugawa Bakufu's Policies for the National Production of Medicines and Dodonaeus' Cruijdeboeck." In *Dodonaeus in Japan*, edited by W. F. Vande Walle and Kazuhiko Kasaya, 167–186. Leuven: Leuven University Press, 2001.

Kawamura Norihisa. "Wagakuni no inyō-zu fu ni tsuite." *Yokusai kenkyūkai dayori* 59 (1992): 2–8.

———. "Itō Keisuke to inyō-zu-hō." In *Seitan nihyakunen kinen Itō Keisuke no shōgai to sono jisseki*, edited by Higashiyama Botanical Garden, 104–106. Nagoya: Higashiyama Botanical Garden, 2003.

Kimura Yōjirō. "Nachurarisuto Iinuma Yokusai." In *Iinuma Yokusai*, edited by Iinuma Yokusai seitan nihyakunen kinenshi henshū i'inkai, 90–100. Gifu: Iinuma Yokusai seitan nihyakunen kinen jigyōkai, 1984.

———. *Edoki no nachurarisuto*. Tokyo: Asahi shinbunsha, 1988.

———. "Shokubutsu no shoku to shu ni tsuite." In *Higashi ajia no honzō to hakubutsugaku no sekai*, vol. 1, edited by Yamada Keiji. Kyoto: Shibunkaku shuppan, 1995.

Kinoshita Naoyuki. "Shashin no hajimari" and "Shashin wo meguru kotoba." In *Nihon no shashinka, Ueno Hikoma to Bakumatsu no shashinkatachi*, vol. 1, edited by Nagano Shigeichi, Iizawa Kōtarō, and Kinoshita Naoyuki, 3–8, 57–65. Tokyo: Iwanami shoten, 1997.

Kishi Masahiro. "Owari igakukan yakuhine mokuroku—'zappin' tenji ni miru hakubutsukan no genten." *Nagoyashi hakubutsukan kiyō* 17 (1994): 19–26.

Kishino Toshihiko, ed. *Owarihan shakai no bunka jōhō gakumon*. Osaka: Seibundō shuppan, 2002.

———. *Owarihan shakai no sōgō kenkyū*. 3 vols. Osaka: Seibundō shuppan, 2007.

Kitazawa Noriaki. "Kōgyō nashionarisumu, bijutsu, 'bijutsu' gainen keiseishi sobyō." In *Gakumon no arukeorojī*, edited by University of Tokyo, 272–292. Tokyo: University of Tokyo Press, 1997.

———. *Kyōkai no bijutsushi: 'bijutsu' keiseishi nōto*. Tokyo: Buryukke, 2000.

Kiyono Kenji. *Nihon kōkogaku jinruigakushi*, vol. 1. Tokyo: Iwanami shoten, 1955.

Koerner, Lisbet. *Linnaeus: Nature and Nation*. Cambridge, MA: Harvard University Press, 1999.

Koizumi Sadahiro. "Nihon shashin yōgoshi no kenkyū." *Nihon daigaku geijutugakubu kiyō* 14 (1985): 3–13, 15 (1986): 11–17.

Kojima Kaoru. "Hattori Sessai, hakubutsu zufu no meishu." In *Bakumatsu meiji no gakatachi*, edited by Tsuji Nobuo, 163–196. Tokyo: Perikansha, 2008.

Kōno Motoaki. "Edojidai 'shasei' kō." In *Nihon kaigashi no kenkyū*, edited by Yamamoto Yunezō sensei koki kinenkai, 384–427. Tokyo: Yoshikawa kōbunkan, 1989.

———. "'Shasei' no gensen-chūgoku." In *Akiyama Terukazu hakushi koki kinen bijutsu ronbunshū*, edited by Akiyama Terukazu hakushi koki kinen bijutsu ronbunshū kankōkai, 481–514. Tokyo: Benridō, 1991.

———. "Edo kaiga to kyakkan shugi." In *Edo no nakano kindai: Akita ranga to kaitai shinsho*, edited by Takashina Shūji, 233–288. Tokyo: Chikuma shobō, 1996.

———. "Takahashi Yuichi." In *Bakumatsu Meijiki no gakatachi*, edited by Tsuji Nobuo, 91–129. Tokyo: Perikansha, 2002.

Kōno Norihisa. "Wagakuni no inyō zufu ni tsuite." *Yokusai kenkyūkai dayori* 59 (1992): 2–8.

Kornicki, Peter. "Public Display and Changing Values: Early Meiji Exhibitions and Their Historical Precursors." *Monumenta Nipponica* 49, no.2 (1994): 167–196.

Kumamoto Kenjirō. "Takahashi Yuichi no shōgai to sakuhin." *Bijutsu kenkyū* 59 (1936): 16–30.

Kurasawa Takashi. *Bakumatsu kyōikushi no kenkyū*. vols. 1–3. Tokyo: Yoshikawa kōbunkan, 1983.

Lacan, Jacques. "Touché and Automaton." In *The Four Fundamental Concepts of Psycho-Analysis*, translated by Alan Sheridan. New York: Norton, 1978.

Leung, Angela Ki Che. "The Business of Vaccination in Nineteenth-Century Canton." *Late Imperial China* 29, no. 1 (2008): 7–39.

Li, Shizen. *Chinese Medicinal Herbs: A Modern Edition of a Classic Sixteenth-Century Manual.* Translated by F. Porter Smith and G. A. Stuart. Mineola, NY: Dover, 2003.

Lindberg, David. *Theories of Vision from Al-kindi to Kepler*. Chicago: University of Chicago Press, 1976.

Liu, Lydia H. *Translingual Practice*. Stanford: Stanford University Press, 1995.

Looser, Tom. trans. "Seiyōga dan." In *Readings in Tokugawa Thought*, 3rd edition, edited by Tetsuo Najita. Chicago: Center for East Asian Studies, 1998.

Low, Morris. *Building a Modern Japan*. New York: Palgrave MacMillan, 2005.

MacGregor, Arthur. *Curiosity and Enlightenment: Collectors and Collections from the Sixteenth to the Nineteenth Century*. New Haven: Yale University Press, 2007.

Machida shiritsu kokusai hanga bijutsukan, ed. *Kinsei nihonkaiga to gafu, e-tehon ten*. Machida: Machida shiritsu kokusai hanga bijutsukan, 1990.

Markus, Andrew. "Shogakai: Celebrity Banquets of the Late Edo Period." *Harvard Journal of Asiatic Studies* 53, no.1 (1993): 135–167.

Maruyama Masao. *Studies in the Intellectual History of Tokugawa Japan*. Translated by Mikiso Hane. Princeton: Princeton University Press, 1974.

Minamoto Ryōen. *Jitsugaku shisō no keifu*. Tokyo: Kōdansha, 1986.

Mitchell, W.J.T. *What Do Pictures Want?* Chicago: University of Chicago Press, 2005.

Miyachi Masao. "Kinsei gazō no shokinō to shashin no shutsugen." In *Bakumatsu maboroshi no aburaeshi Shima Kakoku*, edited by Matsudoshi tojō rekishikan, 166–175. Matsudo: Matsudoshi tojō rekishikan, 1996.

Miyagawa Toshio, Shirai Yasuo, Danki Reiji, Kashimoto Akihiko, and Hiraki Osamu. "Udagawa Kōsai no 'potokarahi' ni tsuite—komonjo no kaidoku to

ranpakushippan no shashin tokusei." *Tōkyo kōgei daigaku kōgakubu kiyō* 11, no. 1 (1989): 14–25.
Miyanaga Takashi. "Bakumatsu roshia ryūgakusei Ichikawa Bunkichi ni kansuru ichishiryō." *Shakai rōdō kenkyū* 39, no.4 (1991): 129–188.
Morris, Rosalind C., ed. *Photographies East: The Camera and Its Histories in East and Southeast Asia*. Durham, NC: Duke University Press, 2009.
Motoyasu Hiroshi. "Shoki shashinshi no chiikiteki shozentei." *Ishikawa kenritsu rekishi hakubutsukan kiyō* 10 (1997): 23–48.
Murakami Yōjirō. *Nihonjin to kindaikagaku*. Tokyo: Shinyōsha, 1979.
———. *Bunmei no naka no kagaku*. Tokyo: Seidosha, 1994.
Naganoken shinano bijutsukan, ed. *Kawakami Tōgai to sono shūhen -Bakumatsu kara Meiji e*. Nagano: Nagano ken shinano bijutsukan, 1990.
Nagasaki City Museum, ed. *A Very Unique Collection of Historical Significance: The Kaptain* (The Dutch Chief) *Collection from the Edo Period—The Dutch Fascination with Japan*. Nagasaki: Nagasaki City Museum, 2000.
Nagata Seiji. "Kaisetsu." In *Hokusai no edehon*, edited by Nagata Seiji, 271–273. Tokyo: Iwanami bijutsusha, 1994.
Nagoya City Museum, ed. *Nagoya no hakurankai*. Nagoya: Nagoya City Museum, 1982.
———. *Tokubetsuten sakariba: Matsuri, misemono, daidōgei Tokubetsuten*. Nagoya: Nagoya City Museum, 2002.
Nagoyashi kyōiku i'inkai, ed. *Nagoya sōsho vol. 19, Zuihitsuhen, Kankyō manhitsu*. Nagoya: Nagoya kyōiku i'inkai, 1960.
Nakamura, Ellen Gardner. *Practical Pursuits: Takano Chōei, Takahashi Keisaku, and Western Medicine in Nineteenth-Century Japan*. Cambridge, MA: Harvard University Press, 2005.
Nakayama Shigeru, David L. Swain, and Yagi Eri, eds. *Science and Society in Modern Japan*. Tokyo: University of Tokyo Press, 1974.
Nappi, Carla. *The Monkey and the Inkpot: Natural History and Its Transformations in Early Modern China*. Cambridge: Harvard University Press, 2009.
Naruse Fujio. "Edojidai no seiyōgaron ni tsuite-sono toyōshisō tono kankei." *Bijutsushi* 85 (1972): 1–17.
Nihon gakushi'in and Nihon kagakushi kankōkai, eds. *Meijizen nihon igakushi*. 5 vols. Tokyo: Nihon gakujutsu shinkōkai, 1955–1964.
Nishimura Tei. *Nihon dōbangashi*. Tokyo: Shomotsu tenbōsha, 1941.
Nosco, Peter. *Remembering Paradise: Nativism and Nostalgia in Eighteenth Century Japan*. Cambridge, MA: Council on East Asian Studies, Harvard University, 1990.
Numata Jirō. "The Introduction of Dutch Language to Japan." *Monumenta Nipponica* 19, no. 3 (1964): 243–253.
Ōba Hideaki, ed. *Nihon shokubutsu kenkyū no rekishi—Koishikawa shokubutsuen sanbyakunen no ayumi*. Tokyo: University of Tokyo Press, 1996.
Ōgaki-city, ed. *Ōgaki shishi*, vol. 3, 410–450 Ōgaki: Ōgaki-city, 1930.
Osaka Municipal Museum, ed. *Rūtsu Nihon no hakubutsukan*. Osaka: Osaka Municipal Museum, 1979.

Osaka rekishi hakubutsukan, ed. *Kimura Kenkadō: Naniwa chi no kyōjin*. Osaka: Osaka rekishi hakubutsukan, 2003.
Osatake Takeki. "Yanagawa Shunsan den." In *Meiji bunka zenshū*, vol. 4, edited by Meiji bunka kenkyūkai, 21–26. Tokyo: Nihon hyōron shinsha, 1955.
Ozawa Kenji. *Nihon no shashinshi—bakumatsu no denhan kara meijiki made*. Tokyo: Nikkor club, 1986.
———. *Bakumatsu Meiji no shashin*. Tokyo: Chikuma shobō, 1997.
Peirce, Charles Sanders. "Logic as Semiotics: The Theory of Signs." In *Philosophical Writings of Peirce*, edited by Justus Buchler, 98–119. New York: Dover, 1955.
Pinney, Christopher, and Nicolas Peterson, eds. *Photography's Other Histories*. Durham, NC: Duke University Press, 2003.
Pomian, Krzystof. "Vision and Cognition." In *Picturing Science Producing Art*, edited by Caroline A. Jones and Peter Galison, 211–231. New York: Routledge, 1998.
Rudolph, Richard. "Thunberg in Japan and His Flora Japonica in Japanese." *Monumenta Nipponica* 29, no. 2 (1974): 163–179.
Sakai Shizu. *Nihon no iryō-shi*. Tokyo: Tokyo shoseki, 1982.
Sakakibara Yoshirō. "Juhasseiki no shokubutsu shasei." In *Higashi ajia no honzō to hakubutsugaku no sekai*, vol. 1, edited by Yamada Keigo, 198–273. Kyoto: Shibunkaku shuppan, 1995.
Sakamoto Kazumichi. *Takahashi Yuichi*. Tokyo: Shinchōsha, 1998.
Sasaki Takami. "Kaisetsu." In *Nagoya sōsho sanpen*, edited by Nagoyashi hōsa bunko, 5–19. Nagoya: Nagoyashi kyōiku i'inkai, 1982.
Satō Dōshin. *Meiji kokka to kindai bijutsu*. Tokyo: Yoshikawa kōbunkan, 1999.
———. *"Nihonbijutsu" tanjō: Kindai nihon no "kotoba" to senryaku*. Tokyo: Kōdansha, 2006.
Screech, Timon. *The Western Scientific Gaze and Popular Imagery in Later Edo Japan: The Lens Within the Heart*. Cambridge: Cambridge University Press, 1996.
———. *The Shogun's Painted Culture: Fear and Creativity in the Japanese States 1760–1829*. London: Reaktion books, 2000.
———. *Japan Extolled and Decried: Carl Peter Thunberg and the Shogun's Realm, 1775–1796*. London: Routledge, 2005.
Seibu Museum of Art, ed. *Kawahara Keiga ten*. Tokyo: Seibu Museum of Art, 1987.
Shimonaka Hiroshi, ed. *Saishoku edo hakubutsugaku shūsei*. Tokyo: Heibonsha, 1994.
Snyder, Joel. "Picturing Vision." In *The Languages of Images*, edited by W.J.T. Mitchell. Chicago: University of Chicago Press, 1980.
———. "Enabling Confusion." *History of Photography* 26, no. 2 (2002): 154–160.
———. "Res Ipsa Loquitur." In *Things That Talk: Object Lessons from Art and Science*, edited by Lorraine Daston, 195–221. New York: Zone Book, 2004.
———. "Pointless." In *Photography Theory*, edited by James Elkins, 369–385. London: Routledge, 2007.
Sōda Hajime. *Zusetsu nihon iryō bunkashi*. Kyoto: Shibunkaku shuppan, 1989.
Strassler, Karen. *Refracted Visions: Popular Photography and National Modernity in Java*. Durham, NC: Duke University Press, 2010.

Sugano Yō. *Edo no dōbanga*. Tokyo: Shinchōsha, 1983.

Sugimoto Isao. *Itō Keisuke*. Tokyo: Yoshikawa Kōbunkan, 1960.

———. "Itō Keisuke no jitsugaku no seikaku." *Nihondaigaku jinbunkagaku kenkyūjo kenkyū kiyō* 2 (1960): 39–70.

Sugimoto Tsutomu. *Edo yōgaku jijō*. Tokyo: Yasaka shobō, 1990.

———. *Edo jidai seiyō hyakkajiten: "kōseishinpen" no kenkyū*. Tokyo: Yūzankaku shuppan, 1999.

———. *Edo no hakubutsugakushatachi*. Tokyo: Kōdansha, 2006.

Suzuki Hiroyuki. *Kōkokatachi no jūkyūseiki*. Tokyo: Yoshikawa kōbunkan, 2003.

Suzuki Susumu. "Kinsei shaseiga no keifu—nihonga no shasei shugi." In *Kinsei shaseiga no keifu*, edited by Kanagawa Prefectural Museum, 5–8. Yokohama: Kanagawa Prefectural Museum, 1971.

Tabei Ryūtarō. *Kokon chūkyōgadan*. Nagoya: Kōfū shoin, 1911.

Takashina Shūji. *Nihon kindai bijutsushiron*. Tokyo: Kōdansha, 1977.

Takeda Koichi. "Chūgoku gafu to nihon nanga no kankei." In *Kinsei Nihonkaiga to gafu/e-tehon ten*, edited by Machida shiritsu kokusai hanga bijutsukan, 147–158. Machida: Machida shiritsu kokusai hanga bijutsukan, 1990.

Takeuchi, Melinda. *Taiga's True Views: The Languages of Landscape Painting in Eighteenth-Century Japan*. Stanford: Stanford University Press, 1992.

Tetsuo Najita. *Visions of Virtue in Tokugawa Japan: The Kaitokudō Merchant Academy of Osaka*. Honolulu: University of Hawai'i Press, 1987.

———. "History and Nature in Eighteenth-Century Tokugawa Thought." In *The Cambridge History of Japan*, vol. 4, *Early Modern Japan*, edited by John Whitney Hall, 596–659. Cambridge, UK: Cambridge University Press, 1991.

———. "Ambiguous Encounters: Ogata Kōan and International Studies in Late Tokugawa Osaka." In *Osaka: The Merchants' Capital of Early Modern Japan*, edited by James McClain and Wakita Osamu, 213–240. Ithaca, NY: Cornell University Press, 1999.

Toby, Ronald. *State and Diplomacy in Early Modern Japan: Asia in the Development of the Tokugawa Bakufu*. Stanford: Stanford University Press, 1984.

Tokyo Metropolitan Museum of Photography. *The Advent of Photography in Japan*. Tokyo: Tokyo Metropolitan Museum of Photography, 1997.

———. *Nihon no dandizumu*. Tokyo: Nigensha, 2003.

Tsukahara Akira, ed. *Shiba Kōkan hyakka jiten* (Shiba Kōkan His Versatile Life). Kobe: Kobe Municipal Museum, 1996.

Tucker, Anne, ed. *The History of Japanese Photography*. New Haven: Yale University Press, in association with Museum of Fine Arts, Houston, 2003.

Tucker, John A., ed. and trans. *Ogyū Sorai's Philosophical Masterworks: The Bendō and Benmei*. Honolulu: University of Hawai'i Press, 2006.

Tucker, Mary Evelyn. *Moral and Spiritual Cultivation in Japanese Neo-Confucianism: The Life and Thought of Kaibara Ekken 1630–1740*. Albany: State University of New York Press, 1989.

Uchiyama Jun'ichi. *Edo no kōkishin: bijutsu to kagaku no de'a'i*. Tokyo: Kōdansha, 1996.

Ueda Kyōichiirō. "Kawahara Keiga." In *Saishoku edo hakubutsugaku shūsei*, edited by Shimonaka Hiroshi, 293–308. Tokyo: Heibonsha, 1994.

Ueno Masuzō. *Kusawo tenishita shōzōga*. Tokyo: Yasaka shobō, 1986.

———. *Nihon hakubutsugakushi*. Tokyo: Kōdansha, 1989.

University of Tokyo, ed. *Gakumon no arukeoroji*. Tokyo: University of Tokyo Press, 1997.

Vande Walle, W. F., and Kazuhiko Kasaya, eds. *Dodonaeus in Japan*. Leuven: Leuven University Press, 2001.

Watanabe Masao. *The Japanese and Western Science*. Translated by Otto Theodor Benfey. Philadelphia: University of Pennsylvania Press, 1976.

Yabe Ichirō. *Edo no honzō*. Tokyo: Saiensusha, 1984.

Yamada Keiji. "Honzō ni okeru bunrui no shisō." In *Higashiajia no honzō to honzōgaku no sekai*, vol. 1, edited by Yamada Keiji, 3–42. Kyoto: Shibunkaku shuppan, 1995.

Yomiuri shinbun, ed. *Words in Motion: Modern Japanese Calligraphy*. Tokyo: Yomiuri shinbun, 1984.

Yokoe Fuminori. "Kamera obusukyura no denrai kara 'shashinjutsu' no toraimade (Between the Arrival of the Camera Obscura and the Daguerreotype in Japan)." In *The Advent of Photography in Japan*, edited by Tokyo Metropolitan Museum of Photography and Hakodade Museum of Art, 14–15, 163–165. Tokyo: Tokyo Metropolitan Foundation for History and Culture, Tokyo Metropolitan Museum of Photography, 1997.

Yoshida Susugu. "Chūkyō no senkusha Maki Bokusen." *Hokusai kenkyū* 9 (1975): 16–35.

Yoshida Toshihide. "Maki Bokusen saikenshō." *Nagoyashi bijutsukan kenkyū kiyō* 1 (1991): 1–30.

Yoshikawa Yoshiaki. *Ikgaku yōgaku honzōgakusha no kenkyū*. Tokyo: Yasaka shobō, 1993.

Yōrō Takeshi. "Edo no kaibōzu." In *Edo no naka no kindai*, edited by Takashina Shūji, 189–228. Tokyo: Chikuma shobō, 1996.

Index

Italicized page numbers refer to illustrations.

Kolibri 360 GmbH
Ferdinand-Jühlke-Straße 7
99095 Erfurt, Deutschland
produktsicherheit@kolibri360.de